ST(P) MATHEMATICS 5A

ST(P) MATHEMATICS will be completed as follows:

Published 1984
- **ST(P) 1**
- **ST(P) 1** Teacher's Notes and Answers
- **ST(P) 2**

Published 1985
- **ST(P) 2** Teacher's Notes and Answers
- **ST(P) 3A**
- **ST(P) 3B**
- **ST(P) 3A** Teacher's Notes and Answers
- **ST(P) 3B** Teacher's Notes and Answers

Published 1986
- **ST(P) 4A**
- **ST(P) 4B**
- **ST(P) 4A** Teacher's Notes and Answers
- **ST(P) 4B** Teacher's Notes and Answers

Published 1987
- **ST(P) 5A** (with answers)
- **ST(P) 5B** (with answers)

In preparation
- **ST(P) 5C** (with answers)
- **ST(P)** Resource Book

ST(P) MATHEMATICS 5A

L. Bostock, B.Sc.
formerly Senior Mathematics Lecturer, Southgate Technical College

S. Chandler, B.Sc.
formerly of the Godolphin and Latymer School

A. Shepherd, B.Sc.
Head of Mathematics, Redland High School for Girls

E. Smith, M.Sc.
Head of Mathematics, Tredegar Comprehensive School

Stanley Thornes (Publishers) Ltd

First Published 1987 by
Stanley Thornes (Publishers) Ltd,
Old Station Drive,
Leckhampton,
CHELTENHAM GL53 0DN

British Library Cataloguing in Publication Data

ST(P) mathematics 5A
   1. Mathematics—1961–
   I. Bostock, L.
   510   QA39.2

   ISBN 0-85950-254-6

Typeset by Tech-Set, Gateshead, Tyne & Wear
Printed and bound in Great Britain at The Bath Press, Avon

CONTENTS

INTRODUCTION

This book completes the work necessary for the written papers in mathematics at the top level of the GCSE.

Apart from the topics needed to complete the syllabus for each Examining Group, the book contains a good deal of material for revision and consolidation. Since no one syllabus requires all the new work included here a careful check is recommended of the particular syllabus being followed.

There are plenty of straightforward questions and exercises are divided into two types.

The first type, identified by plain numbers, e.g. **12,** are considered necessary for consolidation.

The second type, identified by a double underline, e.g. <u><u>**12,**</u></u> are more demanding questions.

ACKNOWLEDGEMENTS

The authors and publishers would like to thank the following for permission to include material.

Esso, France: for the French map on page 100

Ordnance Survey: for the English map on page 100

British Rail: for the timetable on pages 112, 114 and 115

The Post Office: for the table on page 116

British Telecom: for the table on page 118

1 THE LANGUAGE OF MATHEMATICS

The further anyone progresses in mathematics, the more vital it becomes to express ideas precisely and accurately. You may so far have taken the view that if *you* know what you mean, and your answer is right, then that is all that matters. Sooner or later, however, some of you will need to communicate more advanced ideas to other people so that they know *exactly* what you mean.

The ability to use clear, correct, unambiguous mathematical language cannot be acquired overnight. It depends, first of all, on a frame of mind which sees the necessity for meticulous care in expression, and then it must be developed steadily and consistently until it becomes second nature.

For those who are thinking of taking mathematics further, now is the time to begin to develop the skill of using mathematical language correctly. Because this is done by starting at the beginning, inevitably we shall initially be looking at examples that seem almost too trivial to bother with. This should not deter a student with real mathematical potential, for the introduction to rigour which these simple cases give can create the attitude of mind required to use the language fully.

The language of mathematics is a combination of words and symbols, each symbol being the shorthand form for a word or phrase. When the words and symbols are correctly used a piece of mathematical reasoning can be read, as prose can, in properly constructed sentences.

You have already used a fair number of symbols but not, perhaps, always with enough care for their precise meaning. As we now take a look at some familiar symbols we find that some can be translated correctly in more than one way.

THE USE AND MISUSE OF SYMBOLS AND WORDS

First consider the elementary symbol $+$.
This can be read as 'plus' or 'and' or 'together with' or 'positive'.

e.g.　　$3 + 2$　means　3 plus 2　or　3 and 2

　　　　$\mathbf{a} + \mathbf{b}$　means　vector $\mathbf{a}$ together with vector $\mathbf{b}$

1

The symbol $-$ has a similar variety of translations.

e.g. $5 - 4$ means   5 minus 4

$5 - (-4)$ means   5 minus 'negative 4'

Now consider $\times$ which can be read as 'multiplied by' or 'times' or 'of'.

e.g. 7×5 means   7 multiplied by 5   or   7 times 5

$\frac{1}{100} \times x$ means   one hundredth of x

Note.   When 'times' is used for $\times$, it really means 'lots of', e.g. 3×8 means 3 lots of 8. It is quite incorrect, therefore, to say, 'times 3 by 8' a phrase which teachers often hear, because, clearly it is nonsense to write 'lots of 3 by 8'. (This emphasises that 'times' is not a verb.) If we want to use $\times$ as an instruction, the corresponding word must be 'multiply', i.e. '3×8' can read 'multiply 3 by 8'.

The next sign we consider is $=$ which means 'is equal to'. This symbol should be used *only* to link two quantities that are equal in value. Used in this way a short complete sentence is formed, e.g.

$x = 3$ is read as 'x is equal to 3'.

It is very easy to slip into the habit of saying 'equals' or 'equal' instead of 'is equal to'. For instance it is not good English to write 'Let $x = 3$' because this really translates to 'Let x is equal to 3'. While this sort of misuse may seem (and, up to this level, is) trivial it can be serious at a more advanced level and is better avoided altogether.

It is also bad practice to use $=$ in place of the word 'is'. For instance, when defining a symbol such as the radius of a circle we should say 'the radius of the given circle is r cm' and *not* 'the radius of the given circle $= r$ cm', because r cm and the radius are not two separate quantities of equal value; r *stands for* the radius.

Of course, if the radius is later found to be 4 cm, it is then correct to say '$r = 4$'.

EXERCISE 1a

In this exercise we give some problems followed by solutions which, although ending with the correct answer, contain nonsense on the way. These solutions have been taken from actual students' work and are examples of very common misuses of language.

In each case criticise the solution, and write a correct version.

1. Simplify $2\frac{1}{2} + 1\frac{1}{4} - 2\frac{1}{3}$.

$$2\tfrac{1}{2} + 1\tfrac{1}{4} = 3\tfrac{3}{4} - 2\tfrac{1}{3}$$
$$= 1\tfrac{5}{12}$$

2. Solve the equation $6x + 5 = 3x + 11$.

$$6x + 5 = 3x + 11$$
$$= 3x + 5 = 11$$
$$= \quad 3x = 6$$
$$= \quad x = 2$$

3. Find $\widehat{A}$ when $\sin A = 0.5$

$$\sin A = 0.5$$
$$= 30°$$

4. Write down the formula for the circumference of a circle.

The formula for the circumference of a circle is $2\pi r$.

5. Two angles of an isosceles triangle each measure $70°$. Find the size of the third angle.

$$\text{Third angle} = 70° + 70°$$
$$= 180° - 140°$$
$$= 40°$$

6. Three buns and two cakes cost $54\,p$ and five buns and one cake cost $62\,p$. Find the cost of one bun and of one cake.

$$\text{Buns} = x\,p \qquad \text{Cakes} = y\,p$$
$$3x + 2y = 54\,p$$
$$5x + y = 62\,p$$
$$10x + 2y = 124\,p$$
$$7x = 70\,p$$
$$x = 10\,p \quad \text{and} \quad y = 12\,p$$

7. $A = \begin{pmatrix} 1 & 2 \\ 3 & 4 \end{pmatrix}$, $B = \begin{pmatrix} 4 & -3 \\ 2 & 1 \end{pmatrix}$. Find $A + B$.

$$A = \begin{pmatrix} 1 & 2 \\ 3 & 4 \end{pmatrix} + B = \begin{pmatrix} 4 & -3 \\ 2 & 1 \end{pmatrix}$$
$$= \begin{pmatrix} 5 & -1 \\ 5 & 5 \end{pmatrix}$$

SYMBOLS THAT CONNECT STATEMENTS

The symbol $\therefore$, meaning 'therefore', introduces a fact, complete in itself, which follows from a previous complete fact. It is correct to write

$$x^2 = 9$$

$$\therefore \qquad x = \pm 3$$

It is not correct, however, to use $\therefore$ to link the next two lines

$$3x - 4y - 2x + y$$

$$x - 3y$$

Each of these lines is simply an expression, not a complete fact, and in this case we link the lines by the symbol $\Rightarrow$. This means 'giving' or 'which gives', i.e.

$$3x - 4y - 2x + y$$

$$\Rightarrow \qquad x - 3y$$

Note that $\Rightarrow$ can correctly be used as an alternative to $\therefore$, e.g.

$$x^2 = 9 \quad \Rightarrow \quad x = \pm 3$$

but the converse is not usually true.

Note also that in the context of Mathematical Logic, the symbol $\Rightarrow$ means 'implies that'.

There are occasions when none of these symbols is absolutely correct, a very simple example being

$$3 = x$$

$$x = 3$$

These two statements give exactly the same information, so neither $\therefore$ nor $\Rightarrow$ is quite right. In this situation it is best to use 'i.e.' ('that is'):

$$3 = x$$

i.e. $\qquad x = 3$

It is quite common to see the word 'or' where we have used 'i.e.'. Although this is not actually wrong it should be treated with caution because 'or' strongly suggests that an *alternative result* is being given, and not just the same result rearranged.

A similar criticism of bad practice can be levelled at the way some people try to give their reasons for steps in a solution. In the solution of a pair of simultaneous equations, for example, we sometimes see

$$3x - 4y = 5 \qquad \times 2$$
$$2x + 7y = 13 \qquad \times 3$$

$$6x - 8y = 10$$
$$6x + 21y = 39$$

These four lines are disjointed, do not really explain what is happening and cannot be read sensibly in words. It is much better to present this piece of work in one of the following ways:

a)

$$3x - 4y = 5 \qquad [1]$$
$$2x + 7y = 13 \qquad [2]$$

$$2 \times [1] \quad \Rightarrow \quad 6x - 8y = 10$$
$$3 \times [2] \quad \Rightarrow \quad 6x + 21y = 39$$

After the initial definition of equations [1] and [2] this now reads '2 times equation [1] gives $6x - 8y = 10$' and '3 times equation [2] gives $6x + 21y = 39$'.

b)

$$3x - 4y = 5 \qquad [1]$$
$$2x + 7y = 13 \qquad [2]$$

$$6x - 8y = 10 \qquad (\text{multiplying } [1] \text{ by } 2)$$
$$6x + 21y = 39 \qquad (\text{multiplying } [2] \text{ by } 3)$$

This version too can be read:
$6x - 8y = 10$; multiplying equation [1] by 2, etc.

Note. In (a) an instruction was given first, describing the operation to be carried out and leading to an equation, whereas in (b) the operation was carried out and then explained.

Either of these approaches can be extended satisfactorily to more advanced mathematics.

MORE USEFUL LINK WORDS

The words and symbols mentioned so far do, in fact, provide sufficient vocabulary to write and read most mathematics at this level, and will continue to be used as the work develops, supplemented by the extra symbols needed for each new area of study.

Even at present, however, variety can be added by a few more words that link or introduce facts.

A traditional, but still useful, one is 'hence' which means 'from this'. It fits nicely into the following type of situation:

Circle A has a radius of 4 cm and circle B has a radius of 2 cm.

Hence the area of circle A is four times that of circle B.

It sometimes happens that, in a solution, one line of thought is pursued for a few steps and then a new idea is introduced. This situation is clearly expressed by the word 'now', as in the following example:

$$2x + 3y = 7 \qquad [1]$$
$$8x - 5y = 11 \qquad [2]$$
$$4 \times [1] \quad \Rightarrow \qquad 8x + 12y = 28 \qquad [3]$$
$$[3] - [2] \quad \Rightarrow \qquad 17y = 17$$
$$\therefore \qquad y = 1$$
$$\text{In } [1], y = 1 \quad \Rightarrow \qquad 2x + 3 = 7$$
$$\Rightarrow \qquad x = 2$$

$\therefore$ the solution of the equations is $x = 2, y = 1$.

Now we know that if two lines, whose equations are given, are plotted on the same axes, the coordinates of their point of intersection satisfy both equations.

Hence the lines with equations $2x + 3y = 7$ and $8x - 5y = 11$, meet at the point $(2, 1)$.

Here is an example showing the use of some of the words and symbols discussed in this chapter.

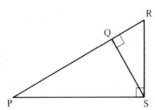

The area of $\triangle PQS$ is $16\,cm^2$ and the area of $\triangle QRS$ is $9\,cm^2$. Find the ratio of PQ to QR. Find also the value of $PR : PS$, using the ratio of the areas of triangles PQS and PSR.

The area of $\triangle PQS$ is given by $\frac{1}{2}PQ \times QS$

$\therefore$ $$\frac{1}{2}PQ \times QS = 16 \qquad [1]$$

The area of $\triangle QRS$ is given by $\frac{1}{2}QR \times QS$

$\therefore$ $$\frac{1}{2}QR \times QS = 9 \qquad [2]$$

$[1] \div [2] \quad \Rightarrow$ $$\frac{\frac{1}{2}PQ \times QS}{\frac{1}{2}QR \times QS} = \frac{16}{9}$$

i.e. $$PQ : QR = 16 : 9$$

In triangles PQS and PSR

$$P\widehat{Q}S = P\widehat{S}R \qquad (\text{both } 90°)$$

$$Q\widehat{P}S \text{ is common}$$

$$P\widehat{S}Q = P\widehat{R}S \qquad (\text{third angles equal})$$

Hence $$\triangle s \frac{PQS}{PSR} \text{ are similar.} \quad (3 \text{ pairs of equal angles})$$

Now the ratio of the areas of similar figures is equal to the ratio of the squares of corresponding sides.

$\therefore$ $$\frac{\text{Area } \triangle PSR}{\text{Area } \triangle PQS} = \frac{(PR)^2}{(PS)^2}$$

$\Rightarrow$ $$\left(\frac{PR}{PS}\right)^2 = \frac{\text{Area } \triangle PQS + \text{Area } \triangle QRS}{\text{Area } \triangle PQS}$$

$$= \frac{16 + 9}{16}$$

Hence $$\frac{PR}{PS} = \sqrt{\frac{25}{16}}$$

i.e. $$PR : PS = 5 : 4.$$

Now try writing out this solution in grammatical prose. (In lines 10, 11 and 13 you will need to translate the first bracket of the 'reason' as 'because' or 'because there are'.)

SOME MEDIA MISUSES

Sometimes when items appearing in the press or on radio or television contain an element of mathematics, insufficient care is taken to use the correct terminology. What is worse is that there are cases of deliberate misrepresentation in an attempt to make an argument appear to be stronger. It is therefore vital that we do not accept everything we read or hear without carefully considering its true meaning and implications, its accuracy and its honesty.

Here are two examples of careless terminology:

a) In mathematics the word 'plus' is used only when collecting two quantities together, one on either side of 'plus'. However, advertisements often include phrases such as 'Send in your order today and you will have the chance to win £1000. PLUS Steve Stardust will present the cheque to the women.' (To express this correctly, although in a less eye-catching way, we could say 'Send in your order today and you will have the chance to win £1000 plus the chance to have it presented by Steve Stardust'.)

This particular misuse is both grammatically and mathematically wrong. Grammar, however, does tend to change with time and no doubt 'plus' will continue to be used in this way. This does *not* mean, though, that the mathematical definition of 'plus' can be changed, so take care.

b) We often hear a news item saying, for instance, that 'the rate of inflation has fallen by one percentage point'. What is meant is that 'the rate of inflation has fallen by one per cent'. The misuse of 'percentage point' can be very confusing, especially when a decimal point is involved, e.g. if the fall in the rate of inflation is from 4.7% to 3.7%.

Keep your eyes and ears open, and see if you can spot more examples of misuse — there are a good many.

MISCONCEPTIONS

The wrong interpretation of information given in the media can often arise, either from misleading presentation or from misuse of words or from lack of care in reading and considering the information.

The problem of misleading presentation usually occurs when graphs or charts are used to illustrate a situation. Enough has been written already in this series about misleading graphs, so we will not labour that point here. The misuse of words has just been discussed so this too we need take no further.

Therefore we will now consider the importance of careful reading and thoughtful weighing up of news items with a mathematical content.

Take, for example, the statement 'the rate of inflation is falling'. Some people think that this means that prices are coming down, but it doesn't. Inflation means rising prices; the rate of inflation measures how fast they are rising. If the *rate* goes down it simply means that prices do not rise so quickly — but they still keep on rising.

There are areas where conflicting figures are given to represent what is apparently the same thing. In wage negotiations, for example, the pay of a 'typical employee' might vary considerably; a union official might choose the basic wage of the lowest-paid after all possible deductions whereas the employer might quote the earnings (including overtime) of a higher-paid employee before any deductions are made. Clearly neither of these is typical and anyone reading only one version would be misled.

These are merely examples of a much wider problem and there are many pitfalls for the unwary in pseudo-mathematical reporting. The best way to avoid any of these is to have a background of precise language, cautious, careful reading and a strong objection to believing everything you hear or read — even in textbooks.

You can now amuse yourself by finding examples of 'bad practice' in the use of symbols in this book!

2 THE BASIC ARITHMETIC PROCESSES

This chapter is in sections. Each section starts with a set of diagnostic tests, followed by revision material and further practice exercises.

The tests will help you to decide whether revision of a topic is necessary (you should get all the questions correct). It is sometimes difficult to know whether mistakes indicate carelessness or lack of understanding. Remember that accuracy matters, so do not rush these tests, and check your answers for reasonableness and for accuracy.

The chapter ends with a set of mixed exercises. These use the basic processes in a variety of problems and should not be omitted.

WHOLE NUMBERS

Do not use a calculator in these tests.

Test 2A   The Four Rules

Calculate

1.	$36 + 27 + 19$	**4.**	215×3	**7.**	199×9
2.	$32 - 18$	**5.**	$1279 + 398$	**8.**	$36 - 51 + 81$
3.	$29 - 4 + 7$	**6.**	$301 \div 7$	**9.**	$2 \times 4 \times 6$

Test 2B   Directed Numbers and Order of Operations

Calculate

1.	$7 - 12$	**7.**	$4 \div (-2)$	**13.**	$(3 - 2)(5 + 1)$
2.	-3×4	**8.**	$4 - (-5)$	**14.**	$6 - 3 \times 4 + 2$
3.	-5×-2	**9.**	$-4 - 6$	**15.**	$4 - 3(2 + 1)$
4.	$-2 - 4 + 3$	**10.**	$1 + 3 \times 2$	**16.**	$-2(76 + 5) \div 9$
5.	6×-2	**11.**	$8 \times 3 - 2$	**17.**	$3 \times 2 + 7 \times (3 - 5)$
6.	$-8 \div 4$	**12.**	$4 - 2(1 - 2)$	**18.**	$4 + 15 \div 3 + 2$

Test 2C   Factors and Multiples

Find the highest number that is a common factor of

1. 4 and 6   **3.** 14 and 21   **5.** 70 and 84

2. 8 and 20   **4.** 35 and 56   **6.** 16, 24 and 40

Find the lowest number that is a common multiple of

7. 2 and 3   **9.** 6 and 5   **11.** 25 and 20

8. 4 and 6   **10.** 12 and 15   **12.** 8, 6 and 15

Express the following numbers as products of their prime factors.

13. 30   **14.** 108   **15.** 3150

THE FOUR RULES

Addition and Subtraction   This is a case in which practice makes perfect. Both speed and accuracy are improved with practice. Check additions by adding in reverse order, i.e. up a column and then down the column, or left to right and then right to left. One way to speed up the addition of a string of numbers is to look for pairs of digits that add up to ten — but be careful that you don't count them in twice.

There are quick ways of adding and subtracting numbers that are close to 100, 200, ..., etc. For example to add 197, you could add 200 and then take 3 away. To subtract 197 you could subtract 200 and then add on 3.

When a calculation involves both addition and subtraction, remember that the additions can be done first. For example, $5 - 6 + 2$ can be done in the order $5 + 2 - 6$.

Multiplication and Division   For speed and accuracy you need to know the multiplication tables.

Remember that multiplication is repeated addition,

e.g. 199×8 means 199 lots of 8.

Hence 199×8 can be calculated by finding 200×8 and subtracting 8, i.e. $199 \times 8 = 1600 - 8 = 1592$.

EXERCISE 2a

Find, without using a calculator,

1. $25 + 14 + 72$	**5.** $89 - 17$	**9.** $86 - 21 + 15$
2. $36 + 56 + 125$	**6.** $31 - 19$	**10.** $72 - 95 + 26$
3. $253 + 891$	**7.** $127 - 59$	**11.** $18 + 3 - 15$
4. $1993 + 827$	**8.** $2561 - 290$	**12.** $42 + 15 - 26 - 18$

13. 94×8	**17.** $152 \div 8$	**21.** $2 \times 5 \times 3$
14. 3×506	**18.** $186 \div 6$	**22.** $9 \times 8 \times 7$
15. 297×6	**19.** $135 \div 5$	**23.** $4 \times 3 \times 2 \times 5$
16. 1009×51	**20.** $882 \div 7$	**24.** $3 \times 2 \times 2 \times 5$

DIRECTED NUMBERS AND ORDER OF OPERATIONS

Remember that

negative multiplied by negative⎫
or   positive multiplied by positive ⎭   gives a positive result;

negative multiplied by positive⎫
or   positive multiplied by negative⎭   gives a negative result.

The same rules apply to division.

Calculations involving mixed operations must be done in the following order: calculations inside brackets first, multiplication and division next and finally addition and subtraction. Remember that the positive sign at the start of a calculation is often omitted. Also remember that it is the sign before the number that tells you what to do with that number, i.e. $6 \div 2 \times 4$ means $(6 \div 2) \times 4$. It does not mean $6 \div (2 \times 4)$.

EXERCISE 2b

Find, without using a calculator,

1. $3 - (-2)$	**5.** $4 - 2(1 - 3)$	**9.** $-2 + 4 - 6$
2. $5 + (-4)$	**6.** $-6 \div -2$	**10.** $-2 \times 4 \times (-6)$
3. $5 \times (-2)$	**7.** $-8 \div 4$	**11.** $(-4)^2$
4. -5×-3	**8.** $-5 - 2$	**12.** $(-1)^2 + (1)^2$

13.	$3 + 2 \times 6$	**17.**	$(3 + 2) \times 5 - 6$	**21.**	$\dfrac{-12}{-3}$
14.	$(3 + 2) \times 6$	**18.**	$3 - 2 \times 4 - 6 \div 2$	**22.**	$-2 \times -3 \times -2$
15.	$8 - 5 \times 2$	**19.**	$3 \times 2 - 4$	**23.**	$5 + 3 - 2 \times 4$
16.	$(8 - 5) \times 2$	**20.**	$3 - 2 + 4$	**24.**	$4 \times (-2)^2$

FACTORS AND MULTIPLES

A *factor* of a number divides exactly into the number. For example, 2, 3, 4 and 6 are factors of 12.

A common factor for two (or more) numbers is a factor of each number. For example, 2, 3 and 6 are common factors of 12 and 18. Further, 6 is the highest common factor of 12 and 18.

To express a number as the product of its *prime factors,* start by trying to divide by 2 and continue to divide the result by 2 until you no longer can exactly. Then try 3 in the same way, then 5 and so on until you are left with 1.

For example, to express 276 as a product of its prime factors, we test the prime numbers in order:

$$
\begin{array}{r}
2\,\overline{)276} \\
2\,\overline{)138} \\
3\,\overline{)\ 69} \\
23\,\overline{)\ 23} \\
1
\end{array}
\qquad
\begin{aligned}
\text{therefore } 276 &= 2 \times 2 \times 3 \times 23 \\
&= 2^2 \times 3 \times 23.
\end{aligned}
$$

A *multiple* of a number has the number as a factor. For example, 6, 9, 12,... are multiples of 3.

A common multiple of two (or more) numbers has each of those numbers as a factor. For example, 24, 48 ... are common multiples of 8 and 6.

EXERCISE 2c

Find the highest number that is a common factor of

1.	18 and 12	**3.**	15 and 20	**5.**	72 and 18
2.	4 and 8	**4.**	36 and 24	**6.**	36 and 48

Find the lowest number that is a multiple of both

7.	3 and 5	**9.**	6 and 15	**11.**	8 and 12
8.	3 and 6	**10.**	4 and 10	**12.**	14 and 21

Express 1764 as a product of its prime factors and hence find $\sqrt{1764}$ without using a calculator.

```
2)1764
2) 882        ∴    1764 = 2 × 2 × 3 × 3 × 7 × 7
3) 441
3) 147                 = (2 × 3 × 7) × (2 × 3 × 7)
7)  49
7)   7        ∴  √1764 = 2 × 3 × 7
     1                 = 42
```

Express 600 as a product of its prime factors and hence find the lowest number by which 600 can be multiplied to make a perfect square.

```
2)600
2)300        ∴    600 = 2 × 2 × 2 × 3 × 5 × 5
2)150        Multiplying by 2 × 3 gives an even number of
3) 75        2s and 3s and 5s as factors.
5) 25        600 × 2 × 3 = 2 × 2 × 2 × 2 × 3 × 3 × 5 × 5
5)  5                    = (2 × 2 × 3 × 5) × (2 × 2 × 3 × 5)
    1
```

∴   multiplying 600 by 6 makes a perfect square.

Express as a product of prime factors:

13. 24   **15.** 588   **17.** 132   **19.** 234

14. 72   **16.** 300   **18.** 78   **20.** 1800

Express each of the following numbers as a product of its prime factors and hence find its square root.

21. 900   **22.** 196   **23.** 3136   **24.** 1521

Express each of the following numbers as a product of its prime factors and hence find the lowest number by which it can be multiplied to give a perfect square.

25. 75   **26.** 162   **27.** 60   **28.** 11 088

FRACTIONS

Do not use a calculator.

Test 2D   The Four Rules

1. $\frac{1}{3} + \frac{3}{4}$ **6.** $1\frac{2}{7} \times \frac{1}{5}$ **11.** $\frac{2}{3}$ of $1\frac{1}{5}$

2. $\frac{1}{3} \times \frac{3}{4}$ **7.** $\frac{3}{4} \times 5$ **12.** $\frac{1}{2} \div \frac{5}{8} + 1\frac{1}{4}$

3. $\frac{3}{4} - \frac{2}{5}$ **8.** $3\frac{1}{8} - 1\frac{3}{5}$ **13.** $\left(\frac{1}{4} + \frac{2}{5}\right) \times \frac{10}{11}$

4. $\frac{3}{4} \div \frac{2}{5}$ **9.** $\frac{5}{12} + \frac{2}{9} \times \frac{5}{6}$ **14.** $\dfrac{\frac{1}{2} + \frac{1}{3}}{\frac{1}{4} - \frac{1}{5}}$

5. $1\frac{2}{7} + \frac{1}{5}$ **10.** $1\frac{1}{3} \div 3\frac{1}{4}$ **15.** $\left(3 - 1\frac{1}{8}\right) \times \left(\frac{1}{2} + 1\frac{1}{3}\right)$

Addition and subtraction   Before fractions can be added or subtracted, they must be expressed with the *same* denominator. When mixed numbers are involved, the whole numbers can be dealt with first, e.g.

$$2\frac{1}{2} - 1\frac{2}{3} = 1\frac{1}{2} - \frac{2}{3} = \frac{9}{6} - \frac{4}{6} = \frac{5}{6}.$$

Alternatively, all mixed fractions can be changed to improper fractions, but this has the disadvantage that the numerators can get rather large.

Multiplication and Division   All mixed numbers must be changed to improper fractions for multiplication and division.

Fractions are multiplied by taking the product of the numerators and the product of the denominators.

Whole numbers can be written as fractions, e.g. $5 = \frac{5}{1}$.

To divide by a fraction, we multiply by its reciprocal.

For example $\frac{1}{2} \div \frac{2}{5} = \frac{1}{2} \times \frac{5}{2}$

$$= \frac{5}{4}$$

Remember to cancel any common factors of numerators and denominators.

For mixed operations, remember that calculations inside brackets must be done first. Sometimes brackets are not there but are implied by the way the calculation is written.

For example, to find $\dfrac{\frac{3}{4}+\frac{1}{9}}{\frac{2}{3}}$, the sum in the numerator must be found first

i.e.

$$\dfrac{\frac{3}{4}+\frac{1}{9}}{\frac{2}{3}} = \left(\tfrac{3}{4}+\tfrac{1}{9}\right) \div \tfrac{2}{3}$$

$$= \left(\tfrac{27+4}{36}\right) \times \tfrac{3}{2}$$

$$= \tfrac{31}{\overset{}{\underset{12}{36}}} \times \tfrac{\cancel{3}^{1}}{2}$$

$$= \tfrac{31}{24} = 1\tfrac{7}{24}$$

EXERCISE 2d

Find:

1. $\frac{5}{8}+\frac{7}{12}$

2. $1\frac{3}{5}+3\frac{1}{7}$

3. $\frac{2}{5}-\frac{1}{4}$

4. $2\frac{1}{3}-1\frac{1}{6}$

5. $\frac{2}{3}\times\frac{9}{10}$

6. $1\frac{1}{2}\times3\frac{1}{3}$

7. $2\frac{5}{11}\div21$

8. $\frac{5}{8}\div1\frac{1}{5}$

9. $\frac{2}{9}-\frac{5}{6}+\frac{3}{4}$

10. $\frac{2}{5}\times\frac{3}{4}\times1\frac{1}{2}$

11. $\left(\frac{3}{4}\div\frac{1}{2}\right)\times1\frac{1}{5}$

12. $1\frac{1}{2}+\frac{2}{5}-1\frac{1}{3}$

13. $\frac{2}{3}\times\left(-\frac{1}{4}\right)$

14. $\frac{2}{5}\div\frac{10}{11}+\frac{4}{25}$

15. $\frac{1}{2}+\frac{3}{4}\times\frac{1}{5}$

16. $\frac{3}{8}$ of $2\frac{1}{2}$

17. $18\times3\frac{2}{9}+4$

18. $27\div2\frac{2}{11}$

19. $\dfrac{\frac{3}{4}-\frac{1}{3}}{\frac{1}{2}\times\frac{1}{4}}$

20. $\dfrac{\frac{1}{4}\times\frac{3}{8}}{\frac{3}{4}\times\frac{1}{12}}$

21. $\dfrac{1\frac{1}{2}+\frac{3}{5}}{\frac{4}{9}-\frac{5}{12}}$

22. $\dfrac{2\frac{1}{2}+\frac{4}{5}}{5\frac{1}{2}}$

23. $\dfrac{\frac{1}{2}}{\frac{3}{4}+\frac{1}{3}}\times\dfrac{\frac{2}{5}}{\frac{1}{3}-\frac{1}{4}}$

24. $\dfrac{1\frac{1}{2}+\frac{1}{8}}{13}\div\dfrac{26}{2\frac{1}{6}\times\frac{3}{4}}$

DECIMALS

Test 2E   The Four Rules

Find, without using a calculator,

1. $1.2 - 0.9$
2. $0.03 + 1.2$
3. $5 - 0.79$
4. $1.3 - 2.9 + 0.21$

5. 1.2×0.6
6. 0.01×1.7
7. 15×0.03
8. 0.02×0.8

9. $15 \div 0.3$
10. $2.7 \div 9$
11. $3.2 \div 0.8$
12. $1.2 \div 0.03$

Test 2F   Calculator Work

Use a calculator and give the answer correct to the number of significant figures given in brackets.

1. 1.275×0.37 (2)
2. 25.9×362 (2)
3. 589×1902 (3)

4. $270 \div 15.2$ (3)
5. $0.098 \div 1.39$ (3)
6. $\dfrac{0.25 + 1.092}{0.36 \times 1.33}$ (2)

7. 0.1291×1.992 (4)
8. $3.8752 \div 0.77$ (4)
9. $\dfrac{1.37 - 0.0025}{5.89}$ (3)

The four rules without a calculator When adding or subtracting decimals, make sure that tenths are added to tenths etc.

To multiply decimals, first ignore the decimal point and multiply the numbers. Then the number of decimal places in the answer is the sum of the number of decimal places in the original numbers.

To divide by a decimal, that decimal and the number it is being divided into must both be multiplied by the power of the ten that makes the divisor into a whole number. For example $8.1 \div 0.09 = 810 \div 9 = 90$.

EXERCISE 2e

Find, without using a calculator,

1. $1.8 \div 0.03$
2. 0.27×0.02
3. 0.8×0.08
4. $3.7 - 2.91$

5. $0.15 + 1.273$
6. $16 \div 0.02$
7. $(1.2)^2$
8. $9 - 12.4$

9. $12 + 0.58$
10. $(0.5)^3$
11. $1.72 - 2.45$
12. $0.53 - 0.072$

13. $\dfrac{2.5 \times 0.3}{0.15 \times 5}$ **15.** $\dfrac{2.7 - 1.9}{4 \times 0.2}$ **17.** $(1.8 + 3.6) \div 80$

14. $\dfrac{1.9 + 0.1}{4.1 - 3.6}$ **16.** $\dfrac{5.3 + 0.7}{0.3 \times 0.4}$ **18.** $(6.9 \div 115) \times 15$

Significant figures and decimal places   Many calculations do not have exact answers, and even if they do it is not always necessary to give the exact value. It is usually sufficient to give values correct to the nearest tenth, hundredth, or whatever is required by the context of a problem.

Most problems specify the degree of accuracy required, as, for example, correct to 3 significant figures or correct to 1 decimal place.

Decimal places are easy to identify; the first decimal place is the first digit to the right of the point, and so on. The first significant figure in a number is the first *non-zero* digit reading from left to right.

For example, in both 26.4 and 0.0264 the first significant figure is 2.

The second significant figure is the digit (including zero this time) on the right of the first, and so on for the third and fourth significant figures etc.

For example, 0 is the second significant figure in 2.05 and the third significant figure is 5.

To correct a number to a given number of figures, look at the next figure: if this is 5 or more, add 1 to the previous figure (i.e. round up); if this is less than 5, leave the previous figure alone (i.e. round down).

For example,   3.897 is 3.90 correct to 3 s.f.
and   3.894 is 3.89 correct to 3 s.f.

Using a calculator   It is important to have a rough idea of the size of an answer before using a calculator. This can be found by approximating each number in the calculation to one significant figure,

e.g. $\dfrac{1.802 + 9.75}{22.3} \approx \dfrac{2 + 10}{20} \approx \dfrac{10}{20} = 0.5$

Remember that it is not necessary to write down all the figures in the display. If an answer is required *correct* to three significant figures, then five significant figures can be written down for any intermediate steps and four significant figures only need be written down before rounding up or down.

EXERCISE 2f

Find, without using a calculator, an approximation (to 1 s.f.) of each of the following.

1. 1.752×93.64

2. $\dfrac{0.85 \times 3.24}{0.00165}$

3. $(0.0735)^2$

4. $(0.492)^3$

5. $\dfrac{3.82 \times 9.75}{12.6}$

6. $\dfrac{0.19 + 0.0752}{0.00029}$

Use a calculator to evaluate the following expressions, correcting your answers to three significant figures,

7. $0.0086 \div 0.0000925$

8. 3500×0.00101

9. $(0.792)^2$

10. $1.827 \times 3.6 - 1.325$

11. $(0.0185)^3$

12. $(0.527 \times 18) \div 52$

13. Use a calculator to give the values of questions 1 to 6 correct to 2 decimal places.

PERCENTAGES

Test 2G Percentages, Fractions and Decimals

1. Express as percentages

a) $\frac{6}{25}$ b) 0.29   c) $1\frac{3}{4}$ d) 0.015

2. Express as fractions

a) 42%   b) 0.28   c) $2\frac{1}{2}$%   d) 1.2

3. Express as decimals

a) 12%   b) 115%   c) $6\frac{2}{5}$ d) $\frac{9}{20}$

4. Find 15% of £380.

5. Find $\frac{2}{9}$ of 153 m³.

6. In a school with 640 pupils, $\frac{3}{8}$ are boys. How many girls are there ?

7. When discs are cut from a sheet of card, 28% of the card is wasted. What area is used from a sheet of 400 cm² ?

8. A boy ate 8 sweets from a bag of 20 sweets. What fraction of the sweets did he leave ?

9. A girl gets 26 out of 40 in a test. What is her percentage mark ?

Test 2H   Increase and Decrease

1. A standard pack of detergent holds 250 g. A promotional pack contains an extra 20 % by weight. What is the weight in the promotional pack ?

2. An uncooked chicken weighs 5 lb. It loses 12 % of its weight during cooking. How much does it weigh when cooked ?

3. A curtain was 3.5 m long when new. After washing it was 2.8 m long. Find the percentage change in length.

4. After a price increase of 20 %, a unit of electricity costs 6 p. Find the cost before the increase.

5. After public examinations in June, the number of pupils attending a school falls by 20 % to 640. How many pupils attend before the examinations ?

FRACTIONS, DECIMALS AND PERCENTAGES

To express a fraction as a decimal, divide the denominator into the numerator,

e.g. $\frac{7}{8} = 7 \div 8 = 0.875$

'Per cent' means 'per hundred', so 12 % means 12 per 100. Hence to express a percentage as a fraction or a decimal, divide the percentage by 100,

e.g. $12\% = \dfrac{12}{100}$

$= 0.12$

To express a fraction or a decimal as a percentage, the operation is reversed, i.e. multiply by 100,

e.g. $\frac{2}{9} = \frac{2}{9} \times 100\% = \dfrac{200}{9}\% = 22\frac{2}{9}\%$

and $0.136 = 0.136 \times 100\% = 13.6\%$

Finding a percentage of a given quantity is therefore the same as finding the equivalent fraction of that quantity,

e.g. $12\% \text{ of } £40 = \dfrac{12}{100} \text{ of } £40$

$= £\dfrac{12 \times 40}{100} = £4.80$

EXERCISE 2g

Give all answers that are not exact correct to 3 s.f.

1. Express the following as fractions.

a) 30 %   b) 0.82   c) 130 %   d) 1.25

2. Express the following as percentages.

a) 0.93   b) 0.095   c) $\frac{3}{8}$ d) $\frac{7}{15}$

3. Express the following as decimals.

a) 54 %   b) $2\frac{3}{5}$ c) 138 %   d) $\frac{4}{11}$

4. Arrange the following in order of size with the smallest first.

$$\frac{2}{7}, \qquad 28\%, \qquad 0.3, \qquad \frac{1}{3}$$

5. Arrange the following in order of size with the largest first.

$$\frac{5}{11}, \qquad 55\%, \qquad \frac{4}{7}, \qquad 0.48$$

6. Find   a) 22 % of £ 10   b) 140 % of 500 g.

7. Find   a) $\frac{3}{8}$ of £ 24   b) $\frac{5}{9}$ of 63 miles.

8. Find   a) $12\frac{1}{2}$ % of 65 m   b) 105 % of £ 52.

9. Find   a) $\frac{4}{11}$ of 132 litres   b) $\frac{5}{8}$ of 0.96 m.

10. Express 20 out of 65 as   a) a fraction   b) a percentage.

11. Ten of a box of 25 oranges were bad. What percentage of the oranges were bad ?

12. In a class of 40 children, 5 % were away from school. How many were at school ?

13. Peter got 15 out of 26 for a test. What percentage did he get ?

14. For a particular football match, 23 % of the available 3700 tickets were unsold. How many tickets were sold ?

15. A public library lost 260 books from a total of 6500 books. What percentage of the books were lost ?

PERCENTAGE CHANGE

When a quantity is increased or decreased by a given percentage, the change is always calculated as a percentage of the original quantity (i.e. the quantity before the increase or decrease).

Thus, if £20 is increased by 2%

then   the increase is 2% of £20

and   the increased sum is (100% + 2%) of £20

$$= 102\% \text{ of } £20$$

If 20 m is increased to 25 m,

then   the change is an increase of 5 m

and   the percentage increase is 5 m as a percentage of 20 m,

i.e. $\dfrac{5}{20} \times 100\% = 25\%$

Similarly if £450 is decreased by 15%

then   the decrease is 15% of £450

and   the decreased quantity is (100% − 15%) of £450

$$= 85\% \text{ of } £450$$

When a quantity is known after a percentage change has taken place, then

the changed quantity = (100% ± % change) of the original

Thus if a sum of money is increased by 15% to £260

then   £260 = 115% of sum before the increase

i.e. $£260 = \dfrac{115}{100} \times (\text{original sum})$

Similarly if a marked price is decreased by 25% to £60

then   £60 = 75% of marked price

i.e. $£60 = \dfrac{75}{100} \times (\text{marked price})$

EXERCISE 2h

The bus fare from Petts Corner to the Old Oak was increased by 5% to 42 p. Find the fare before the increase.

Let the original fare be x pence,

then $\qquad\qquad\qquad$ 42 p $= 105\%$ of x pence

i.e. $\qquad\qquad\qquad 42 = \dfrac{105}{100} \times x$

giving $\qquad\qquad\qquad \dfrac{\overset{2}{\cancel{42}} \times \overset{20}{\cancel{100}}}{\underset{21}{\cancel{105}}} = x$

$\qquad\qquad\qquad\qquad 40 = x$

The fare before the increase was 40 p.

---

1. Increase £ 80 by $\quad$ a) 10% $\qquad$ b) 15% $\qquad$ c) $7\frac{1}{2}\%$

2. Decrease £ 150 by $\quad$ a) 7% $\qquad$ b) 30% $\qquad$ c) $12\frac{1}{2}\%$

3. Find the percentage change when
a) £ 80 increases to £ 100
b) 20 m decreases to 16 m
c) 5 m increases to 5.5 m
d) 16 litres decreases to 14 litres.

4. After an increase of 8%, a car costs £ 6048. Find the cost before the increase.

5. A builders' merchant marks prices net of VAT. Find the price to be paid for an item marked at £ 36 if VAT is charged at 15%.

6. A DIY shop marks prices inclusive of VAT at 15%. Find the price, exclusive of VAT, of a tin of paint marked at £ 5.20.

7. The label on a roll of fabric says "allow for 5% shrinkage". A dress pattern requires 2.5 m of fabric. How much should be bought to allow for shrinkage ?

8. A chicken weighs $4\frac{1}{2}$ lb before cooking and 4 lbs after cooking. Find the percentage change in its weight.

The questions in the remainder of this exercise are mixed percentage problems.

9. A house bought for £ 87 000 is sold at a profit of 15%. Find the selling price of the house.

10. In a box containing two dozen eggs, it is found that nine are broken. What percentage of the eggs are broken ?

11. A girl is 8% taller this year than she was last year. Her height is now 175 cm. How tall was she last year ?

12. In a sale a jacket was marked down from £58.00 to £43.50. What percentage reduction was this ?

13. In an examination a boy got a mark of 49 and was told this was 70%. What was the maximum possible mark ?

14. Mr Evans paid £448 income tax in one year. The tax was charged at a standard rate of 32%. Find Mr Evans' taxable income for that year.

15. If bread dough loses 8% of its weight when cooked, find the cooked weight of a loaf made from 500 g of dough.

16. It is found that 12% of the wheat crop from a field is not saleable. How much is saleable from a crop of 750 tons ?

17. After a wage increase of $6\frac{1}{2}$%, Brenda Adstore's weekly pay was £159.75. What was her weekly pay before the increase ?

18. Maya Jacobs invested £750 in a deposit account which paid interest at 6.5% p.a. She did not withdraw the interest. How much was in the account after two years ?

19. A machine automatically cuts rolls of paper into short lengths. One day it is incorrectly set so that it cuts pieces 31.5 cm long instead of 30 cm long. What is the percentage error on

a) one piece of paper,   b) 1000 pieces of paper ?

RATIO

Test 2I

1. One square has sides of length 5 metres and a second square has sides of length 7.5 metres. What is the ratio of these lengths ?

2. Express the ratio $2:7$ in the form $1:n$.

3. A map ratio is $1:100\,000$. How long is a road which on the map measures 2.5 cm ?

4. Simplify the ratio $50:125$.

5. Divide £36 into two parts in the ratio $7:2$.

6. Divide a length of 48 m into three parts in the ratio 1 : 3 : 4.

7. The lengths of two lines are in the ratio 3 : 5. The shorter line is 18 cm long. How long is the other line ?

RATIO

Ratio is a way of comparing sizes.

For example, if one line is 8 cm long and another line is 12 cm long, then

the length of the shorter line compared with the length of the longer line is 8 : 12

But 8 : 12 is the same as 2 : 3,

i.e.   for every 2 units of length in the shorter line, there are 3 units of length in the longer line.

Thus a ratio can be simplified (by dividing or by multiplying both parts of the ratio by the same number).

Ratios can also be expressed in the form 1 : n by performing the appropriate divisions,

i.e. $2 : 3 = 1 : \frac{3}{2} = 1 : 1.5$

Map ratios are usually given in this form, i.e. 1 : n.

If the ratio of two quantities is known and the size of one quantity is known, the size of the other can be found.

For example, if two sums of money are in the ratio 3 : 5 and the smaller sum is £1.80, then for every 3 portions in the smaller sum there are 5 portions in the larger sum.

i.e.   the larger sum $= £\dfrac{1.80 \times 5}{3} = £3.00$.

Similarly, if we have a total sum of money, say £10, and need to divide it into three amounts in the ratio 2 : 7 : 11 then for every 2 portions in the smallest sum there are 7 portions in the middle sum and 11 portions in the larger sum.

Hence we divide £10 into (2 + 7 + 11) portions.

Each portion is $£10 \times \frac{1}{20}$.

Therefore the three sums are

$$£10 \times \tfrac{2}{20}, \quad £10 \times \tfrac{7}{20}, \quad £10 \times \tfrac{11}{20}$$

i.e.   £1,   £3.50,   £5.50

EXERCISE 2i

Simplify:

1. $400 : 75$ **2.** $5 : \frac{2}{3}$ **3.** $14 : 28 : 49$

Find in the form $1 : n$

4. $4 : 7$ **5.** $8 : 25$ **6.** $3 : 12\,000$

Find the ratio of the two given quantities:

7. £2, £3.50   **8.** 2 cm, 1 m   **9.** $1\,cm^2 : 1\,m^2$

Find, in the form $1 : n$, the ratio of the given quantities:

10. 1 cm, 2 km   **11.** 2 mm, 5 m   **12.** 3 inches, 2 yards

13. Two lines have lengths in the ratio $2 : 5$. If the longer line is 15 cm long, find the length of the other line.

14. A road on a map measures 2.5 cm. The actual road is 1 km long. Find the map ratio in the form $1 : n$.

15. A map ratio is $1 : 100\,000$. Find the length of a road which on the map measures 2.6 cm.

16. Divide £45 into 2 parts in the ratio $4 : 5$.

17. Divide £56 into 3 parts in the ratio $2 : 3 : 3$.

18. A design team makes a model of a table, using a scale of $1 : 20$. The actual table is intended to be 1 metre long. How long is the model?

19. A school decides to give 20 % of the proceeds of a jumble sale to charity and the rest to the school fund. In what ratio are the proceeds to be divided?

20. A path on a map measures 2.5 cm. The actual path is 650 m long. Find the map scale in the form $1 : n$.

MIXED EXERCISES

Use your calculator where necessary and give all answers that are not exact correct to 3 s.f.

EXERCISE 2j

1. Ice cubes are made by freezing water in cuboid containers so that the frozen ice cube measures 2 cm by 3 cm by 2.5 cm. The water increases in volume by 5% when frozen.

a) What volume of water is needed to make one ice cube ?

b) How many ice cubes can be made from 1 litre of water ?

2. Three people stake £10 on a lottery and win £850. Peter paid £2, John paid £4.50 and Clare paid the rest towards the stake. The winnings are shared in the ratio of the contributions. How much does Clare get ?

3. What is the largest odd number that is a factor of 860 ?

4. Find the exact value of $\dfrac{15.2 \times 8.6}{0.05}$

5. From the set of numbers $\{1, 4, 5, 7, 9, 12, 15\}$ write down

a) the prime numbers

b) multiples of 3

c) factors of 60

d) perfect squares (square numbers).

6. A school fête raised £4800 and one quarter of this sum was paid out for expenses. Three-quarters of the remainder was given to the school fund and the rest was donated to a local charity. How much did the charity receive ?

EXERCISE 2k

1. Express 441 as a product of its prime factors and hence find $\sqrt{441}$ without using a calculator.

2. A motorist is asked to pay a net premium of £280.50 for car insurance. This is after a discount of 45% on the gross premium. Find the gross premium.

3. A bottle of concentrated orange drink holds $32\frac{1}{2}$ fluid ounces. The recommended dilution is one capful to a glass of water. The cap holds $\frac{1}{3}$ gill and a gill is taken as 5 fluid ounces. How many glasses of orange drink can be obtained from one bottle ?

4. Find the exact value of $\dfrac{2\frac{1}{5} - 1\frac{3}{4}}{1\frac{1}{2} \times 2\frac{8}{11}}$

5. When Lord Worth died his estate was valued at £150 000. His will stated that 20% of his estate was to be given to charity and 5% of what was left was to be given to his housekeeper. The remainder was to be given to his only son. How much did his son receive ?

6. A road on a map is 2.5 cm long while the actual road is 4.8 km long. Find the map ratio in the form $1 : n$.

EXERCISE 2I

1. Find
   a) the lowest number that is a common multiple of 9 and 12
   b) the highest number that is a common factor of 72 and 48
   c) the lowest number which when multiplied by 12 makes it into a perfect square.

2. The Red Brick Building Society states that it will give individuals a maximum mortgage of $2\frac{1}{2}$ times gross income. In certain circumstances it will increase the amount by up to $1\frac{1}{4}$ times the partner's income. The maximum mortgage the society will give on any property is 85 % of its own valuation.

   Mr and Mrs Hope see a house which the society values at £68 000. Mr Hope has a gross income of £15 000 and his wife has an income of £4500.

   Find the difference between the maximum mortgage available on the house and the maximum mortgage the society will give Mr and Mrs Hope.

3. Don Lucky won £50 at bingo and decided to give it to his three children, in amounts that are in the same ratio as their ages. The children were 10, 12 and 18 years old respectively. How much did they each get ?

4. Tom Peterson invested £500 in a deposit account which paid interest at 7.25 % p.a. He left the interest in the account. How much was in the account at the end of two years ?

5. Find the exact value of $2\frac{5}{11} \times 1\frac{2}{9} + \frac{5}{6}$

6. Mr Ford's gross motor insurance premium this year is £280. This represents an increase of 5 % on last year's gross premium. In calculating the net premium that Mr Ford had to pay last year, a 30 % no claims bonus was given. Find the net premium paid last year.

EXERCISE 2m

Do not use a calculator.

Each question is followed by several possible answers. Write down the letter that corresponds to the correct answer.

1. Correct to 3 s.f., the value of $\dfrac{1.87 \times 0.0073}{0.019}$ is

   A 7.18   **B** 0.0718   **C** 0.718   **D** 71.8

2. The third significant figure in 0.001 020 5 is
   A 0   **B** 2   **C** 5   **D** 1

3. The highest factor of 60 apart from itself is
   A 2   **B** 5   **C** 10   **D** 30

4. In a sale, a store offers 15% off marked prices. When the marked price is £25, the sale price is

A £3.75   **B** £21.25   **C** £25   **D** £28.75

5. A common multiple of 6 and 9 is

A 18   **B** 3   **C** 15   **D** 12

6. The exact value of $(0.105)^2$ is

A 0.110 25   **B** 0.011 025   **C** 11 025   **D** 0.011

7. Two-thirds of a class of children had school dinners. The remaining 8 children went home for lunch. The number who stayed for school dinner was

A 16   **B** 12   **C** 24   **D** 4

8. Correct to three significant figures, 87 059.3479 is

A 87 059.348   **B** 87 100   **C** 87 060   **D** 871

9. A good approximation for $\dfrac{89.37 \times 3.05}{0.92}$ is

A 300   **B** 30   **C** 0.3   **D** 3000

10. The value of 199×53 is

A 5353   **B** 10 653   **C** 10 547   **D** 5247

3 THE STANDARD ALGEBRAIC PROCESSES

In this chapter we revise the standard processes and methods of solving equations.

SIMPLIFYING EXPRESSIONS

Reminder x^2, $2x$ and 6 are *unlike* terms, so $x^2 + 2x + 6$ cannot be written more simply. $3x^2$ and $4x^2$ are *like* terms, so $3x^2 + 4x^2$ can be written as $7x^2$.

EXERCISE 3a

Simplify the following expressions.

1. $7x + 2 - 9x - 5 + 5x^2$

2. $2x \times 3x$

3. $x \times x \times x$

4. $x + x + x$

5. $2x^2 - 3x - x^2$

6. $3a - 7b - 4a - 5c + 10b + 6c$

7. $6x \times 7y$

8. $x \times x - x$

9. $x^2 \times x$

10. $x^2 \div x$

Expand and simplify, where possible, the following expressions.

11. $6(x - 2) + 4$

12. $3(2a + 2b) - 2(b - 4)$

13. $4(3x - 1) - (2x - 3)$

14. $x(x - 2) - 2(x + 4)$

15. $c(a - b) + b(c - a)$

16. $3(2x + 4) + x(3 - 4x)$

Expand the following expressions.

17. $(x + 3)(x + 4)$

18. $(x - 3)(x + 4)$

19. $(x - 5)(x + 5)$

20. $(x - 5)^2$

21. $(2x + 3)(x - 4)$

22. $(a - b)(c - d)$

23. $(2x + 5)^2$

24. $(1 - 2x)(1 + 3x)$

25. $(x + y)(x - 2y)$

26. $(2 - x)(3 + x)$

Simplify:

27. $(2x-3)^2 - 4x(x-4)$ **30.** $(x+3)^2 + (x-1)^2$

28. $(a+b)^2 - 2ab$ **31.** $4-(x-2)^2$

29. $4(x-2)(x+1)$ **32.** $6(2x-1)(3-x)$

FACTORS

The factors of 10 are 2 and 5, i.e. $2 \times 5 = 10$. We need to be aware of this when we are, say, simplifying fractions.

In the same way we often need to know the factors of algebraic expressions in order to simplify the expressions and solve problems.

$3x^2 - 6x$ is made up of the factors $3, x$ and $(x-2)$ i.e. $3x^2 - 6x = 3x(x-2)$.

When you have decided on the factors, you should check, by multiplying out, that you have the correct ones.

Always start by looking for any common factors.

EXERCISE 3b

Factorise $6x^2 - 15x$.

$$6x^2 - 15x = 3x(2x-5)$$

Factorise the following expressions.

1. $2x-4$ **5.** $x^2y + xy^2$

2. $6a - 9c$ **6.** $x^2 - x$

3. $2x^2 - 6x$ **7.** $4x + 20y$

4. $12a + 18b - 24c$ **8.** $9x^2y + 18xy^2$

Factorise $3ax - 6ay - 4bx + 8by$.

$$3ax - 6ay - 4bx + 8by = 3a(x-2y) - 4b(x-2y)$$
$$= (x-2y)(3a - 4b)$$

Factorise the following four-term expressions.

9. $ax + bx + ay + by$

10. $pq - qr + ps - rs$

11. $2mn + pq - pn - 2mq$

12. $x^3 + x^2 + x + 1$

13. $4uv + 3v - 12u - 9$

14. $ab + 4xy - 2bx - 2ay$

15. $ax^2 + a^2x - ay - xy$

16. $ab + 6cd - 2bc - 3ad$

FACTORS OF QUADRATIC EXPRESSIONS

Quadratic expressions will sometimes factorise into two brackets. Remember that the x term in each bracket must multiply to give the x^2 term and the product of the number terms (including the signs) from the brackets gives the number term in the quadratic expression.

Always check your factors by multiplying out

EXERCISE 3c

Factorise   a) $x^2 - 10x + 16$ b) $x^2 - 16$

a) $x^2 - 10x + 16 = (x - 2)(x - 8)$ $-2x - 8x = -10x$

b) ($x^2 - 16$ is the difference of two squares)

$x^2 - 16 = (x - 4)(x + 4)$

Factorise the following expressions.

1. $x^2 + 12x + 32$

2. $x^2 - 7x + 12$

3. $x^2 - 9$

4. $x^2 - 6x + 9$

5. $49 + a^2 - 14a$

6. $y^2 - 18 + 7y$

7. $x^2 + 2x - 15$

8. $x^2 - 4x - 77$

9. $30 + 11x + x^2$

10. $x^2 - 9x$

Factorise $5x^2 - 15x - 20$.

(Look for the common factor first.)

$$5x^2 - 15x - 20 = 5(x^2 - 3x - 4)$$
$$= 5(x - 4)(x + 1) \qquad -4x + 1x = -3x$$

Factorise the following expressions by taking out the common factor first, then factorising the remaining expression if possible.

11. $3x^2 + 6x + 3$

12. $4x^2 + 8x - 12$

13. $9y^2 - 36$

14. $5y^2 - 40y + 35$

15. $12x^2 - 18x$

16. $2b^2 - 14b + 20$

17. $8x^2 + 8x - 48$

18. $3x^2 - 48$

Factorise $2x^2 - x - 1$.

(There is no common factor.)

$$2x^2 - x - 1 = (2x + 1)(x - 1) \qquad +x - 2x = -x$$

Factorise the following expressions.

19. $2x^2 + 7x + 3$

20. $3x^2 + 11x - 4$

21. $9x^2 - 4$

22. $6x^2 + 11x + 3$

23. $12x^2 - 29x + 15$

24. $5x^2 + 3x - 2$

25. $4x^2 + 4x - 3$

26. $12x^2 - 28x + 15$

27. $12x^2 - 41x + 15$

28. $25x^2 - 9$

29. $x^2 - 25x$

30. $25 - x^2$

31. $x^2 - 10x + 25$

32. $x^2 - 26x + 25$

33. $x^2 + 5ax + 5x + 25a$

34. $1 - x - 12x^2$

35. $10x^2 - 25x + 15$

36. $5x^2 + 50x + 125$

> Factorise $2x^2 - 18$ and hence find the prime factors of 182.
>
> $$2x^2 - 18 = 2(x^2 - 9)$$
> $$= 2(x - 3)(x + 3)$$
> $$182 = 2(91)$$
> $$= 2(100 - 9)$$
> $$= 2(10^2 - 3^2)$$
> $$= 2(10 - 3)(10 + 3)$$
> $$= 2 \times 7 \times 13$$

37. Factorise $101^2 - 99^2$ and hence work out its value.

38. Factorise $4x^2 - 9$ and hence find the prime factors of 391.

39. Factorise $3x^2 - 48$ and hence find the prime factors of 252.

40. Given that $8.12^2 - 1.88^2 = 10k$, find k.

41. Find the value of $7.86^2 - 2.14^2$.

42. By expressing 899 as the difference between two perfect squares, find the prime factors of 899.

FINDING THE VALUE OF AN EXPRESSION

Remember that if $y^2 = 9$ then y can have two possible values i.e. 3 or -3.

We write $y = \pm 3$.

On the other hand, $y = \sqrt{9}$ means the *positive* square root of 9 i.e. $+3$.

EXERCISE 3d

Find the value of each of the following expressions, given that $a = 6$, $b = -3$ and $c = 2$.

1. $2a + 3b + 4c$

2. $a^2 + b^2$

3. $2a^2$

4. $\sqrt{a + b + 3c}$

5. $\dfrac{a}{b + c}$

6. $(a + 3c)(b + 2c)$

7. $\dfrac{c}{a} + \dfrac{a}{c}$ **9.** $\dfrac{1}{a} + \dfrac{1}{b}$ **11.** $\sqrt{a^2 + b^2 + c^2}$

8. $\dfrac{ab}{c}$ **10.** $\sqrt{a - c}$ **12.** $a + bc$

13. Given that $z^2 = x^2 + y^2$, find the two possible values of z if
  a) $x = 5$ and $y = 12$
  b) $x = \sqrt{2}$ and $y = \sqrt{7}$

14. Given that $a = 3$, $b = -4$ and $c = -1$, evaluate
  a) $2a^2$ b) $a - c^2$ c) $a(b - c)$

15. If $y^2 = 4a(x + 5)$ find the two possible values of y when $a = \frac{1}{25}$ and $x = 4$.

INDICES

x^4 means $x \times x \times x \times x$

x^{-4} means the reciprocal of x^4, i.e. $\dfrac{1}{x^4}$

x^0 means 1

$x^{1/4}$ means the fourth root of x

MULTIPLICATION AND DIVISION

To *multiply* numbers in the same base we *add* the indices.

For example, $x^4 \times x^3 = x^7$

To *divide* numbers in the same base we *subtract* the indices.

For example, $x^6 \div x^4 = x^2$

EXERCISE 3e

If $x = 2$, $y = 3$ and $z = 5$, find the value of each of the following expressions.

1. x^5 **3.** $5x$ **5.** $\dfrac{1}{z^2}$ **7.** xy^2

2. z^0 **4.** y^3 **6.** y^{-2} **8.** $\left(\dfrac{y}{z}\right)^{-1}$

Complete the following statements.

9. $64 = 2^{\square}$ **10.** $125 = \square^3$ **11.** $4^{\square} = 1$ **12.** $\frac{1}{4} = 2^{\square}$

Simplify the following expressions where possible.

13. $a^5 \times a^2$ **16.** $a^5 \times a^{-2}$ **19.** $a^5 \div a^{-2}$

14. $a^5 \div a^2$ **17.** $a^5 + a^2$ **20.** $a^5 \div a^5$

15. $(a^5)^2$ **18.** $5a + 2a$ **21.** $(a^2)^{-5}$

Find the value of each of the following expressions.

22. $36^{1/2}$ **24.** $125^{1/3}$ **26.** $\left(\dfrac{9}{4}\right)^{1/2}$

23. $16^{1/4}$ **25.** $\left(\dfrac{27}{8}\right)^{1/3}$ **27.** $27^{1/3}$

Simplify the following expressions where possible.

28. $x^{1/2} \times x^{1/2}$ **32.** $6x^2 \times 3x^4$ **36.** $(x^3)^{1/3}$

29. $x^{1/3} \div x^{1/3}$ **33.** $\dfrac{x^2 \times x^4}{x^3 \times x}$ **37.** $a^{1/2} \times b^{1/3}$

30. $x^0 \times x^{1/4}$ **34.** $\dfrac{x^3 \times x^{-2}}{x^{1/2}}$ **38.** $x^{1/2} \div x^{1/4}$

31. $x^{1/2} + x^{1/3}$ **35.** $(x^{1/2})^4$ **39.** $x^{1/2} \times x^{1/2} \times x^{1/2} \times x^{1/2}$

Solve the equation $3^x = 9^3$.

(Before we can compare indices, the base must be the same on both sides.)

$$3^x = 9^3$$
$$3^x = (3^2)^3$$
$$3^x = 3^6$$
$$\therefore \quad x = 6$$

Find the value of x in each of the following equations.

40. $3^x = 81$ **42.** $5^x = 25^3$ **44.** $4^x = 2^6$

41. $2^x = 4^2$ **43.** $(2^x)^2 = 16$ **45.** $2^x \times 3^x = 6^3$

EXERCISE 3f

Evaluate   a) $27^{2/3}$ b) $4^{-1/2}$ c) $\left(\dfrac{81}{16}\right)^{-1/4}$

a) $(27^{2/3}) = (27^{1/3})^2$ b) $4^{-1/2} = \dfrac{1}{4^{1/2}}$

$\qquad\qquad = 3^2$ $= \dfrac{1}{2}$

$\qquad\qquad = 9$

c) $\left(\dfrac{81}{16}\right)^{-1/4} = \left(\dfrac{16}{81}\right)^{1/4} = \dfrac{2}{3}$

Find the value of each of the following expressions.

1. $4^{3/2}$ **4.** $8^{-1/3}$ **7.** $\left(\dfrac{4}{9}\right)^{-1/2}$

2. $81^{3/4}$ **5.** $125^{2/3}$ **8.** $1000^{2/3}$

3. $1^{1/3}$ **6.** $27^{4/3}$ **9.** $1000^{-4/3}$

Simplify the following expressions where possible.

10. $x^{3/2} \times x^{1/2}$ **13.** $x^{3/2} \times x^{-1/2}$ **16.** $\dfrac{1}{x^{-1/2}}$

11. $x^{3/2} + x^{1/2}$ **14.** $x^{3/2} \div x^{-1/2}$ **17.** $x \div x^{1/2}$

12. $x^{3/2} \div x^{1/2}$ **15.** $x^{-1/2} \div x^{3/2}$ **18.** $x \times x^{1/2}$

19. If $a = 2 \times 10^4$ and $b = 3 \times 10^6$, find the value of each of the following expressions, giving your answers in standard form.

a) ab b) $\dfrac{b}{a}$ c) $a + b$ d) $b - a$

20. Given that $x = 2.4 \times 10^{-3}$ and $y = 1.6 \times 10^{-4}$, find in standard form, the value of

a) xy b) $\dfrac{x}{y}$ c) $x + y$ d) $x - y$

EQUATIONS

When dealing with the simpler equations remember to tidy them first; then decide which side has the most xs.

EXERCISE 3g

Solve the following equations.

1. $2x + 4 = 8$ **4.** $7 - 2x = 5$

2. $6x - 3 = 7$ **5.** $5 - 6x = 2 - 2x$

3. $3x - 4 = 6 + x$ **6.** $4x - 8 = 2 - 6x$

7. $5x - 2 + 6x + 7 = 38$ **10.** $3(x - 1) + 5(x - 2) = x + 1$

8. $3(2 + y) = -18$ **11.** $x - 3 - 6(x - 1) = 2(6 - x)$

9. $5(2x - 1) = 4(4 - x)$ **12.** $7 = x - (4 - x)$

13. I think of a number, double it and add 7. I take the same number and subtract it from 19. The two answers are the same. What is the number?

14. John is x years old now and his father is three times as old. In 12 years time, his father will be twice as old as John will be. Find x.

15. Two angles of a triangle are $x°$ and $2x°$ and the third angle is $42°$ more than the first angle. Find the sizes of the three angles.

SIMULTANEOUS EQUATIONS

Elimination method   Remember that if the signs are different we add the equations, and if the signs are the same we subtract.

EXERCISE 3h

Solve the equations $x + y = 10$
$$x - y = 6$$

$$x + y = 10 \qquad [1]$$
$$x - y = 6 \qquad [2]$$

[1] + [2] gives $2x = 16$

$$x = 8$$

In [1] $8 + y = 10$

$$y = 2$$

Check in [2] $x - y = 8 - 2 = 6$

The solution is $x = 8,\ y = 2$.

Solve the following pairs of equations.

1. $2x + y = 5$
$3x - y = 5$

4. $3x - 2y = 2$
$5x + 2y = -2$

2. $4x + y = 0$
$2x + y = 2$

5. $x - y = 7$
$3x - y = 31$

3. $x + 3y = -7$
$x + y = -1$

6. $4x - y = 8$
$6x - y = 0$

Sometimes one, or both, of the equations must be changed before adding or subtracting.

Solve the equations $6x - 5y = 8$
$$4x - 3y = 6$$

$$6x - 5y = 8 \qquad [1]$$
$$4x - 3y = 6 \qquad [2]$$

[1] × 3 gives $18x - 15y = 24 \qquad [3]$

[2] × 5 gives $20x - 15y = 30 \qquad [4]$
$$18x - 15y = 24 \qquad [3]$$

[4] − [3] gives $2x = 6$
$$x = 3$$

In [1] $18 - 5y = 8$
$$18 = 8 + 5y$$
$$10 = 5y$$
$$y = 2$$

Check in [2] $4x - 3y = 12 - 6 = 6$

The solution is $x = 3, y = 2$.

Solve the following pairs of equations.

7. $4x + 6y = 50$
   $3x + 2y = 30$

10. $3x - 5y = 13$
   $2x - 3y = 8$

8. $2x + y = 7$
   $5x - 3y = 12$

11. $x + 5y = 9$
   $2x + 3y = 11$

9. $6x + 7y = 12$
   $4x + 3y = 18$

12. $5x + 11y = -2$
   $2x + 9y = 13$

In questions 13 to 18, some of the equations need rearranging first.

13. $x = 3 + y$

$x + y = 6$

16. $x + 3 = 2x + y$

$4x - y = 7$

14. $2y + x = 1$

$2x + y = -1$

17. $2x - y = -8$

$2y - x = 10$

15. $9 = 2x - 3y$

$3y + x = 9$

18. $4 + x = 9 + y$

$4 - x = 3 + y$

In each of the following questions form two equations. Remember to state first the meaning of the two letters that you use.

19. The sum of two numbers is 25 and the difference between them is 13. What are the numbers ?

20. In a two-digit number the sum of the digits is 9 and the units digit is 5 more than the tens digit. What is the number ?

21. The perimeter of a rectangle is 52 cm. The difference between the length and the width is 8 cm. Find the length and the width.

22. Three apples and five oranges cost 85 p; five apples and three oranges cost 83 p. How much does one apple cost ?

QUADRATIC EQUATIONS

Some quadratic equations can be solved by factorising. If this is not possible then we can use the method of completing the square or we can use the formula.

For the equation $ax^2 + bx + c = 0$

$$x = \frac{-b \pm \sqrt{b^2 - 4ac}}{2a}$$

Whether factorisation or the formula is used it is necessary to arrange the equation with the terms in the correct order on one side and 0 on the other.

For example, $2x - 3 = 4 - x^2$ is rearranged as $x^2 + 2x - 7 = 0$.

Remember that   the sum of the roots is $\dfrac{-b}{a}$

the product of the roots is $\dfrac{c}{a}$

EXERCISE 3i

Solve the equation $(x+4)(x-1) = 0$.

$$(x+4)(x-1) = 0$$

Either $\qquad x+4 = 0 \qquad$ or $\qquad x-1 = 0$

$$x = -4 \qquad\qquad x = 1$$

Solve the following equations.

1. $(x-2)(x+3) = 0$

2. $x(x-4) = 0$

3. $2(x-1)(x-5) = 0$

4. $(1+4x)(7x-6) = 0$

Solve the equation $x^2 - 6x - 7 = 0$.

$$x^2 - 6x - 7 = 0$$
$$(x-7)(x+1) = 0$$

Either $\qquad x-7 = 0 \qquad$ or $\qquad x+1 = 0$

$$x = 7 \qquad\qquad x = -1$$

Check   Sum of the roots $= 7+(-1) = 6, \qquad \dfrac{-b}{a} = \dfrac{6}{1} = 6$

Product of the roots $= 7\times(-1) = -7, \qquad \dfrac{c}{a} = \dfrac{-7}{1} = -7$

$\therefore \quad$ the solution is $x = 7 \quad$ or $\quad x = -1$.

Solve the following equations by factorising.

5. $x^2 - 7x + 12 = 0$

6. $x^2 - 25 = 0$

7. $x^2 - 6x = 0$

8. $2x^2 + 2x - 12 = 0$

9. $x^2 - 14x + 49 = 0$

10. $3x^2 + 9x = 0$

11. $x^2 - 2x = 8$

12. $7x - x^2 = 10$

13. $9x^2 = 9$

14. $4x^2 + 4x - 15 = 0$

15. $5x^2 - 15x + 10 = 0$

16. $2 - 4x^2 = 2x$

Solve the equation $2x^2 - 3x - 3 = 0$ by completing the square.

$$2x^2 - 3x - 3 = 0$$

$$x^2 - \tfrac{3}{2}x = \tfrac{3}{2}$$

$$x^2 - \tfrac{3}{2}x + \tfrac{3}{4} \times \tfrac{3}{4} = \tfrac{3}{2} + \tfrac{3}{4} \times \tfrac{3}{4}$$

$$\left(x - \tfrac{3}{4}\right)^2 = \tfrac{33}{16}$$

$$x - \tfrac{3}{4} = \pm\sqrt{\tfrac{33}{16}}$$

Either $\quad x = \tfrac{3}{4} + \tfrac{\sqrt{33}}{4} \qquad$ or $\qquad x = \tfrac{3}{4} - \tfrac{\sqrt{33}}{4}$

$$= \tfrac{3}{4} + \tfrac{5.744}{4} \qquad\qquad\qquad = \tfrac{3}{4} - \tfrac{5.744}{4}$$

$$= 2.186 \qquad\qquad\qquad\qquad = -0.686$$

Check $\quad 2.186 + (-0.686) = 1.50$

$$-\frac{b}{a} = 1.5$$

The solution is $x = 2.19$ or $x = -0.69$ correct to 2 d.p.

Solve the following equations by completing the square.

Check your solution by comparing the sum of the roots with the value of $-\dfrac{b}{a}$

17. $x^2 - 2x - 7 = 0$

18. $2x^2 + x = 1$

19. $x^2 + 3x + 1 = 0$

20. $3x^2 - x = 3$

21. $x^2 = x + 4$

22. $7x = 2x^2 - 1$

Solve the equation $2x^2 - 3x - 3 = 0$ by using the formula.

$$2x^2 - 3x - 3 = 0$$

$$a = 2, \quad b = -3, \quad c = -3$$

$$x = \frac{-b \pm \sqrt{b^2 - 4ac}}{2a}$$

$$= \frac{3 \pm \sqrt{9 - (-24)}}{4}$$

$$= \frac{3 \pm \sqrt{33}}{4}$$

$$= \frac{3 \pm 5.744}{4}$$

Either $\quad x = \dfrac{8.744}{4} \qquad$ or $\quad x = \dfrac{-2.744}{4}$

$$= 2.186 \qquad\qquad\qquad = -0.686$$

Check   Sum of roots $= 2.186 - 0.686 = 1.5$

$$\frac{-b}{a} = \frac{-(-3)}{2} = 1.5$$

Product of roots $= 2.186 \times (-0.686) = -1.499$

$$\frac{c}{a} = \frac{-3}{2} = -1.5$$

The solution is $x = 2.19$ or $x = -0.69$ correct to 2 d.p.

Solve the equations in questions 15 to 24 by completing the square or by using the formula. Give your answers correct to 2 decimal places.

23. $x^2 + 2x - 5 = 0$ **28.** $x^2 = x + 1$

24. $2x^2 + 4x + 1 = 0$ **29.** $x^2 - 20 = 0$

25. $x^2 - 3x - 1 = 0$ **30.** $2x^2 - 7x + 2 = 0$

26. $3x^2 - 5x + 1 = 0$ **31.** $6 - 4x - x^2 = 0$

27. $x^2 + 4x = 6$ **32.** $3x^2 = x + 6$

33. Find x when $x^2 - 7 = 1 - x^2$.

34. The roots of the equation $x^2 + px + q = 0$ are both 5. Find the values of p and q.

35. Form the equation whose roots are 3 and 2.

36. The length of a rectangular carpet is 1 m more than the width, and its area is 10 m². Find its width correct to 3 significant figures.

37.

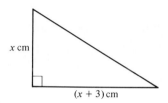

x cm

$(x + 3)$ cm

The area of the triangle is 14 cm². Find x.

38. The product of two consecutive numbers is eleven more than the sum of the two numbers. Find the two numbers.

39. When a number is subtracted from its square, the remainder is 72. Find the two possible values of the number.

40. I travel 90 km at an average speed of x km/h.
   a) In terms of x how long does the journey take ?
   b) If the speed were 5 km/h greater, how long would the journey take ?
   c) By travelling at the faster speed I would take a quarter of an hour less than at the slower speed. Form an equation and find x.

41. I could save half an hour on a 45 mile journey if I travelled on average 15 m.p.h. faster than I had planned to do. Find the speed at which I had planned to travel.

42. Mary intends to spend £3.20 on balloons for a party, the balloons costing x pence each. However, she finds that the balloons each cost 2p more than she expected, so she will have to buy 8 fewer for her £3.20. Find x.

ALGEBRAIC FRACTIONS

Algebraic fractions are simplified by following the same rules as for arithmetic fractions: if we wish to add or subtract we look for a common denominator; if we wish to multiply or divide we look for factors to cancel.

In every case we first find the factors, if any.

MULTIPLICATION AND DIVISION

EXERCISE 3j

Simplify the following expressions.

1. $\dfrac{x}{2} \times \dfrac{y}{5}$

4. $1 \div \dfrac{a}{b}$

2. $\dfrac{5a}{c} \times \dfrac{c}{3a}$

5. $\dfrac{8pq}{r^2} \times \dfrac{r}{q^2}$

3. $\dfrac{1}{a} \times \dfrac{1}{b}$

6. $\dfrac{4a}{bc} \times \dfrac{3b}{a}$

7. $\dfrac{(x-1)}{4} \times \dfrac{3}{(x-1)}$

9. $\dfrac{x(x-1)}{x+2} \times \dfrac{3(x+2)}{x}$

8. $\dfrac{3(x+1)}{5} \div \dfrac{2(x+1)}{3}$

10. $\dfrac{x^2}{x-5} \div \dfrac{2x}{x-5}$

In the following questions factorise first, then cancel if possible. Remember that $(x-a) = -1(a-x)$.

11. $\dfrac{2x+4}{x^2+2x}$

15. $\dfrac{x^2-9}{3x-9}$

12. $\dfrac{x-2}{2-x}$

16. $\dfrac{ax+bx+ay+by}{ax-bx+ay-by}$

13. $\dfrac{x^2-4}{x^2+4x+4}$

17. $\dfrac{x^2-1}{2-2x}$

14. $\dfrac{x^2-5x+6}{x^2-4x+3}$

18. $\dfrac{2x+4}{3x-9} \times \dfrac{2x-6}{3x+6}$

19. $\dfrac{x^2 - 16}{x + 4} \div \dfrac{x - 4}{2}$

21. $\dfrac{x^2 + 5x + 6}{x^2 - 5x + 6} \div \dfrac{x^2 + 6x + 8}{x^2 - 6x + 8}$

20. $\dfrac{x - a}{x + a} \times \dfrac{1}{x^2 - a^2}$

22. $\dfrac{a^2 - b^2}{a^2 + b^2} \times \dfrac{a + b}{a - b}$

FRACTIONAL EXPRESSIONS AND EQUATIONS

$\dfrac{x + 1}{6} - \dfrac{2x + 1}{3} = 1$ is an *equation*. If we multiply every term by 6, the balance is not upset and the equation is still valid.

On the other hand, $\dfrac{x + 1}{6} - \dfrac{2x + 1}{3}$ is an *expression* and if we are simplifying it we must retain the common denominator, 6. We do *not* multiply it by 6 as this would make it six times as large, i.e. it would be changed.

EXERCISE 3k

Simplify the expression $\dfrac{4x + 1}{2} - \dfrac{2(x - 2)}{3}.$

$$\dfrac{(4x + 1)}{2} - \dfrac{2(x - 2)}{3} = \dfrac{3(4x + 1) - 4(x - 2)}{6}$$

$$= \dfrac{12x + 3 - 4x + 8}{6}$$

$$= \dfrac{8x + 11}{6}$$

Simplify the following expressions.

1. $\dfrac{x}{3} + \dfrac{2x}{5}$

4. $\dfrac{2(x - 1)}{5} - \dfrac{3(1 - x)}{2}$

7. $\dfrac{4}{x} + \dfrac{3}{2x}$

2. $\dfrac{x}{7} + \dfrac{1}{3} - \dfrac{2}{7}$

5. $\dfrac{x}{3} - \dfrac{x - 1}{4}$

8. $\dfrac{1}{x + 2} - \dfrac{1}{x - 3}$

3. $\dfrac{x + 1}{4} + \dfrac{x - 1}{3}$

6. $\dfrac{1}{x} + \dfrac{x}{2}$

9. $\dfrac{x + 2}{4} - \dfrac{2x - 4}{5}$

Solve the equation $\dfrac{4x+1}{2} - \dfrac{2(x-2)}{3} = \dfrac{9}{2}.$

$$\frac{4x+1}{2} - \frac{2(x-2)}{3} = \frac{9}{2}$$

$$^3\!\!\not{6} \times \frac{(4x+1)}{\not{2}_1} - ^2\!\!\not{6} \times \frac{2(x-2)}{\not{3}_1} = ^3\!\!\not{6} \times \frac{9}{\not{2}_1}$$

$$3(4x+1) - 4(x-2) = 27$$

$$12x + 3 - 4x + 8 = 27$$

$$8x + 11 = 27$$

$$8x = 16$$

$$x = 2$$

The solution is $x = 2.$

Solve the following equations.

10. $\dfrac{x}{3} + \dfrac{2x}{5} = 1$

11. $\dfrac{x}{7} + \dfrac{1}{3} - \dfrac{2}{7} = -\dfrac{2}{3}$

12. $\dfrac{x+1}{4} + \dfrac{x-1}{3} = \dfrac{17}{12}$

13. $\dfrac{2(x-1)}{5} - \dfrac{3(1-x)}{2} = \dfrac{19}{10}$

14. $\dfrac{x}{3} - \dfrac{x-1}{4} = 7$

15. $\dfrac{1}{x} + \dfrac{x}{6} = \dfrac{7}{6}$

16. $\dfrac{4}{x} + \dfrac{3}{2x} = \dfrac{11}{4}$

17. $(x+2) = \dfrac{(x+2)}{(x-3)}$

18. Express as a single fraction a) $\dfrac{3}{x-y} + \dfrac{2}{x+y}$ b) $\dfrac{a+b}{2} + \dfrac{2}{a-b}$

19. Solve the equation a) $\dfrac{x+1}{4} - \dfrac{x-1}{5} = 1$ b) $\dfrac{2}{3x} - \dfrac{6}{5x} = -8$

20. Find x if a) $\dfrac{1}{x} + \dfrac{x+1}{9} = \dfrac{11}{9}$ b) $\dfrac{3}{x-1} + \dfrac{4}{1-x} = 2$

21. Simplify a) $\dfrac{3}{x-1} + \dfrac{4}{1-x}$ b) $\dfrac{2}{x-y} + \dfrac{3}{y-x}$

SIMULTANEOUS EQUATIONS, ONE LINEAR, ONE QUADRATIC

To solve a pair of equations one of which contains terms in, say, x^2 or y^2 or xy, we need to use the method of *substitution* rather than elimination.

From the simpler equation (i.e. the linear equation), we find one letter in terms of the other and make a substitution in the more complicated equation.

EXERCISE 3I

Solve the pair of equations $x^2 + y^2 = 17$, $x + y = 3$.

$$x^2 + y^2 = 17 \qquad [1]$$
$$x + y = 3 \qquad [2]$$

(x and y are equally convenient to use.)

From [2] $\qquad x = 3 - y \qquad [3]$

$$x^2 = (3 - y)(3 - y)$$
$$= 9 - 6y + y^2$$

Substitute for x^2 in [1]

$$9 - 6y + y^2 + y^2 = 17$$
$$2y^2 - 6y + 9 = 17$$
$$2y^2 - 6y - 8 = 0$$
$$y^2 - 3y - 4 = 0$$
$$(y - 4)(y + 1) = 0$$

Either $y = 4$ or $y = -1$.

From [3], when $y = 4$, $x = -1$

when $y = -1$, $x = 4$

Check in [1]: when $x = 4$ and $y = -1$, $x^2 + y^2 = 16 + 1 = 17$

when $x = -1$ and $y = 4$, $x^2 + y^2 = 1 + 16 = 17$

The solution is $x = 4$, $y = -1$ or $x = -1$, $y = 4$.

Solve the following equations.

1. $x^2 + y^2 = 5$

$\quad x + y = 3$

2. $x^2 + y^2 = 10$

$\quad x - y = 4$

3. $2xy + y = 10$

$\quad x + y = 4$

4. $x^2 - 2xy = 32$

$\quad y = 2 - x$

5. $9 - x^2 = y^2$

$\quad x + y = 3$

6. $\quad xy = 12$

$\quad x + y = 7$

In the previous equations either the choice of letter was obvious or it did not matter which was chosen. In the following equations, one letter is easier to use than the other, but it may not be obvious.

Solve the pair of equations $3xy + y^2 = 13$, $y = x - 3$.

$$3xy + y^2 = 13 \qquad [1]$$

$$y = x - 3 \qquad [2]$$

(It is simpler to substitute for x in [1], to avoid squaring.)

From [2] $\qquad\qquad x = y + 3 \qquad [3]$

In [1] $\qquad\qquad 3y(y + 3) + y^2 = 13$

$\qquad\qquad\qquad 3y^2 + 9y + y^2 = 13$

$\qquad\qquad\qquad 4y^2 + 9y - 13 = 0$

$\qquad\qquad\qquad (4y + 13)(y - 1) = 0$

$\therefore$ Either $4y + 13 = 0 \qquad$ or $y = 1$.

$\qquad\qquad\qquad 4y = -13$

$\qquad\qquad\qquad y = -\frac{13}{4}$

From [3], when $y = -3\frac{1}{4}$, $x = -3\frac{1}{4} + 3 = -\frac{1}{4}$

$\qquad$ when $y = 1$, $x = 4$.

Check when $x = 4$ and $y = 1$, $3xy + y^2 = 12 + 1 = 13$

$\qquad$ when $x = -\frac{1}{4}$ and $y = -3\frac{1}{4}$, $3xy + y^2 = \frac{39}{16} + \frac{169}{16} = 13$

The solution is $x = 4$, $y = 1$ or $x = -\frac{1}{4}$, $y = -3\frac{1}{4}$.

Solve the following pairs of equations.

7. $xy - x = -10$

$2x + y = 2$

10. $x + 3y = 14$

$xy = 8$

8. $xy = x$

$2x - y = 3$

11. $13 - x^2 = y^2$

$x + y = 5$

9. $y^2 = 4x$

$x - y = 0$

12. $x^2 - y^2 = 24$

$x - 2y = 3$

Solve the equations $x^2 + y^2 = 13,\ 2x + 3y = 13.$

$$x^2 + y^2 = 13 \qquad\qquad [1]$$

$$2x + 3y = 13 \qquad\qquad [2]$$

From [2] $2x = 13 - 3y$

$$x = \frac{13 - 3y}{2} \qquad\qquad [3]$$

$$x^2 = \frac{(13 - 3y)(13 - 3y)}{4}$$

$$= \frac{169 - 78y + 9y^2}{4}$$

In [1] $\dfrac{169 - 78y + 9y^2}{4} + y^2 = 13$

Multiply both sides by 4 $169 - 78y + 9y^2 + 4y^2 = 52$

$$13y^2 - 78y + 117 = 0$$

Divide both sides by 13 $y^2 - 6y + 9 = 0$

$$(y - 3)(y - 3) = 0$$

$$\therefore\quad y = 3$$

From [3], when $y = 3,\ x = 2.$

Solve the following pairs of equations.

13. $x^2 + y^2 = 2$

$3x - 2y = 5$

14. $x^2 - y^2 = 0$

$3x + 2y = 5$

15. $xy = 6$

$4x - 3y = 6$

16. $xy + x = -3$

$2x + 5y = 8$

17. $x^2 = y + 3$

$2x - 3y = 8$

18. $\dfrac{3}{x} + \dfrac{6}{y} = 4$

$x + y = 5$

19. $x(x - y) = 6$

$x + y = 4$

20. $x^2 + xy + y^2 = 7$

$2x + y = 4$

MIXED EXERCISES

EXERCISE 3m

1. Simplify   a) $\dfrac{6a}{b} \times \dfrac{3a}{c}$ b) $\dfrac{6a}{b} \div \dfrac{3a}{c}$ c) $\dfrac{b}{6a} + \dfrac{c}{3a}$

2. Simplify $\dfrac{6x - 3x^2}{x}$.

3. Solve the equations   a) $x^2 = 5x$ b) $\dfrac{3}{x} = 12$

4. Express $\dfrac{1}{x - 2} - \dfrac{1}{x + 3}$ as a single fraction.

5. Find the value of   a) 4^0 b) $4^{1/2}$ c) 4^{-3}

6. Solve the equation $2x^2 + 3x - 12 = 0$ giving your answers correct to 2 d.p.

EXERCISE 3n

1. Find a value of x such that $\dfrac{3}{x} = \dfrac{x}{27}$.

2. Simplify   a) $x^2 \times x^3$ b) $x^2 \div x^3$ c) $(x^2)^3$

3. Solve the equations $x + 2y = 12,\ 3x - 4y = 1$.

4. Factorise   a) $x^2 - x - 6$ b) $x^3 - x^2 + x - 1$

5. Express $\dfrac{4(x-1)}{3} - \dfrac{2(x+1)}{5}$ as a single fraction.

6. Solve the equation $x^2 - x - 42 = 0$.

EXERCISE 3p

1. Solve the equation $\dfrac{1}{2x} + \dfrac{x+1}{2} = 2\frac{1}{6}$.

2. Factorise   a) $4x^2 - 16$ b) $4x^2 - 16x$ c) $4x^2 - 5x + 1$

3. Solve the equation $4(x - 1) - 3(2x + 1) = 3(x + 4)$.

4. Simplify   a) $2^4 \div 2^2$ b) $2^0 \div 2^3$ c) $(2^3)^2$

5. Solve the equations $2x - 3y = 2,\ 3x - 2y = -7$.

6. Solve the equation $x^2 - 6x + 1 = 0$, giving your solutions correct to 2 d.p.

4 AREAS AND VOLUMES

Apart from revising previous work on areas and volumes this chapter considers the problems of errors and estimating.

We begin by summarising the most important results. Some of you may prefer to remember certain of these results in words rather than as formulae.

AREAS

Rectangle: Area = length × breadth

i.e. $A = lb$

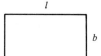

Triangle: Area = half base × perpendicular height

i.e. $A = \frac{1}{2}bh$

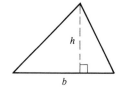

Parallelogram: Area = base × perpendicular height

i.e. $A = bh$

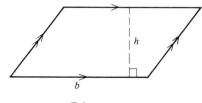

Trapezium: Area = half sum of parallel sides × distance between them

i.e. $A = \frac{1}{2}(a+b)h$

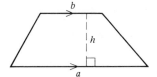

Circle: Area $= \pi r^2$

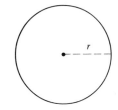

Sector of circle: Area $= \pi r^2 \times \dfrac{x}{360}$

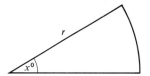

We can also find the area of any compound shape that is made up of a combination of the basic shapes

e.g.

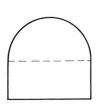

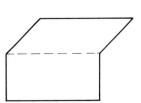

The surface area of a solid can be found if its plane faces are shapes whose areas are already known.

We can also find the surface area of a few curved surfaces.

Cylinder: Curved surface area $= 2\pi rh$

Therefore   total surface area $= 2\pi r^2 + 2\pi rh$

$$= 2\pi r(r+h)$$

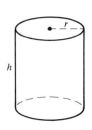

Sphere: Total surface area $= 4\pi r^2$

Cone: Curved surface area $= \pi rl$ where l is the slant height

Therefore   total surface area $= \pi r^2 + \pi rl$

$$= \pi r(r+l)$$

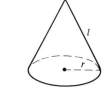

In the exercises that follow give any answers that are not exact, correct to three significant figures. Use the value of π on your calculator.

EXERCISE 4a

Find the areas of the shapes given in questions 1 to 12. All measurements are given in centimetres.

1.

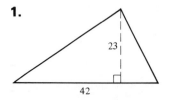

2.

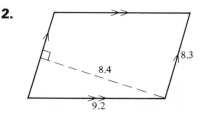

3.

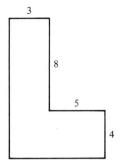

7.

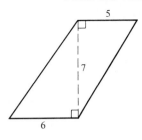

4.

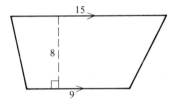

8.

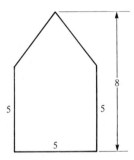

5.

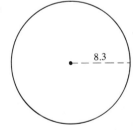

9.

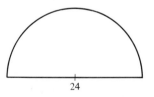

6.

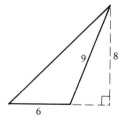

10.

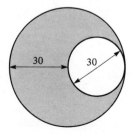

11.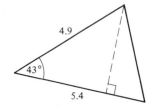

12.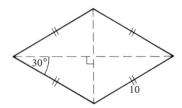

13. The hands of a clock are 8 cm and 5 cm long.
   a) What area does the minute hand pass over in half an hour ?
   b) What area does the hour hand pass over in half an hour ?

14. Find the curved surface area of a solid cylinder of base radius 8 cm and height 12 cm. What is the total surface area of this cylinder ?

15. Leanda uses her ruler to measure the sides of a page of a book and finds them to be 294 mm by 216 mm.
   a) Express these measurements   i) in centimetres   ii) in metres.
   b) Find the area of the page in square centimetres
      i) correct to one decimal place
      ii) correct to three significant figures.

The total surface area of a closed cylindrical metal can is 200 cm². If the diameter of the base of the can is 5 cm find its height.

$$A = 200 \qquad r = 2.5$$

$$A = 2\pi r(r + h)$$

$$200 = 2\pi \times 2.5(2.5 + h)$$

where h is the height in centimetres

$$2.5 + h = \frac{200}{2\pi \times 2.5}$$

$$2.5 + h = 12.73$$

$$h = 10.23$$

∴   the height of the can is 10.2 cm correct to 3 s.f.

---

16. Find the curved surface area of a right circular cone that has a base radius of 6 cm and a slant height of 10 cm. How high is the cone ?

17. The total surface area of a sphere is 100 cm². Find its radius.

18. The curved surface area of a cylinder is 150 cm². If the cylinder is twice as high as it is wide, find its radius.

19.

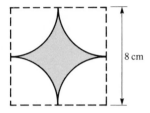

8 cm

The sketch shows the pressing, from a square sheet of metal of side 8 cm, required in the manufacture of a piece of machinery. The perimeter is formed from four quarter circles with equal radii.

a) Calculate   i) the perimeter of the pressing
               ii) its area
               iii) the percentage of the metal sheet wasted.

b) Several of these pressings have to be made. Can you suggest arrangements, using large sheets of metal, so that less metal is wasted ?

20. The scale of a map is 1:50 000. Find the actual area   a) in hectares, b) in square kilometres, of a lake represented by an area of 3.8 square centimetres on the map.
(1 hectare = 10 000 m².)

21.

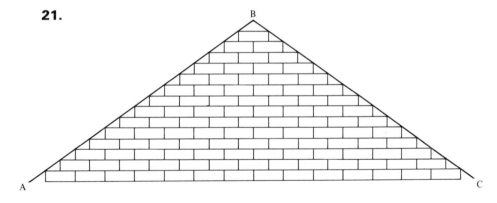

The sketch shows the gable end of a garage where bricks are used to fill the gable. Each brick measures 210 mm by 70 mm. If there are fourteen bricks in the first row and there are fourteen rows of bricks, find

a) the inclination of AB and BC to the horizontal

b) the distance AC, in metres

c) the height of B above the level of AC

d) the area of triangle ABC.

22.

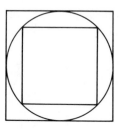

A square is inscribed in a circle of radius 10 cm and a second square is circumscribed to the circle. Find:

a) the area inside the larger square but outside the circle

b) the area outside the smaller square but inside the circle.

Express your answer to (a) as a percentage of your answer to (b).

23.

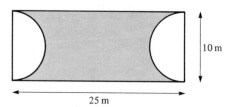

10 m

25 m

The diagram shows a rectangular garden, 25 m by 10 m. A semicircular flowerbed has been prepared at each end, and the remainder (the shaded area) put down to lawn. Using $\pi \approx 3$, calculate:

a) the combined area of the two flowerbeds

b) the area of the lawn

c) the perimeter of the lawn.

24. Two teapot stands have exactly the same area. If the square stand has a side of 120 mm, what is the radius of the circular stand ?

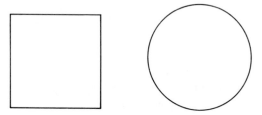

25.

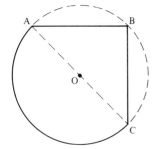

The shape of the top of a kitchen unit is part of a circle, centre O.
Angle $ABC = 90°$ and $AB = BC = 60$ cm. Find:

a) the area of the top in cm²

b) the perimeter of the top in centimetres

c) the cost of covering the top with a laminate costing £8 per square metre

d) the cost of plastic strip for the curved edge of the top at 24 p per metre.

26. Two similar triangles are such that every linear dimension in one triangle is twice the corresponding linear dimension in the other. If their combined areas amount to 500 cm², find the area of the larger triangle.

VOLUMES

1 litre $= 1000$ cubic centimetres.

Cuboid: Volume $=$ length $\times$ breadth $\times$ height

i.e. $V = lbh$

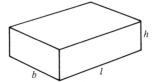

Pyramid: Volume $= \frac{1}{3}$ area of base $\times$ perpendicular height

i.e. $V = \frac{1}{3}Ah$

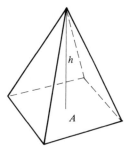

Cylinder: Volume = area of base × height

i.e. $V = \pi r^2 h$

Cone: Volume = $\frac{1}{3}$ area of base × perpendicular height

i.e. $V = \frac{1}{3}\pi r^2 h$

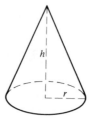

Sphere: $V = \frac{4}{3}\pi r^3$

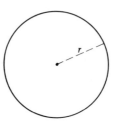

Prism: (solid with constant cross-section):

Volume = area of cross-section × length

i.e. $V = Al$

It follows that we can find the volume of any solid that is made by combining the basic solids

e.g.

this salt cellar combines a cylinder with half a sphere (i.e. a hemisphere),

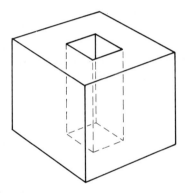

this solid wooden cube has had a cuboid removed from it.

EXERCISE 4b

1. Find the volume of a brick measuring 210 mm by 100 mm by 70 mm. Give your answer in cubic centimetres. How many such bricks are required to give a total volume that exceeds one cubic metre ?

2. A pyramid has a square base of side 12 cm and is 15 cm high. Find its volume.

3. A cone has a base radius of 3.8 cm and is 8 cm high. Find its volume.

4. Find the volume of a sphere of radius 15 cm.

5. The area of the end of a pencil with uniform cross section is $\frac{1}{5}$ cm². If the pencil is 17 cm long find its volume   a) in cubic centimetres b) in cubic millimetres.

6. A cylindrical can is 12 cm high and has a base radius of 25.8 mm. Find its capacity in millilitres.

7. The radius of a sphere and the radius of the circular base of a cylinder are each 5 cm. If both solids have the same volume, how high is the cylinder?

8. The volume of a circular cylinder is 500 ml. If its height is equal to twice its radius, find its radius.

9.

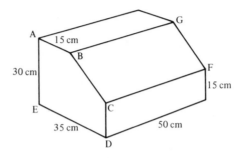

The sketch shows a closed wooden box of uniform cross-section, with a sloping front. The dimensions of the box are given on the sketch. Calculate:

a) the length of BC

b) the area of the sloping front BCFG

c) the area of the end ABCDE

d) the total surface area of the box

e) the volume of the box.

10. The volume of a right circular cone is 1000 cm³. If its height is three times its radius, find its height.

11. The volume of a pyramid with a square base is 2187 cm³. If its height is 9 cm, find the length of a side of the base.

12. A cylindrical jam jar has a diameter of 70 mm, is 120 mm high, and is filled with water to a depth of 80 mm. 1200 identical spherical ball-bearings of radius 2.5 mm are placed in the jar and are completely immersed. By how much does the surface of the water rise?

13.

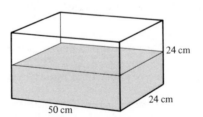

Water stands in a rectangular tank, measuring 50 cm × 24 cm × 24 cm, to a depth of 12 cm. Five cubical heavy metal blocks, each of side 9 cm, are placed on the bottom of the tank. By how much does the water level rise?

14. A hemispherical cut-glass bowl has an internal diameter of 22 cm and is 7.5 mm thick throughout. If 5% of the original glass has been removed by the cutting of the design, calculate the volume of glass in the resulting bowl, giving your answer in cubic centimetres.

15.

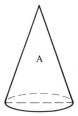 

A cone (A) and a cylinder (B) have equal bases and equal heights.
a) If the volume of the cylinder is 36 cm³, find the volume of the cone.
b) If the volume of the cone is 18.6 cm³, find the volume of the cylinder.
c) If the combined volume of the two solids is 64 cm³ find the volume of the cylinder.

16. The diagram shows the goalmouth and netting prepared for a football match. The netting is attached to the goal posts and crossbar, and also to two metal tubes, each in the form of a quarter of a circle.

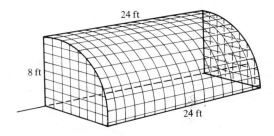

a) Use the dimensions given in the diagram to calculate the total area of netting required, in square feet.
b) If 1 foot = 0.305 metres, give your answer to (a) in square metres.
c) Find, in cubic metres, the volume of the space contained within the netting that is behind the face of the goal.

17. A solid metal cone of height 20 cm and radius 12 cm is melted down to form a cylinder of the same height. What is the radius of the cylinder?

18. A solid metal cylinder with diameter 2 cm and height 4 cm stands inside an empty cylindrical measuring jar with a diameter of 4 cm. Water is poured into the jar to a depth of 3 cm.

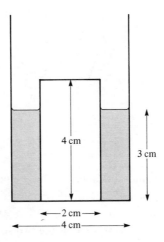

Find:

a) the volume of water poured into the cylinder

b) the additional volume of water required just to cover the metal cylinder

c) the total volume of water required if the depth of water in the cylinder at its deepest point is 6 cm.

(Give all your answers as a multiple of π.)

19. A salt cellar is in the form of a cylinder, of diameter 3 cm and height 3 cm, topped by a hemisphere. It contains salt to a depth of 2.5 cm. It is inverted with its outlet covered. Find the distance from the new level of the salt (assumed horizontal) to the flat base of the salt cellar.

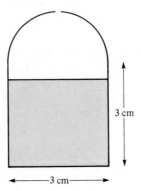

20.

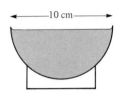 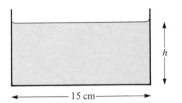

Soup placed in a bowl forms a hemisphere of diameter 10 cm. Six such bowls are filled with soup that has been warmed in a cylindrical saucepan with diameter 15 cm.

a) Find the depth of soup in the saucepan if there is just sufficient to fill all six bowls.

b) By what factor is the area of soup open to the atmosphere increased when the soup has been poured from the saucepan into the bowls ?

21.

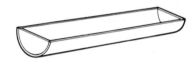

The diagram shows a 2 metre length of plastic gutter with semicircular end-stops of diameter 7 cm. The gutter is made by bending sheet plastic. Find:

a) the maximum amount of water the gutter will hold
    i) in cubic centimetres   ii) in litres

b) the area, in square metres, of sheet plastic required to make the gutter, including the endstops. Assume that there is no overlap.

22.

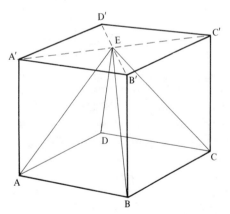

ABCD is the base of a wooden cube of side 6 cm. The diagonals of the opposite face A'B'C'D' intersect at E. Wood is removed from the cube to form the pyramid ABCDE.

a) Find the volume of
    i) the cube   ii) the pyramid   iii) the wood wasted.

b) Find the surface area of
    i) the cube   ii) the pyramid.

23.

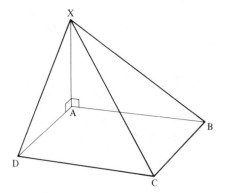

The diagram shows a pyramid which has a rectangular base ABCD, and a vertex X which is vertically above A. If AB = 40 cm, BC = 30 cm and $\widehat{ABX} = 32°$, find

a) the height of X above A

b) the length of AC

c) the angle ACX

d) the combined area of the *sloping* faces of the pyramid.

24.

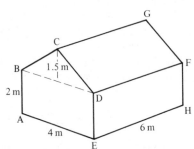

The diagram shows a workshop which stands on a rectangular base, and has a ridged roof. Its dimensions in metres are given on the diagram. Find:

a) the area of the end ABCDE

b) the length CD, one of the sloping edges of the roof

c) the area of CDFG, one of the sloping faces of the roof

d) the capacity of the shed

e) the ground area on which the workshop stands

f) the volume of rain falling vertically on the roof of the shed during a storm of rainfall 2 cm.

25. Liquid is discharged from a pipe at a rate of 44.4 kilograms per minute. Express this rate in grams per second.

If the density of the liquid is 0.925 grams per cubic centimetre, calculate the volume of liquid discharged per second.

The liquid is discharged into cylindrical containers of height 72 cm and base radius 20 cm. Find the time taken, in seconds, to fill each container. (Give your answer as a multiple of π.)

26. Two cola cans of circular cross-section are similar in every respect. One is 8 cm high while the other is 12 cm high.

a) Find the diameter of the larger can if the diameter of the smaller can is 6 cm.

b) The cans are to be placed in a cardboard box, which is to be twice as long as it is wide, with large cans on the bottom and small cans on the top.

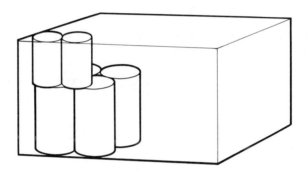

i) Find the minimum length and width of the box if the cans at each level have no room to move about.

ii) How many large cans are required for the bottom layer in the box?

iii) How many small cans are required for the top layer in the box?

iv) How deep is the box?

v) Find the percentage of space wasted in each layer. How does the percentage of space wasted in the upper layer compare with the percentage of space wasted in the bottom layer?

27.

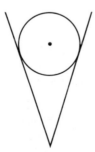

The sketch shows a cone of radius 4 cm and depth 12 cm on which rests a sphere of ice-cream of radius 2.5 cm. The ice-cream melts, and, without any loss or change in volume, runs into the cone. Find the depth of liquid in the cone when the ice-cream has melted completely.

ROUGH ESTIMATES

There are many everyday situations in which we need to estimate. How long will it take to drive to London ? How many tiles will we need to cover the bathroom walls ? How many litres of paint will we require to paint the lounge ? For most estimates of this kind we use our inbuilt knowledge of shape, size and units, or take advantage of the experience of other people. We also need to develop a feel for the size of an answer to a calculation, e.g. what, roughly, is $8.926 \times 0.042\,34$?

One way of estimating the answer to a calculation is to write each number correct to 1 significant figure e.g.

$$8.926 \times 0.04234 \approx 9 \times 0.04 = 0.36$$

EXERCISE 4c

Write down an estimate for each of the following quantities. Where possible check your estimates by measuring or weighing or working them out.

1. How high is the door you used to come into the room ?

2. How wide is the blackboard ?

3. What are the dimensions of the room ?

4. What are the dimensions of this page ?

5. How high is the desk ?

6. How long is a new stick of chalk ?

7. How long is the school hall ?

8. Two hundred 1 kg bags of sugar are laid end to end on their sides. How far will they stretch ?

9. How heavy is   a) a pencil,   b) a shoe,   c) a 50 p coin,   d) an elephant,   e) a coach,   f) a large ship ?

10. How high is a telegraph pole ? How could you calculate its height without climbing to the top ?

Correct each number to 1 s.f. and hence give a rough answer to

a) $\dfrac{6.623 \times 18.7}{93.4}$
b) $\dfrac{0.534}{6.242 \times 0.819}$
c) $\dfrac{3.72^2 + 8.64^2}{3.72 \times 8.64}$

a) $\dfrac{6.623 \times 18.7}{93.4} \approx \dfrac{7 \times 20}{90} = \dfrac{140}{90} \approx 1.5$

b) $\dfrac{0.534}{6.242 \times 0.819} \approx \dfrac{0.5}{6 \times 0.8} = \dfrac{5}{6 \times 8} = \dfrac{5}{48} \approx \dfrac{5}{50} = 0.1$

c) $\dfrac{3.72^2 + 8.64^2}{3.72 \times 8.64} \approx \dfrac{4^2 + 9^2}{4 \times 9} = \dfrac{16 + 81}{36} = \dfrac{97}{36} \approx \dfrac{100}{40} = 2.5$

Correct each number to 1 s.f. and hence give a rough answer to each of the following calculations.

11. 16.21×0.937

12. $40.34 \div 7.626$

13. $\dfrac{86.04 \times 0.07824}{5.95}$

14. $\left(\dfrac{0.7921}{0.4105} \right)^2$

15. $\sqrt{\dfrac{44.75}{14.78}}$

16. 0.0664×81.7

17. $8933 \div 47$

18. $\dfrac{5.95 \times 0.1932}{0.613}$

19. $\left(\dfrac{2.285}{7.732} \right)^2$

20. $\sqrt{\dfrac{0.8877}{0.0093}}$

21. $\dfrac{9.26^2 + 2.964^2}{9.26 \times 2.964}$

22. $\sqrt{\dfrac{17.26^2 - 9.64^2}{2 \times 17.26^2 + 4 \times 9.64^2}}$

23. $\dfrac{1}{84} + \dfrac{1}{38}$

24. $\sqrt{\dfrac{0.7842}{0.0517}}$

25. $\dfrac{5.2^2 - 0.98 \times 7.47}{3.14 \times 3.97 - 5.2 \times 1.24}$

26. $\dfrac{1}{36} - \dfrac{1}{58}$

The value of T is given by the formula

$$T = 2\pi \sqrt{\frac{h^2 + k^2}{hg}}$$

Without using a calculator estimate the value for T when $h = 8.92$, $k = 3.22$, $g = 9.81$ and $\pi = 3.14$. Show clearly the approximations made at each stage of your working.

Writing each value correct to 1 significant figure, $h \approx 9$, $k \approx 3$, $g \approx 10$ and $\pi \approx 3$.

$$\therefore \qquad T \approx 2 \times 3 \sqrt{\frac{9^2 + 3^2}{9 \times 10}}$$

$$= 6 \times \sqrt{\frac{81 + 9}{90}}$$

$$= 6 \times \sqrt{\frac{90}{90}}$$

$$= 6 \times \sqrt{1}$$

$$= 6$$

Therefore the approximate value of T is 6.
(The calculator value is 6.370 correct to 4 s.f.)

In questions 27 to 31 do not use a calculator.

27. The value of E is given by the formula $E = \frac{1}{2}mu^2$.
Estimate E when $m = 12.6$ and $u = 9.21$.

28. The value of I is given by the formula $I = \dfrac{m(a^2 + b^2)}{3}$.
Estimate I when $m = 5.92$, $a = 21.4$ and $b = 9.72$.

29. The value of T is given by the formula $T = 2\pi \sqrt{\dfrac{l}{g}}$.
Estimate T when $l = 41.3$, $g = 9.81$ and $\pi = 3.142$.

30. The value of V is given by the formula $V = (u^2 + v^2)^{1/2}$.
Estimate V when $u = 8.24$ and $v = 5.72$.

31. Estimate the value of $\sin^2\alpha + \cos^2\alpha$ when $\sin\alpha = 0.3907$ and $\cos\alpha = 0.8746$ [$\sin^2\alpha$ means $(\sin\alpha) \times (\sin\alpha)$].

EVERYDAY ESTIMATES

We are used to numbers bombarding us from all sides:

80 000 at Wembley Stadium Last Night

Ruritania to spend £ 3.46m on defence next year

Population of China reaches 2000m

300 feared lost in air disaster.

These are typical headlines from our daily newspapers, but they do not mean exactly what they say! They are approximations or 'guesstimates'. Perhaps they are correct to the first figure. For example, '3000 feared dead in an earthquake' could mean anything between 2500 and 3499 if the figure is correct to the nearest thousand. If the number had been thought to be a few less than 2500 the estimate would have been 2000, whereas if the number had been thought to be 3500, or a few more, the estimate would have been 4000.

When we quantify something our value is often approximate. The length of our lounge is perhaps 4 m, correct to the nearest metre, but 426 cm correct to the nearest centimetre. We estimate we spend £ 30 each week at the supermarket. Perhaps this means £ 27.40 one week but £ 34.72 the following week.

The next exercise considers the range between which given numbers lie.

EXERCISE 4d

The attendance at a pop concert is given as 65 000, correct to the nearest thousand. Find:

a) the maximum number that could have been present

b) the minimum number that could have been present.

a) The largest number that could have been present is 65 499 (This number correct to the nearest thousand is 65 000, whereas 65 500 correct to the nearest thousand would be 66 000.)

b) The smallest number that could have been present is 64 500.

1. The attendance at a football match is given as 46 000 correct to the nearest thousand. Find:

a) the greatest number of spectators that could have been present

b) the smallest number of spectators that could have been present.

2. It is reported that 1500 people have lost their lives in an earthquake. If this is correct to the nearest 100 what is the smallest possible number of deaths ?

3. A sports' club has 600 members, correct to the nearest 100. Between what numbers must the membership lie ?

4. The population of China is 2000 million, correct to the nearest 100 million.

a) What is the largest possible population ?

b) What is the smallest possible population ?

Give each answer correct to the nearest million.

5. Jenny spends £45 each week at the supermarket. If this is correct to the nearest £5, find

a) the largest amount she could spend

b) the smallest amount she could spend.

6. A local authority is to spend £750 000, correct to the nearest £10 000, on a new road project. What are the upper and lower limits within which they expect the expenditure to be ?

7. A motor manufacturer aims to produce 2000 cars a week, correct to the nearest 100. What are the upper and lower limits that satisfy this target ?

8. The Australian government decides that the number of immigrants next year, to the nearest thousand, is to be restricted to 35 000. What is the difference between the largest number that will be allowed and the smallest number that will be allowed ?

9. Copy and complete the following table, which shows corrected numbers and the smallest and largest values from which they can arise.

Number	Correct to nearest	Smallest possible value	Largest possible
4 560	10	4555	4564
1 800	100	1750	1849
5 000	1000		
80 000	10 000		
30 000	1000		
66 700	100		
	100		4549
	1000	3500	

FIRST ORDER APPROXIMATIONS

The approximations and estimates we have considered in the previous exercise have all concerned whole numbers. However, many of the measurements we make can involve decimal or fractional numbers. For example, the length of a room may be 3 m, correct to the nearest metre, but its length could be 3.24 m or 3.243 m, depending on the degree of accuracy possible in measuring.

If a length is given as 3 m, correct to the nearest whole number, then it is possible that the length could really be as low as 2.5 m or nearly as high as 3.5 m. So if the length is l cm

$$2.5 \leqslant l < 3.5$$

Similarly if the mass, m kg, of a bag of potatoes is 5.5 kg, correct to two significant figures, then

$$5.45 \leqslant m < 5.55$$

EXERCISE 4e

In questions 1 to 8 write down the values between which each of the given quantities lies.

1. The width of my protractor is 9.6 cm, correct to 2 s.f.

2. The height of my protractor is 5.6 cm, correct to 2 s.f.

3. The length of my ruler is 31 cm, correct to the nearest cm.

4. The length of my lounge is 8200 mm, correct to the nearest 100 mm.

5. The length of a rectangular field is 126 m, correct to the nearest metre.

6. The weight of a bag of cement is 50 kg, correct to the nearest kg.

7. The distance between Blackborough and Woodside is 300 miles, correct to the nearest 100 miles.

8. John's speed, correct to 3 s.f., was 8.35 m/s.

9. The length of my lounge is 4 m, correct to the nearest metre. If l metres is the actual length between what values must l lie ?

10. Ravi estimates that it is 3 miles from his home to school, correct to the nearest mile. If the actual distance is D miles, between what values must D lie ?

11. Shan is 166 cm tall, correct to the nearest cm. If Shan's actual height is h cm, between what values must h lie ?

12. The diameter of a spindle is given as within twelve thousandths of 0.5 cm. If the actual value is D cm use inequalities to show the range of possible diameters.

COMPOUND APPROXIMATIONS

If the side of a square field is given as 90 m, correct to the nearest 10 m, the area of the field could be anything within a fairly wide range of values.

The smallest value for the length of the side of the field is 85 m, which gives an area of 7225 m² for the field.

The largest value is the value just below 95 m, which gives an area just less than 9025 m². Thus if A m² is the area of the field

$$7225 \leqslant A < 9025$$

The upper limit is 25 % more than the lower limit, and clearly indicates why surveyors calculate their figures with great accuracy.

Similarly, if r is the radius of a sphere, and $r = 3$ cm, correct to the nearest cm, then

$$2.5 \leqslant r < 3.5$$

and the volume of the sphere, V cm³ ($V = \frac{4}{3}\pi r^3$), is such that

$$65 \leqslant V < 180$$

correct to the nearest cubic centimetre.

EXERCISE 4f

The sides of a rectangle are given as 5.2 cm and 3.8 cm, each correct to one decimal place.
Find   a) the upper limit of the possible area
       b) the minimum area
of the rectangle with these dimensions.

By what percentage does the upper limit of the area exceed the minimum area ?

The maximum length that gives a length of 5.2 cm, correct to the nearest tenth of a centimetre, is the length just less than 5.25 cm. Similarly the maximum width is just less than 3.85 cm.

Therefore the upper limit of the area is

$$5.25 \times 3.85 \, \text{cm}^2 = 20.21 \, \text{cm}^2$$

The minimum length that gives a length of 5.2 cm correct to the nearest tenth of a centimetre is exactly 5.15 cm. Similarly the minimum width is exactly 3.75 cm.

The minimum area is therefore $5.15 \times 3.75 \, \text{cm}^2$

$$= 19.31 \, \text{cm}^2$$

$$\text{Difference in areas} = (20.21 - 19.31) \, \text{cm}^2$$

$$= 0.90 \, \text{cm}^2$$

$$\text{Percentage difference} = \frac{0.90}{19.31} \times 100\,\%$$

$$= 4.7\% \quad (\text{correct to 1 d.p.})$$

1. The side of a square, correct to the nearest whole number, is 10 cm. For this square write down
a) the upper limit for the length of a side of this square
b) the smallest possible value for the length of a side
c) the upper limit for the area of this square
d) the smallest possible area.

2. A rectangle measures 20 cm by 15 cm, each measurement being given correct to the nearest whole number. Find   a) the upper and lower limits for its dimensions   b) the upper and lower limits for its area.

3. A rectangle measures 8.3 cm by 6.9 cm, each measurement being correct to the nearest tenth of a centimetre. Find the upper and lower limits for its area.

4. The diameter of a circular flower bed is 8 m, correct to the nearest metre. Find the upper and lower limits for its area.

5. The side of a solid metal cube is 20 cm, correct to the nearest whole number. Find the upper and lower limits of the volume of metal used to make this cube.

6. The dimensions of a rectangular box are 10 cm by 6 cm by 8 cm. Find the upper and lower limits for the capacity of the box, if each dimension is given correct to the nearest centimetre.

In questions 7 to 15 all values are given correct to the nearest whole number. Use the given formula to find the upper and lower limits for the values for the letter given on the left-hand side.

7. $A = l \times b$ if $l = 10$ and $b = 6$.

8. $A = 2(a+b)h$ if $a = 5$, $b = 7$ and $h = 6$.

9. $V = \frac{4}{3}\pi r^3$ if $r = 12$.

10. $s = \frac{1}{2}(a+b+c)$ if $a = 12$, $b = 15$ and $c = 17$.

11. $V = lbh$ if $l = 3$, $b = 4$ and $h = 5$.

12. $f = \frac{u+v}{2}$ if $u = 12$ and $v = 15$.

13. $C = 2\pi r$ if $r = 22$.

14. $A = \pi r^2$ if $r = 17$.

15. $z = \frac{y^2}{x}$ if $x = 10$ and $y = 12$.

16. Given $V = \frac{1}{3}\pi r^2 h$, $r = 1.25$ and $h = 4.93$, each correct to 2 d.p., find the maximum and minimum values for V. Give these values correct to 2 d.p.

17. If $x = 9$ $y = 6$ correct to the nearest whole number, find the upper and lower limits for the value of z given that $z = (x-y)^2$.
(Hint: z is largest when x is biggest and y is smallest)

18. If $x = 1.4$ and $y = 2.8$, correct to one decimal place, find the largest and smallest values of   i) $2x+3y$ ii) $3y-x$ iii) $2xy$.

19. Two variables x and y are such that $7.5 \leqslant x \leqslant 10.5$ and $4 \leqslant y \leqslant 6$. Calculate the smallest and largest values of

a) $x+y$ b) $x-y$ c) xy d) $\frac{x}{y}$.

20. Two variables y and z are such that $1.2 \leqslant y \leqslant 3.4$ and $4.3 \leqslant z \leqslant 6.8$.

Calculate the smallest and largest values of

a) $y+z$ b) yz c) $\frac{z}{y}$ d) y^2+z^2

21. The length of an edge of a cube is 5 cm, correct to the nearest whole number. Calculate the smallest and largest values of
a) an edge of the cube
b) the surface area of the cube
c) the volume of the cube.

A weights and measures inspector finds that the petrol pumps at a particular garage give 5% less petrol than is indicated on the meter.

a) How much petrol goes into the tank when a motorist thinks she has bought 5 gallons ?

b) If the marked price of the petrol is £1.80 per gallon, how much does the motorist actually pay per gallon ?

c) By how much was the motorist overcharged for the petrol she bought ?

a) For each gallon paid for, the motorist receives $(100 - 5)\%$ of a gallon.

$$1 \text{ gallon registered} \equiv 95\% \text{ of 1 gallon received}$$

$$5 \text{ gallons registered} \equiv 5 \times \frac{95}{100} \text{ gallons received}$$

$\therefore$ the motorist receives 4.75 gallons.

b) Price paid for 4.75 gallons is £5 × 1.80.

Therefore the price per gallon is

$$£\frac{5 \times 1.80}{4.75} = £1.89$$

c) The motorist should have paid

$$4.75 \times £1.80 = £8.55$$

Payment actually made $= 5 \times £1.80$

$$= £9$$

She was therefore overcharged by £9 − £8.55, i.e. by 45 p.

---

22. A 30 cm wooden ruler is checked and found to be inaccurate. A line which is known to be exactly 10 cm long measures 10.1 cm according to this ruler.

a) What percentage error will result from measuring distances with this ruler ?

b) What measurement will this ruler give for the length of a box known to be 4 metres long ?

c) What is the true length of a distance measured as 3700 cm with this ruler ?

23. Tim measures the space to replace a broken window using his mother's tape measure. He thinks that the piece of glass he requires measures 900 mm by 500 mm, but when he brings his glass home finds that it is too large. On checking the tape against a metal tape he finds that every 10 cm on the tape measure is actually 9.7 cm. What measurements should he have given for the sheet of glass ?

24. Three liquids *A*, *B* and *C* are mixed together to make an elixir. They are mixed in the proportions $1:2:3$ using a ladle that holds 4% less liquid than its stated capacity. The most important ingredient is liquid *A*. If Zoe must take one teaspoonful of the elixir will she be taking

a) the exact amount of liquid *A*

b) too much of liquid *A*

c) too little of liquid *A* ?

Give a reason for your answer.

25. A faulty electricity meter results in every 100 units of electricity I actually use being recorded as 106. My bill for the units I used in a quarter, excluding the standing charge, was £90.10. By how much did I overpay ?

26. A faulty gas meter records 10% fewer units than I actually use. My bill last quarter charged £129.60 for the 324 therms recorded as having been used. Find:

a) the number of therms actually used

b) the cost of gas per therm

c) the amount by which I underpaid for the gas I had used.

27. The reading on a kitchen weighing machine is 20% more than the actual weight placed on it. For a particular recipe the requirements are

125 g butter
50 g sugar
175 g flour

How much of each ingredient should be indicated on these scales to give the correct amounts for the recipe ?

28. The sides of a rectangular piece of cardboard are each 10% too long. By what percentage is the area of the sheet too large ?

29. Use this temperature conversion table to complete the statements that follow.

°C	0	10	20	30	40	50	60	70	80	90	100
°F	32	50	68	86	104	122	140	158	176	194	212

a) 56 °C is approximately   °F.

b) 100 °F is approximately   °C.

c) A rise in temperature of 143 °F is approximately a rise of   °C.

d) A fall in temperature of 53 °C is approximately a fall of   °F.

30.

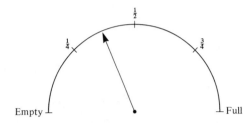

Empty   Full

The petrol tank on my car holds 16 gallons when full. The needle is in the position shown in the sketch. Approximately how far can I go if my car does, on average, 36 miles to the gallon ?

31. The table shows the area of a circle, A cm^2, for particular values of its radius, r cm.

r	3	5	10	15	20	25	30
A	28	79	310	710	1260	1960	2830

a) What is the approximate area of a circle with radius   i) 7 cm   ii) 23 cm ?

b) What is the approximate radius of a circle whose area is   i) 50 cm^2 ii) 2500 cm^2 ?

32. On the continent petrol consumption is quoted in litres/100 km. In the United Kingdom it is given as miles per gallon. The table shows the connection between the two methods.

Petrol	in m.p.g.	26.4	32.8	39.2	45.5	52.3
consumption	in litres/100 km	10.7	8.6	7.2	6.2	5.4

a) Approximately what is the consumption in m.p.g. if the consumption in litres/100 km is
i) 9.5   ii) 7.9   iii) 5 ?

b) Approximately what is the consumption in litres/100 km if the consumption in m.p.g. is
i) 30   ii) 40   iii) 50 ?

EXERCISE 4g

In this exercise several alternative answers are given. Write down the letter that corresponds to the correct answer.

1. A circular garden pond has a perimeter of 12 metres. An approximate value for the area of its surface is

A   45 m^2 **B**   12 m^2 **C**   100 m^2 **D**   75 m^2

2. The area of a trapezium is 32 cm^2. If the parallel sides are of length 5.90 cm and 9.89 cm, the approximate distance between them is

A   4 cm   **B**   6 cm   **C**   8 cm   **D**   10 cm

3. The perimeter of the sector of a circle of radius 7.8 cm is 23.4 cm. The angle that the arc subtends at the centre is

  A about 60°   **B** less than 45°

  C between 90° and 180°   **D** more than 180°

4.

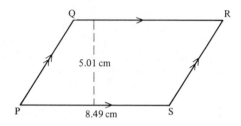

The area of parallelogram PQRS is approximately

  A 21 cm²   **B** 52 cm²   **C** 30 cm²   **D** 43 cm²

Questions 5 to 8 refer to the following diagram.

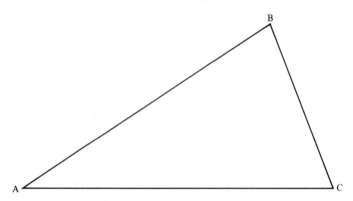

5. The length of AB is about

  A 5 cm   **B** 8 cm   **C** 12 cm   **D** 4 cm

6. The perimeter of the triangle ABC is about

  A 12 cm   **B** 32 cm   **C** 21 cm   **D** 40 cm

7. The value of $A\widehat{C}B$ is about

  A 70°   **B** 45°   **C** 60°   **D** 50°

8. The area of △ABC is about

  A 37 cm²   **B** 9 cm²   **C** 18 cm²   **D** 27 cm²

5 GEOMETRY AND DRAWING

This chapter revises the main facts and techniques. Remember that not all the work in this chapter is required by each examination board.

PARALLEL LINES

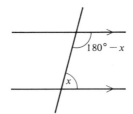

TRIANGLES AND POLYGONS

In any triangle,

the sum of the interior angles is $180\,^\circ$

an exterior angle is equal to the sum of the interior opposite angles.

In a right-angled triangle,

the square on the hypotenuse is equal to the sum of the squares on the other two sides.

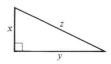

$z^2 = x^2 + y^2$

For any polygon,

the sum of the exterior angles is 360°.

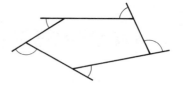

For a polygon with n sides,
the sum of the interior angles is

$$(180n - 360)°$$
$$= (n - 2)180°$$

A regular polygon has all its sides equal
and all its angles equal.

Two figures are *congruent* if they are
exactly the same shape and size.

To prove that two triangles are congruent, show that

either   the three sides of one triangle are equal to the three sides of the
other triangle.

or   two angles and one side of one triangle are equal to two angles
and the *corresponding* side of the other triangle.

or   two sides and the included angle of one triangle are equal to two
sides and the included angle of the other triangle.

or   both triangles have a right angle and two sides of one triangle are
equal to the two corresponding sides of the other triangle.

i.e.

Two figures are *similar* if they are the same shape, but not necessarily the same size (i.e. one figure is an enlargement of the other).

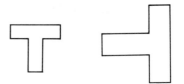

To prove that two triangles are similar, show that

either the three angles of one triangle are equal to the three angles of the other triangle.

or the corresponding sides of each triangle are in the same ratio.

or there is one pair of equal angles and the sides containing these angles are in the same ratio.

i.e.

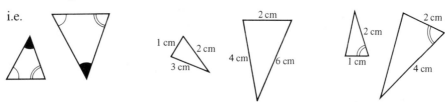

The ratio of the *areas* of similar figures is equal to the ratio of the *squares* of corresponding lengths.

The ratio of the *volumes* of similar objects is equal to the ratio of the *cubes* of corresponding lengths.

EXERCISE 5a

1. a)

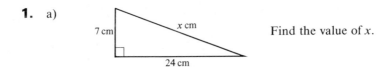

Find the value of x.

b) x cm is the length of the hypotenuse and p cm and q cm are the lengths of the other two sides of a right-angled triangle. Copy and complete the table.

p	q	x
3	4	
5		13
7	24	

c) The two shorter sides of a right-angled triangle are 70 mm and 240 mm. Find the length of the other side.

2. Find the sizes of the angles marked *p*, *q* and *r*.

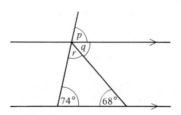

3. These polygons are regular. Find the sizes of the marked angles.

a)

b)

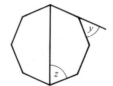

c)

4. Each of the following diagrams contains at least two triangles which are similar and/or congruent. Name these triangles and state, with reasons, which type they are.

a)

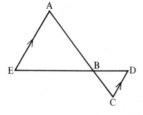

b)

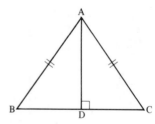

c)

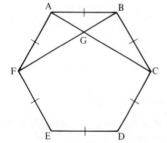

d)

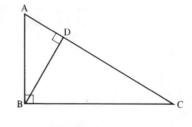

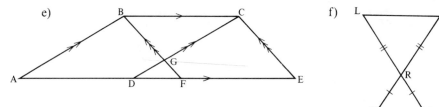

e)

f)

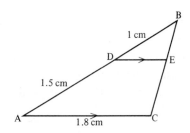

5.

$AD = 1.5\,\text{cm}$, $DB = 1\,\text{cm}$ and $AC = 1.8\,\text{cm}$.

Find:

a) BE : BC

b) the length of DE

c) the ratio of area $\triangle ABC$ to area $\triangle DBE$.

6. In triangle ABC, AB = AC. L and M are the midpoints of AB and AC respectively. Show, with reasons, that $\widehat{BLC} = \widehat{CMB}$.

7. ABCDE is a regular pentagon. Sides AB and DC are extended to meet at F. Find the size of $\widehat{BFC}$.

8. Two jugs are mathematically similar. The smaller jug holds 0.8 litres and the larger jug holds 2.7 litres.

a) The smaller jug is 10 cm high. Find the height of the larger jug.

b) The lid on the larger jug has a surface area of 36 cm². Find the surface area of the lid for the smaller jug.

9. ABCDEF is a regular hexagon. Sides AB and DC are produced to meet at X. Sides BA and EF are produced to meet at Y and sides FE and CD are produced to meet at Z.

a) Prove that $\triangle XYZ$ is equilateral.

b) If AB is 8 cm long, find the length of XY.

10.

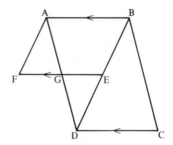

AB, EF and CD are parallel and equal. G is the midpoint of AD.

a) Show that △AFG is similar to △BDC.

b) Show that △AFG is congruent with △DEG.

c) Find the ratio of the area of △AFG to the area of △BDC.

d) Find the ratio of the area of △AFG to the area of ABCD.

CIRCLES

This section revises the angle and tangent properties of a circle.

The angle at the centre of a circle is equal to twice the angle subtended at the circumference by the same arc.

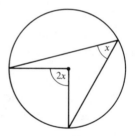

All the angles subtended at the circumference by an arc of a circle are equal.

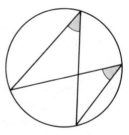

The angle in a semicircle is 90°.

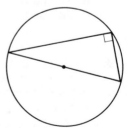

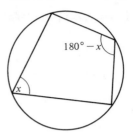

The opposite angles of a cyclic quadrilateral add up to 180°.

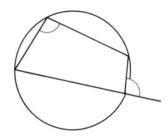

The exterior angle of a cyclic quadrilateral is equal to the interior opposite angle.

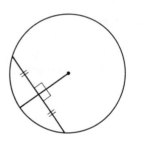

The radius through the midpoint of a chord is perpendicular to the chord.

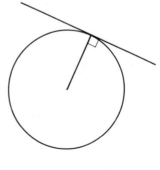

The angle between a tangent and the radius through the point of contact is 90°.

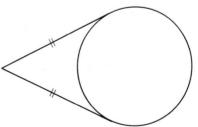

The two tangents drawn from an external point to a circle are of equal length.

THE ALTERNATE SEGMENT THEOREM

The angle between a tangent and a chord drawn from the point of contact is equal to any angle in the alternate segment.

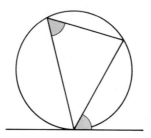

EXERCISE 5b

Questions 1 to 12 can be done without using the Alternate Segment Theorem. Questions 13 to 17 need the Alternate Segment Theorem.

In questions 1 to 6 find the sizes of the marked angles.

1.

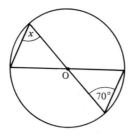

2.

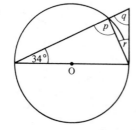

3.

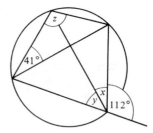

4.

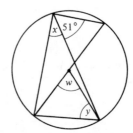

5.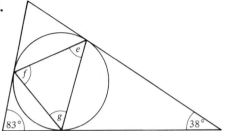

6.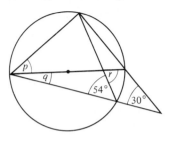

7. A, B, C and D are four points on the circumference of a circle, and $A\widehat{B}C = 90°$. Prove, with reasons, that $A\widehat{D}C = 90°$.

8. ABC is an equilateral triangle inscribed in a circle, centre O. A radius is drawn from O through the midpoint of AB to meet the circumference of the circle at D. Prove that $\triangle$ODA is equilateral.

9.

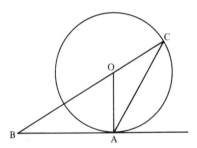

BA is a tangent to the circle at A and O is the centre of the circle. $O\widehat{B}A = 40°$. Find $O\widehat{A}C$.

10.

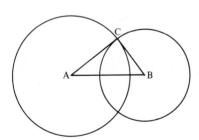

A is the centre of the larger circle whose radius is 4 cm and B is the centre of the smaller circle whose radius is 3 cm. AC is a tangent to the circle with centre B. Find the length of AB.

11.

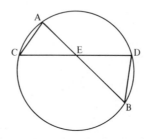

AB and CD are two chords of a circle that cut at E.

a) Prove that △AEC is similar to △DBE.

b) Hence show that AE × EB = CE × ED.

c) Use the result in (b) to find EB when AE = 6 cm, CE = 8 cm and ED = 9 cm.

12. PQ and RS are two chords of a circle that cut inside the circle at M. Use a method similar to that in question 11 to prove that PM × MQ = RM × MS.

In questions 13 to 16, TC is a tangent. Find the sizes of the marked angles.

13.

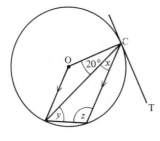

15.

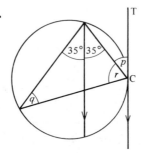

14.

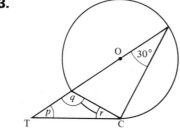

16.

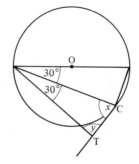

17. ABC is an isosceles triangle inscribed in a circle. AB = AC. Prove that the tangent to the circle at A is parallel to BC.

THE INTERSECTING CHORD THEOREM

The result proved in questions 11 and 12 of the last exercise is known as *the intersecting chord theorem* and can be quoted.

If AB and CD are two chords of a circle that intersect inside the circle at P, then

$$AP \times PB = CP \times PD$$

EXERCISE 5c

Use the intersecting chord theorem to find the value of x in each diagram. All lengths are in centimetres.

1.

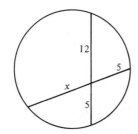

4.

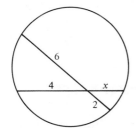

2.

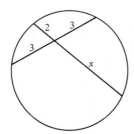

5.

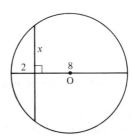

3.

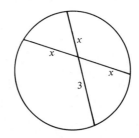

6.

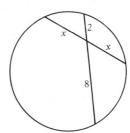

7.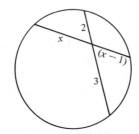

9.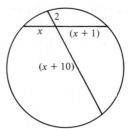

8.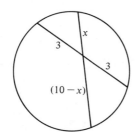

10.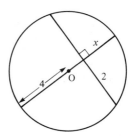

CONSTRUCTIONS USING RULER AND COMPASSES

Sometimes a construction has to be done without the aid of a protractor or a set square. Here is a reminder of the most useful 'ruler and compasses only' constructions.

In these diagrams, the positions where the point of the compasses have to be placed are marked P_1, P_2... for the first, second, ... positions.

To construct an angle of 60° at P_1

This involves constructing an equilateral triangle but drawing only two sides. Keep the radius the same throughout.

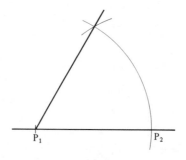

To bisect angle AP₁B

Keep the radius the same throughout.

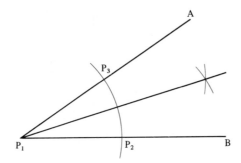

To construct an angle of 30° at P₁

Construct an angle of 60° at P₁ and then bisect it. Keep the radius the same throughout.

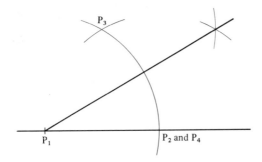

To construct an angle of 90° at P₁

Bisect an angle of 180° at P₁. Enlarge the radius for the arcs drawn from P₂ and P₃.

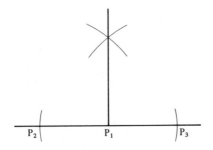

To construct an angle of 45° at P_1

Construct an angle of 90° at P_1 and then bisect it.

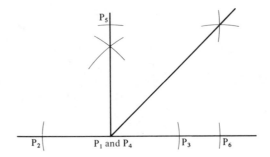

To draw a perpendicular from P_1 to the line AB

Keep the radius the same thoughout.

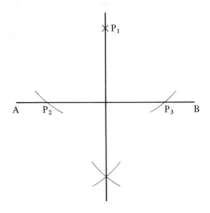

To bisect a line AB

Keep the radius the same throughout.

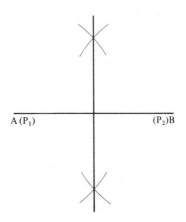

ACCURACY

When they are done properly, ruler and compasses constructions give more accurate results than using the protractors and set squares commonly available. That accuracy depends, however, on

1. using compasses that are reasonably stiff (a loose joint means that the radius will change in use)

2. using a *sharp* pencil, preferably an H, and keeping it sharp

3. making the construction as large as is practical, in particular *not* using a radius on the compasses that is too small. Aim for a minimum radius of about 5 cm, and do not rub out any arcs.

EXERCISE 5d

Construct the following figures using only a ruler and a pair of compasses. Check your construction by measuring the length marked x. Your result should be within 1 mm of the given answer.

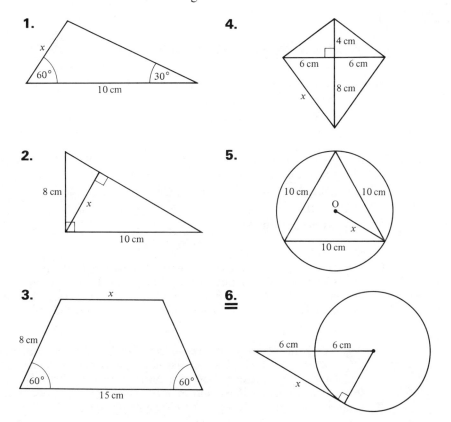

SKETCHES

In the last exercise, a sketch was given of the shape to be drawn accurately. Sometimes, however, only a description of the figure is given. In this case a sketch is essential because it gives some understanding of the shape and properties of the accurate drawing that has to be produced.

A sketch should be drawn freehand, but neatly. It should be big enough for essential information to be marked. Sometimes a first attempt at a sketch does not fit the given information. When this happens make another sketch.

The next exercise gives practice in producing sketches and getting information from them.

EXERCISE 5e

Draw sketches of the figures described below.

1. ABCD is a quadrilateral in which $AB = 10$ cm, $BC = 8$ cm, $BD = 10$ cm and $A\widehat{B}C = B\widehat{C}D = 90°$.
   a) Find CD.   b) Find the area of ABCD.

2. ABC is an equilateral triangle of side 6 cm. The line which bisects $\widehat{B}$ cuts AC at D and BD is produced to E so that $DE = BD$.
   a) What type of quadrilateral is ABCE ? Give reasons.
   b) Find BE.
   c) Find the area of ABCE.

3. A flagpole PQ stands on level ground. Q is the foot of the pole and R and S are two points on the ground such that QRS is a straight line. $RS = 12$ m, the angle of elevation of P from R is $60°$ and of P from S is $30°$. Find
   a) the angles of $\triangle PRS$ b) PR   c) the height of the pole.

4. A garden is in the shape of a trapezium ABCD in which $AB = 50$ m and $DC = 40$ m. AD is the back of the house; it is perpendicular to both AB and DC and is of length 20 m. The owner wants to plant a tree that is at least 40 m from the house and at least 5 m from any fence. Describe the shape of the region in which the tree can be planted.

5. A first-floor window A is 4m above ground level, and overlooks a level garden with a tree PQ standing in it. From A, the angle of depression of Q, the foot of the tree, is $30°$ and the angle of elevation of P, the top of the tree, is $60°$.

Find   a) the angles in $\triangle APQ$
         b) the length AQ
         c) the height of the tree.

SCALES

The construction of a scale drawing usually requires the choice of a scale and the calculation of lengths to be used. The following exercise gives practice in these techniques, and in interpreting information from scale drawings.

EXERCISE 5f

1. A quadrilateral PQRS is to be drawn to scale. PQ = 92 m and QR = 76 m.

a) Find what the lengths of PQ and QR should be if the scale is
   i) 1 cm to 10 m   ii) 1 cm to 8 m

b) On a drawing using a scale of 1 cm to 10 m, the measured length of PR is 6.8 cm. What is the real length of PR ?

c) On a drawing using a scale of 1 cm to 8 m, the measured length of QS is 7.2 cm. What is the real length of QS ?

d) Which scale was easier to use ? Would a scale of 1 cm to 5 m be easier to use than 1 cm to 8 m ? Give your reasons.

2. A map is drawn to a scale of $1 : 50 000$. Two towns A and B are 36 km apart.

a) What real length does 1 cm on the map represent ?

b) How far apart are the points on the map representing A and B ?

c) On the map, town C is 42 cm from A. How far apart are the towns C and A ?

3.

This is an extract from an Esso road map of France. The scale is 1 : 1 000 000.
Find, in kilometres, the straight line distance between

a) Beziers and Narbonne

b) Carcassonne and Narbonne

c) Caunes and Sigean.

4.

This is an extract from an Ordnance Survey road map whose scale is
1 inch : 3 miles. Find, in miles, the straight line distance between

a) junctions 3 and 4 on the M18

b) junctions 1 and 4 on the M18

c) junction 1 on the M18 and Hickleton.

5. A length which represents 800 km is drawn to a scale of 1 cm to 100 km but an error of 1 mm is made.

a) What error, in kilometres, does the error represent ?

b) What is the percentage error ?

The length is redrawn to a different scale but an error of 1 mm is again made. What length, in kilometres, does the error represent if the scale is

c) 1 cm to 200 km

d) 1 cm to 50 km ?

What advice would you give to someone who has to choose the scale for a drawing ?

6. A triangle PQR is to be constructed using a scale of 1 cm to 5 m. PR = 48 m and $\widehat{Q} = 90°$.

a) If $\widehat{R} = 25°$, calculate the length of PQ on the drawing.

b) When drawing the triangle to scale, an error is made in drawing $\widehat{R}$ as 26° instead of 25°. The other measurements are correct. Calculate the drawn length of PQ in this case and hence find the percentage error in PQ.

c) Construct △PQR accurately when $\widehat{R} = 25°$. Comment on any difference between your measured length of PQ and the calculated answer in (a).

SCALE DRAWING

Remember never to draw right angles or parallel lines by eye. Use a protractor, a pair of compasses or a ruler and set square after you have studied the rough sketch and decided on a method.

EXERCISE 5g

The illustrations in this exercise are not drawn to scale.

1.

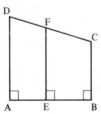

The diagram shows the plan of two gardens. DFC is a straight line. AD = 15 m, AB = 20 m, BC = 11 m and AE = 9 m.

a) Draw the plan accurately, using a scale of 1 cm to 2 m.

b) Use your drawing to find the real lengths of AC, BD and EF.

c) Calculate the areas of the two gardens, and hence find the ratio of the area of ADFE to that of EFCB.

2.

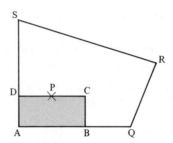

The diagram shows the grounds of a school, with the school building in one corner. There is a door at P, the midpoint of CD, and there are gates at Q, R and S.   AS = 83 m,   AQ = 80 m,   QR = 41 m, $\widehat{Q}$ = 95°,   AB = 42 m   and AD = 20 m.

a) Draw the plan accurately, using a scale of 1 cm to 10 m.

b) Andrew comes out of the door at P. If he can cross the grounds as he pleases, which of the three gates is nearest to P ? (Give reasons for your answer.)

c) Draw a perpendicular from R to ABQ produced. Calculate the true area of AQRS, measuring from the drawing any lengths that you need.

3.

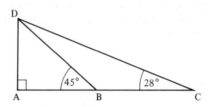

AD is a pole standing on level ground. The angle of elevation of D from B is 45° and of D from C is 28°. BC is 8.4 m.

a) Choose your own scale and draw an accurate diagram (start by drawing △BCD).

b) How high is the flagpole ?

4.

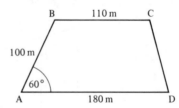

ABCD represents a paddock, which is in the shape of a trapezium with BC parallel to AD. Using a scale of 1 cm for 10 m, make a scale drawing of the paddock.

a) Find the perimeter and area of the paddock.

b) The owner wants to plant some trees in the paddock. They must be at least 70 m from D, and not more than 40 m from BC. Shade the area in which the trees can be planted.

5.

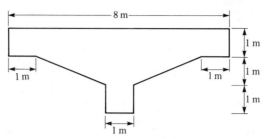

The diagram shows the cross-section of a girder for a motorway bridge.

The girder has one line of symmetry.

Use a scale of 1 cm to 0.5 m to make a scale drawing of the cross-section.

a) Take measurements from your drawing to find the perimeter of this cross-section.

b) Find the volume of concrete needed to make a girder 40 m long.

BEARINGS

Bearings are measured from a north-pointing line in a clockwise direction.

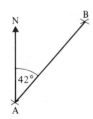

The bearing of B from A is 042 °.

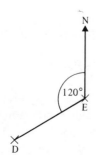

The bearing of D from E is (360 − 120) °, i.e. 240 °.

EXERCISE 5h

1.

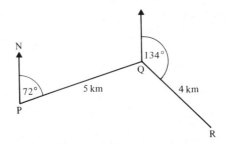

Abel walks 5 km on a bearing of 072° from a point P to point Q and then for 4 km on a bearing of 134° to point R.   a) Find $P\hat{Q}R$.   b) Draw $\triangle PQR$ accurately using a scale of 2 cm to 1 km, and find how far R is from P.

2.   Baxter drives 36 km due South from L to M and then he drives 28 km on a bearing of 292° to K. Choose a suitable scale and use an accurate drawing to find the direct distance from K to L.

3.   Carlos sails from port A for 20 nautical miles on a bearing of 040° to a port B. He then sails for 40 nautical miles on a bearing of 290° to a port C. Choose a suitable scale and use an accurate drawing to find the distance and bearing of C from A.

4.

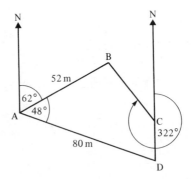

The diagram shows part of a park with two large trees A and B and the positions C and D where new trees are to be planted.

a) Sketch the diagram and calculate the remaining angles of the quadrilateral.

b) What is the bearing of A from D ?

c) Draw the quadrilateral accurately using a scale of 1 cm to 10 m. (Notice that there is no need to draw the north-pointing lines; start with AD.)

d) Use your drawing to find the distance apart of the trees at A and C.

e) Use your drawing to find the bearings of C from A and of B from D.

5. Two people set out from a point P at the same time. The first person, A, walks at 3.5 km/h on a bearing of 025° and the second person, B, walks at 4.5 km/h on a bearing of 065°.

a) Draw a sketch showing the positions of A and B twenty minutes after leaving P.

b) Make an accurate drawing of your sketch, using a scale of 10 cm to 1 km.

c) Use your drawing to find the distance and bearing of A from B twenty minutes after leaving P.

6 SOCIAL ARITHMETIC

This chapter uses arithmetic in a variety of everyday situations.

IMPERIAL UNITS

As Imperial units are still part of our everyday vocabulary it is sensible to be familiar with those in common use. These are given below.

Units of length: the inch (in), the foot (ft), the yard (yd), the furlong and the mile

$$1\,\text{ft} = 12\,\text{inches}$$
$$1\,\text{yd} = 3\,\text{ft}$$
$$1\,\text{furlong} = 220\,\text{yd}$$
$$1\,\text{mile} = 1760\,\text{yd}$$

⊢————————————⊣
1 inch

Units of area: the square yard, the acre, the square mile (occasionally the square inch and square foot)

Units of volume: cubic feet and cubic yards

Units of capacity: pints and gallons
$$1\,\text{gallon} = 8\,\text{pints}$$

Units of weight: the ounce (oz), the pound (lb), and the ton; occasionally the stone and the hundredweight (cwt)

$$1\,\text{lb} = 16\,\text{oz}$$
$$1\,\text{stone} = 14\,\text{lb}$$
$$1\,\text{cwt} = 112\,\text{lb}$$
$$1\,\text{ton} = 2240\,\text{lb}$$

EXERCISE 6a

Express the given quantity in terms of the unit in brackets.

1. 4 ft (in)

2. $\frac{1}{2}$ lb (oz)

3. 36 in (ft)

4. 15 ft (yd)

5. $6\frac{1}{2}$ stones (lb)

6. $2\frac{1}{2}$ gallons (pints)

7. 20 oz (lb)

8. 12 pints (gallons)

9. $\frac{1}{6}$ yd (in)

10. 2 sq ft (sq in)

11. $\frac{3}{4}$ lb (oz)

12. 5280 yd (miles)

13. A dripping tap fills a 1 pint milk bottle in 3 minutes. How many gallons of water drip from the tap in 24 hours ?

14. A rectangular room is 12 ft wide and 15 ft long. What is the floor area in square yards ?

15. A recipe asks for $\frac{1}{2}$ gill of cream. One pint is 20 fluid ounces and 1 gill is $\frac{1}{4}$ pint. How many fluid ounces of cream are required ?

16. One acre is 4840 sq yd. A farmer buys a square field of 40 acres. What is the perimeter of the field ?

17. a) Find out why a furlong is so called.
b) What is the area, in square yards, of a square field of side one furlong ?
c) What is the relationship between an acre and one square furlong ?

18. Find out why a hundredweight is 112 lb and not 100 lb.

EQUIVALENCE BETWEEN METRIC AND IMPERIAL UNITS

In supermarkets it is not unusual to find apples, say, in priced prepacked bags of 500 g or 1 kg but to find loose apples sold by the pound.

This mixture of Imperial and metric units in everyday use means that we should all know rough equivalents for corresponding units.

Length: 8 km ≈ 5 miles

            1 metre ≈ 39 inches   (1 m is roughly 1 yd)

            10 cm ≈ 4 inches

Area: 1 hectare ≈ 2.5 acres

Weight: 1 tonne ≈ 1 ton   (1 tonne is slightly less than 1 ton)

1 kg ≈ 2.2 lb

100 g ≈ 3.5 oz   (100 g is roughly a quarter of a lb)

Capacity: 1 litre ≈ 1.75 pints

1 gallon ≈ 4.5 litres

EXERCISE 6b

The use of rough equivalents cannot give exact answers, so bear this in mind when giving answers.

The cost of posting a first class letter is 19 p for a letter weighing up to 60 g and 26 p for a letter weighing over 60 g and up to 100 g. A letter weighs 2 oz on my kitchen scales. What is the value of the stamps I should put on it ?

$$100\,g \approx 3.5\,oz$$

∴ $$1\,oz \approx \frac{100}{3.5}\,g$$

so $$2\,oz \approx \frac{100 \times 2}{3.5}\,g$$

$$= 57\,g \quad (\text{to the nearest g})$$

A 19 p stamp is enough.

1.   The petrol tank on my car holds 12 gallons. How many litres is this ?

2.   After an excise duty increase, the price of petrol is quoted on the radio as £ 2.10 a gallon. How much is this per litre ?

3.   An old dressmaking pattern states that the seam allowance is $\frac{5}{8}$ inch. How many millimetres is this ?

4. A recipe for jam requires 20 lb of sugar. How many 1 kg bags of sugar are needed ?

5. Sarat knows that he is 5 ft 10 in tall, but he has to give his height in metres on his application form for a passport. What should he enter for his height ?

6. A farm is advertised for sale as 140 hectares. How many acres is this ?

7. Apples are sold in 1 kg bags at 98 p, or loose at 42 p a pound. Which are cheaper ?

8. Floorboards are sold in units of 30 cm. (30 cm is sometimes called a metric foot !) What length of floorboard should be bought to replace one that is 12 ft long ?

9. An old knitting pattern requires twenty 1 oz balls of double knitting yarn. How many 50 g balls are needed to make up this pattern ?

10. A particular brand of carpet is priced at £ 8.75 per square yard in a local shop and at £ 9.60 per m² in a department store. Which is cheaper ?

RATES OF CONSUMPTION

The rate at which quantities are consumed is usually expressed as an amount of one quantity with respect to one unit of another.

Speed is the rate at which a moving object covers distance and it is given as the distance covered in one unit of time.

The common units for speed are kilometres per hour (km/h), miles per hour (m.p.h.) and metres per second (m/s).

Fuel consumption is the rate at which fuel is used. For cars etc. it is usually expressed as distance covered on one unit quantity of fuel, e.g. miles per gallon (m.p.g.), kilometres per litre (km/l). It is also often given as litres per 100 km.

When vast quantities of fuel are used, it is given as the fuel used in covering a unit of distance or time, e.g. gallons per mile or litres per second.

Coverage of paint, fertilizer, spray etc. is the rate at which it covers (or should cover) area, so it is expressed in quantity of liquid per unit of area; for example, litres per square metre (l/m²).

EXERCISE 6c

Express a speed of 60 m.p.h. in m/s using $8 \text{ km} \approx 5 \text{ miles}$.

$$60 \text{ m.p.h.} = \frac{60 \times 8}{5} \text{ km/h}$$

$$= \frac{60 \times 8 \times 1000}{5} \text{ m/h}$$

$$= \frac{60 \times 8 \times 1000}{8 \times 60 \times 60} \text{ m/s}$$

$$= \frac{80}{3} \text{ m/s} = 26.7 \text{ m/s} \quad (\text{correct to 3 s.f.})$$

i.e. $60 \text{ m.p.h.} \approx 27 \text{ m/s}.$

In questions 1–10, give answers correct to 3 s.f. when possible. When that degree of accuracy is not possible, give 2 s.f.

1. Express 80 km/h in m/s.

2. Express 100 m/s in km/h.

3. Express 40 m.p.h. in ft/s.

4. Express 40 km/h in m.p.h.

5. Express 70 m.p.h. in km/h.

6. Express 500 litres/second in litres/hour.

7. Express 50 m.p.g. in kilometres/litre.

8. Express 50 m²/litre in sq ft/gallon.

9. Express 25 litres/second in gallons/minute.

10. Express 4 kg/cm² in lb/sq in.

In questions 11–15, use your own judgement with regard to the context of the question, as to the accuracy required for the answer.

11. The average petrol consumption of a car is 35 m.p.g. The driver puts 25 litres of petrol into the tank. How far can the car be expected to travel ?

12. A bottle of liquid insecticide contains 500 ml. The instructions state this it should be diluted in the ratio 2.5 ml of insecticide to 2 litres of water and the diluted mixture applied at the rate of 250 ml/m². What area will the contents of the bottle treat ?

13. A car manual states that the front tyre pressures should be 3.5 kg/cm², and the pressure gauge at the local garage measures pressure in lb/sq in. What should the gauge read to give the correct pressure ?

14. A petrol pump can deliver petrol at the rate of $\frac{1}{2}$ litre/second. How long does it take to fill a tank that holds 20 gallons ?

15. A firm uses packaging cartons that are cubes of side 2 ft. The full cartons are lifted into containers at the rate of 5 cartons per minute. How long does it take to fill a container measuring 3.5 m by 5 m by 10 m ?

TIME AND TIMETABLES

There are two systems for giving the time of day.

One system uses two periods of twelve hours. The first starts at midnight and continues to noon; times within this period are referred to as 'a.m.' The second period goes from noon to midnight and times are referred to as 'p.m.'

The other system has one period of twenty-four hours, starting at midnight and continuing through to the next midnight.

Thus 1 hour after noon is 1 p.m. in the first system and 1300 hours (read as 'thirteen hundred hours') or 13.00 in the second system.

Most clocks and watches with dials use the a.m./p.m. system whereas most digital clocks and watches, and timetables, use the 24 hour system.

EXERCISE 6d

1. Give the equivalent times on a 24-hour clock.
  a) 8.30 a.m.   b) 8.30 p.m.   c) 5.42 a.m.   d) 2.36 p.m.

2. Give the equivalent time on a 12-hour clock, using a.m. or p.m.
  a) 03.00   b) 19.42   c) 08.51   d) 22.43

Find the elapsed time between 9.20 a.m. and 3.52 p.m.

Method 1

Elapsed time from 9.20 a.m. to noon is

$$12\,h - 9\,h\,20\,min = 2\,h\,40\,min$$

Elapsed time from noon to 3.52 p.m. is 3 h 52 min.

∴   Elapsed time from 9.20 a.m. to 3.52 p.m. is

$$2\,h\,40\,min + 3\,h\,52\,min = 6\,h\,32\,min$$

Method 2

9.20 a.m. = 09.20   and   3.52 p.m. = 15.52.
Elapsed time is from 09.20 to 15.52.

i.e. 6 h 32 min

(Remember that 15.52 means 15 hours 52 minutes.)

---

3. Find the elapsed time between

a) 03.20 and 15.08 hours on the same day

b) 11.30 a.m. and 5.42 p.m. on the same day

c) 20.35 and 09.40 on the next day

d) 10.40 a.m. and 7.52 a.m. on the next day.

Here is part of a timetable for trains from Paddington to Bristol and Weston-super-Mare.

```
Paddington*          d  00 50  06 35  06 55  07 25  07 35  08 00  08 05  08 35  09 00  09 05  09 35
Slough               d   ____  06 48  07 08  07 38  07 48   ____   ____  08 48   ____   ____  09 48
Reading C*           d   ____  07 03  07 23  07 53  08 03  08 24  08 30  09 03  09 24  09 30  10 03
Didcot               d   ____  07 17  07 34  08 06  08 16   ____   ____  09 16   ____   ____  10 16
Swindon              a  02 17  07 37  07 54  08 27  08 37   ____   ____  09 37   ____   ____  10 37
Chippenham           a   ____  07 51   ____  08 40  08 50   ____   ____  09 50   ____   ____   ____
Bath Spa             a   ____  08 09   ____  08 52  09 02   ____   ____  09 16  10 02   ____  10 16  ____
Bristol Parkway      a   ____   ____  08 21   ____   ____  09 12   ____   ____  10 12   ____  11 01
Bristol Temple Meads a   ____  08 25   ____  09 07  09 17   ____   ____  09 31  10 17   ____  10 31  ____
Nailsea & Backwell   a   ____   ____   ____  09 25   ____   ____  10 11   ____   ____   ____   ____
Yatton               a   ____   ____   ____  09 32   ____   ____  10 18   ____   ____   ____   ____
Weston-super-Mare    a   ____   ____   ____  09 41   ____   ____  10 32  10 38   ____  11 12   ____

Paddington*          d  10 05  10 35  11 00  11 05  11 35  12 05  13 00  13 05  13 35  13 35  14 00
Slough               d   ____  10 48   ____   ____  11 48   ____  12 25   ____  13 48  13 48   ____
Reading C*           d   ____  10 03  11 24  11 30  12 03   ____  13 24  13 30  14 03  14 03   ____
Didcot               d   ____  11 16   ____   ____  12 16   ____   ____   ____  14 16  14 16   ____
Swindon              a  10 55  11 37   ____   ____  12 37  12 55   ____  14 00  14 37  14 37  14 49
Chippenham           a   ____  11 50   ____   ____   ____  13 08   ____   ____   ____  14 50
Bath Spa             a  11 17  12 02   ____  12 16   ____  13 20   ____  14 22   ____  15 02   ____
Bristol Parkway      a   ____   ____   2 12   ____  13 01   ____  14 12   ____   ____  15 14
Bristol Temple Meads a  11 32  12 17   ____  12 31   ____  13 35   ____  14 37   ____  15 17   ____
Nailsea & Backwell   a  12 11   ____   ____  13 21   ____  14 21   ____  15 22   ____  16 26   ____
Yatton               a  12 18   ____   ____  13 28   ____  14 28   ____  15 29   ____  16 33   ____
Weston-super-Mare    a  12 34  12 38   ____  13 42   ____  14 42   ____  15 42   ____  16 45   ____
```

Heavy type indicates through trains and light type indicates connecting trains.

If I catch the 11.05 from Paddington, to go to Weston-super-Mare, do I have to change trains, and if so where ? How long should the journey take ? (Use the timetable on p. 114.)

The 11.05 is in heavy type to Bristol Temple Meads and in light type beyond there, so I have to change trains at Bristol Temple Meads.

Arrival time in Weston-super-Mare is 13.42, so the journey should take   13 h 42 min − 11 h 5 min

i.e. 2 h 37 min

Use the timetable on the previous page to answer questions 4 to 9.

4. What is the earliest through train from Paddington to Weston-super-Mare, and how long does it take ?

5. Mrs Angelon has an appointment in Swindon at 2.00 p.m. She wants to travel by train from Slough and has to arrive at Swindon at least 15 minutes before 2.00 p.m. What is the time of the latest train she can catch ? How long will she have in Swindon before her appointment if the train is on time ?

6. Which train from Paddington gives the quickest journey to Weston-super-Mare ? Is this a through train ?

7. How long do the fastest trains take to go from Paddington to Bristol Temple Meads ?

8. How long do the fastest trains take to go from Paddington to Bristol Parkway ?

9. Mr Black wanted to go to Bath Spa but mistakenly caught the 09.00 from Paddington instead of the 09.05.

a) If Mr Black realises his mistake soon enough he can change trains and still arrive in Bath Spa when he originally intended. Where does he have to make the change and how long should he have to wait there to catch the correct train ?

b) If Mr Black does not realise that he is on the wrong train soon enough to make the change in (a), where does he next have the opportunity to get off the train ?

This is an extract from a British Rail timetable for suburban services in the Glasgow area. Use this timetable to answer questions 10 to 15.

							A C				and at the same minutes past						
Lanark ....d														1524				
Carluke ....d														1534				
Wishaw ....d														1539				
Motherwell ....a	0911b		0918c											1545			1548c	
....d	0915		0922			0952								1545			1552	
Bellshill ....225 d	0921													1551				
Uddingston ....225 d	0925													1555				
Hamilton Central ....d		0928				0958								1558			1558	
Hamilton West ....d		0931				1001								1601			1601	
Blantyre ....d		0934				1004								1604			1604	
Newton ....23 d		0938				0958 1008								1608			1608	
Cambuslang ....225 d		0941				1001 1011								1601			1611	
Rutherglen ....225 d		0944				1004 1014								1604			1614	
Dalmarnock ....d		0948				1008 1018								1608			1618	
Brigeton ....d		0950				1010 1020								1610			1620	
Airdrie ....d				0946				1001	1016						1546			
Coatdyke ....d				0948				1003	1018						1548			
Coatbridge Sunnyside ....d				0951				1006	1021						1551			
Blairhill ....d				0953				1008	1023						1553			
Easterhouse ....d				0957				1012	1027						1557			
Garrowhill ....d				0959				1014	1029						1559			
Shettleston ....d				1001				1016	1031						1601			
Carntyne ....d				1003				1018	1033						1603			
Cumbernauld ....d					0920		0950											
Springburn ....a					0936		1006									1550		1606
Barnhill ....d				0939		1009											1609	
Alexandra Parade ....d				0940		1010											1610	
Duke Street ....d					1013	1013											1613	
Bellgrove ....d				0944		1014											1614	
High Street ....d				0946		1016							1559			1606	1616	
Glasgow Queen Street ....a				0948		1018							1601			1611	1618	
Charing Cross ....d				0951		1021							1603			1613	1621	
....d				0953		1023											1623	

Station	—	0951 0953	0959 1001	1011 1013	1021 1023	1029 1031	1041 1043	past each hour until	1601 1603	1611 1613	1621 1623	—
Glasgow Queen Street …d	…	0951	0959	1011	1021	1029	1041		1601	1611	1621	…
Charing Cross …d	…	0953	1001	1013	1023	1031	1043		1603	1613	1623	…
Argyle Street …d	0943	…	…	…	…	…	…		…	…	…	…
Glasgow Central .223. 225 d	0945	…	…	…	…	…	…		1608	…	…	1623
Anderston …d	0947	…	…	…	…	…	…		1610	…	…	1625
Finnieston § …d	0949	…	…	…	…	…	…		1612	…	…	1627
Partick …d	0953	0958	1005	1018	1028	1035	1048		1614	1618	1628	1629
Hyndland …d	0955	1000	1007	1020	1030	1037	1050		1616	1620	1630	1633
Jordanhill …d	0957	…	…	…	…	…	…		1618	…	…	1635
Scotstounhill …d	0959	…	…	…	…	…	…		…	…	…	…
Garscadden …d	1001	…	…	…	…	…	…		1621 →	…	…	…
Yoker …d	1003	…	…	…	…	…	…		…	…	…	…
Clydebank …d	1006	…	…	…	…	…	…		…	…	…	…
Anniesland …d	…	1002	…	1022	1032	…	1052		…	1622	1632	1637
Westerton …d	…	1005	…	…	1035	…	…		…	1625	1635	1640
Bearsden …d	…	1008	…	…	1038	…	…		…	…	1638	…
Hillfoot …d	…	1010	…	…	1040	…	…		…	…	1640	…
Milngavie …a	…	1014	…	…	1044	…	…		…	…	1644	…
Drumchapel …d	…	…	1013	1028	…	1043	1058		1613	1628	…	1638a →
Drumry …d	…	…	1015	1030	…	1045	1100		1615	1630	…	…
Singer …d	…	…	1017	1032	…	1047	1102		1617	1632	…	…
Dalmuir …d	1008a	…	1019a	1034	…	1049a	1104		1619a	1634	1623a	1638a
Kilpatrick …d	…	…	…	1037	…	…	1107		…	1637	…	…
Bowling …d	…	…	…	1040	…	…	1110		…	1640	…	…
Dumbarton East …d	…	…	1024	1045	…	1054	1115		…	1645	…	…
Dumbarton Central …d	…	…	1026	1047	…	1056	1117		…	1647	…	…
Dalreoch …d	…	…	1027	1048	…	1057	1118		…	1648	…	…
Renton …d	…	…	…	1051	…	…	1121		…	1651	…	…
Alexandria …d	…	…	…	1054	…	…	1124		…	1654	…	…
Balloch Central …a	…	…	…	1057	…	…	1127		…	1657	…	…
Balloch Pier ‡ …a	…	…	…	…	…	…	…		…	…	…	…
Cardross …d	…	…	1032	…	…	1102	…		…	…	…	…
Craigendoran …d	…	…	1037	…	…	1107	…		…	…	…	…
Helensburgh Central …a	…	…	1041	…	…	1111	…		…	…	…	…

For general notes see pages 2–4
Voir pages 4–6 de l'indicateur pour les renseignements généraux et l'explication des signes
Allgemeine Bemerkungen und Zeichenerklärung siehe Seiten 7–9 des Kursbuches

§ For Scottish Exhibition Centre
‡ This station may be closed during the currency of the timetable

A 1231 Airdrie to Helensburgh Central stops additionally at Jordanhill dep. 1309 and Yoker dep. 1311
C 1416 Airdrie to Balloch Central is extended to Balloch Pier from 30 June to 29 August
→ continued in a later column

b From Glasgow Central via Hamilton
c From Glasgow Central via Bellshill
→ continued from an earlier column

Use the timetable on the previous pages to answer questions 10 to 15.

10. What is the earliest train from Lanark to Glasgow Central and how long does it take ?

11. What is the departure time from Motherwell of the earliest train to stop at Anniesland and how long does this journey take, assuming the train runs on time ?

12. What time does the first train after midday leave Motherwell for Glasgow Central ?

13. What time does the first train after 2 p.m. leave Glasgow Queen Street for Helensburgh Central ?

14. Is it possible to travel from Airdrie to Clydebank without changing trains ? Mrs Stuart wanted to go to Clydebank and she caught the 0946 train from Airdrie. Assuming that the trains ran on time,
a) where did she have to change trains,
b) how long did she have to wait for the next train,
c) when did she arrive in Clydebank ?

15. Mr Dodge arrived at Glasgow Queen Street station at 3 p.m., wanting to catch a train for Milngavie. How long did he have to wait and when should he have got to Milngavie ?

POSTAGE AND OTHER STEP-FUNCTIONS

This table gives postal rates (Sept 1986) for inland letters.

Rates for letters within the UK and from the UK to
the Isle of Man, the Channel Islands and the Irish Republic

Weight not over	First Class	Second Class	Weight not over	First Class	Second Class
60g	17p	12p	400g	69p	52p
100g	24p	18p	450g	78p	59p
150g	31p	22p	500g	87p	66p
200g	38p	28p	750g	£1.28	98p
250g	45p	34p	1000g	£1.70	Not admissible
300g	53p	40p	Each extra 250g		over 750g
350g	61p	46p	or part thereof	42p	

Considering just first class post, we see that the cost goes up in steps, depending on the weight. For example, a letter weighing up to 60 g costs 17 p, but when the weight goes over 60 g the cost jumps to 24 p and so on.

This graph, showing cost against weight, illustrates the 'step' nature of postal rates.

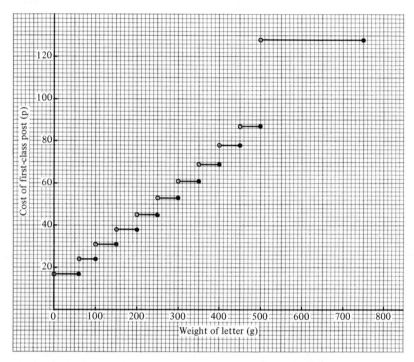

Notice that the 'open' circle at the left-hand end of each line indicates that the point is not included. The closed circle at the right-hand end of each line shows that the point is included.

Many of the charges for services have this 'step' structure. Other examples are parking charges and telephone charges.

EXERCISE 6e

Use the table for postal rates on p. 116 to find the cost of sending a packet weighing $2\frac{1}{2}$ kg by first class letter post.

The cost of the first 1 kg is £1.70.

The cost of the next $1\frac{1}{2}$ kg is $6 \times 42\,p = £2.52$.
($1\frac{1}{2}$ kg $= 6 \times 250\,g$).

The total cost is £4.22.

Use the table on p. 116 to answer questions 1 to 4.

1. Find the cost of first class letter post for a packet weighing
a) 370 g   b) 201 g   c) 100 g   d) 1.7 kg.

2. Find the cost of second class letter post for a letter weighing
a) 40 g   b) 158 g   c) 97 g.

3. What is the maximum weight that can be sent by second class letter post?

4. Copy the graph on p. 119 and use a coloured pencil to draw the lines representing the cost of second class post.

This table gives the charge rates for inland dialled telephone calls. Calls are charged in whole units of 5.2p, e.g. a local call of 3 minutes at standard rate is charged at 2 units of 5.2p each. Use the table to answer questions 5 to 8, giving answers to the nearest penny.

Call Charge Letter	Type of Call	Charge Rate	Time allowed for charge unit of 5.2p on any one call
L	**Local Calls**	**Cheap**	**8 mins**
		Standard	2 mins
		Peak	1 min 30 secs
a	**Calls up to 56 km (35 miles)**	**Cheap**	**144 secs**
		Standard	45 secs
		Peak	30 secs
b	**Calls over 56 km (35 miles) and to the Channel Islands**	**Cheap**	**48 secs**
		Standard	16 secs
		Peak	12 secs

5. Find the cost of a local telephone call lasting 20 minutes at
a) cheap rate   b) standard rate   c) peak rate.

6. Find the cost of a call over 56 km lasting 5 minutes at
a) cheap rate   b) standard rate   c) peak rate.

7. Draw a graph illustrating the cost of local telephone calls of up to 20 minutes duration, at peak rate. Use scales of 1 cm to 1 minute and 1 cm to 5 p.

8. Using the graph drawn for question 7, superimpose the graphs illustrating the cost of local calls at cheap rate and at standard rate. (Use different colours for the different graphs.)

9. A multistorey car-park in the town centre has the following charge tariff:

up to 2 hours	40 p
up to 4 hours	80 p
each additional hour (or part of)	100 p

How much does it cost to park for

a) $1\frac{1}{2}$ hours   b) $2\frac{1}{2}$ hours   c) 6 hours ?

Why do you think there is a steep increase in cost if parking is for more than four hours ?

10. A shop offers a dry cleaning service by weight. The charges are £2 for up to 3 lb and then 30 p per lb or part thereof, up to a maximum weight of 10 lb per load.

a) What is the cost of cleaning a load of clothes weighing $6\frac{1}{2}$ lb ?

b) Why do you think there is a maximum weight of 10 lb ?

c) Find the cost of cleaning a set of 6 curtains, each weighing 3 lb.

INSURANCE

Although all property and people are at risk, relatively few suffer loss. Insurance works by spreading the cost of loss amongst all who are insured.

Some forms of insurance are legal requirements. For example, any driver using a car or motor cycle on a public road must, by law, be insured for third party risks. Third party risk covers damage caused to other people or their property. Other forms of insurance are sensible precautions against possible loss or damage.

The payment for insurance is called the *premium*. The premium may be given in a form such as £2 per £100 insured value, or the premium may be quoted as a total amount for a particular insurance. Premiums are usually payable yearly.

Some insurances give a discount if the insured person bears part of the risk. For example a householder may agree to pay the first £50 of any loss. This has the misleading name of "an *excess*".

Car insurances usually offer discounts for several consecutive years in which no claims have been made. This form of discount is called a *no claims bonus*.

EXERCISE 6f

> The premium for insuring a building worth £86000 is £4.30 per £1000 value. Find the premium to be paid.
>
> $$\text{Premium} = £4.30 \times 86$$
> $$= £369.80$$

The table gives the premiums for insuring buildings and contents as quoted by Northern Star Insurance Co.

	Buildings/£1000	Contents/£1000
Area A	£1.80	£ 6.50
Area B	£2.10	£ 8.00
Area C	£2.20	£10.80
Area D	£2.80	£12.90

Use this table to answer questions 1 to 4.

1. Find the premium for insuring a house worth £150000 (building only) in area D.

2. Find the premium for insuring a house worth £80000 and its contents worth £9000 in area B.

3. Mr and Mrs Hadinsky live in Area A and value the contents of their house at £12000. This includes a piano worth £850, a ring worth £700 and a watch worth £350. One condition of the insurance cover is that all items whose value is greater than 5% of the total insured must be listed.

a) Find the premium to be paid.

b) State the items that have to be listed.

4. A householder living in area C has a house valued at £75000 with contents worth £8000. The discount on the buildings premium is 2% for an excess of £500 and on the contents the discount is 5% for an excess of £200. Find the premium for insuring house and contents if

a) no excess is agreed   b) both excesses are agreed.

5. a) The insurance premium for comprehensive cover on a small family car in Liverpool is £350. Daniel Kirby wants to insure himself to drive such a car but finds that because he is under 25 years of age, there is a 50% surcharge on the premium. What premium does he have to pay?

b) Three years later, when Daniel is over 25 years old, he no longer has to pay the surcharge and because he has made no claims on his insurance, finds that he is entitled to a 40% no claims bonus. However, the basic premium has increased to £420. Find the premium he has to pay now.

Insurance is only one part of the cost of running a car. In addition there are the annual road fund licence, repairs and maintenance, petrol and, lastly, payment for the vehicle itself. Some people also add in the depreciation of the value of the car.

6. Maya Liang wants to take out a loan to buy herself a second-hand car and decides to estimate what it will cost her for a year. The repayments on the loan are £94 a calender month. Maya has been quoted £250 premium for insurance. The road tax is £100 and she estimates that she will do 10000 miles at an average of 40 m.p.g. Petrol is £2.00 a gallon. Maya also estimates that she will need one service costing about £100.

How much does her estimate for 1 year's motoring come to?

How much is this a week?

7. Derek James buys a new car costing £10000 and decides to work out the cost of running the car for its first year. He includes repayments of £350 per month, depreciation of 25% of the purchase price, £500 for insurance, £100 road tax, a service charge of £200 and £14 a week for petrol.

How much does it cost him each week to run his car?

7 TRIGONOMETRY

THE TRIGONOMETRIC RATIOS

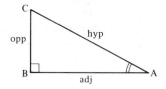

In a right-angled triangle ABC in which $\widehat{B} = 90°$,

$$\sin A = \frac{opp}{hyp} = \frac{BC}{AC} \qquad \cos A = \frac{adj}{hyp} = \frac{AB}{AC} \qquad \tan A = \frac{opp}{adj} = \frac{BC}{AB}$$

To help decide which ratio to use, mark the sides of the triangle as opp, adj, hyp, with respect to the angle to be found or used.

Remember that, if two angles in a triangle are known, the third angle can be found from the angle sum of a triangle.

Remember also that, in a right-angled triangle, if two sides are known, the third side can be calculated using Pythagoras' Theorem.

EXERCISE 7a

In $\triangle ABC$, $\widehat{B} = 90°$, $AB = 7.2\,cm$ and $AC = 9.3\,cm$.
Find   a) BC   b) $\widehat{C}$

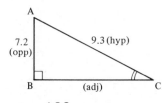

122

a)
$$AC^2 = BC^2 + AB^2 \quad (\text{Pythag. th.})$$
$$9.3^2 = BC^2 + 7.2^2$$
$$86.49 = BC^2 + 51.84$$
$$BC^2 = 86.49 - 51.84$$
$$= 34.65$$
$$BC = 5.886$$
$$\therefore \quad BC = 5.89 \text{ cm} \quad \text{correct to 3 s.f.}$$

b)
$$\sin C = \frac{\text{opp}}{\text{hyp}} = \frac{7.2}{9.3}$$
$$= 0.7742$$
$$\therefore \quad \widehat{C} = 50.7°$$

In $\triangle PQR$, $\widehat{P} = 90°$, $\widehat{R} = 38°$, $RP = 4.3$ cm. Find RQ.

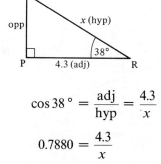

$$\cos 38° = \frac{\text{adj}}{\text{hyp}} = \frac{4.3}{x}$$
$$0.7880 = \frac{4.3}{x}$$
$$x \times 0.7880 = 4.3$$
$$x = \frac{4.3}{0.7880}$$
$$= 5.456$$

$RQ = 5.46$ cm correct to 3 s.f.

In questions 1 to 8, all lengths are in centimetres. Calculate x, giving angles correct to 1 d.p. and lengths correct to 3 s.f.

1.

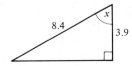

5.

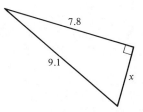

2.

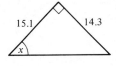

6.

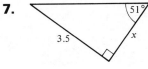

3.

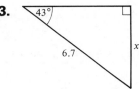

7.

4.

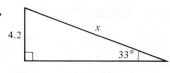

8.

9. In $\triangle PQR$, $\widehat{P} = 90°$, $PQ = 3.2$ cm and $QR = 5.7$ cm. Find $\widehat{Q}$ and $\widehat{R}$.

10. In $\triangle XYZ$, $\widehat{X} = 90°$, $XY = 2.5$ cm, $XZ = 3.2$ cm. Find YZ and $\widehat{Y}$.

11. In $\triangle ABC$, $\widehat{C} = 90°$, $AB = 4.3$ cm and $\widehat{A} = 54°$. Find $\widehat{B}$ and AC.

12. In $\triangle LMN$, $\widehat{M} = 90°$, $LM = 48$ cm, $LN = 64$ cm. Find $\widehat{L}$ and $\widehat{N}$.

PROBLEMS IN TWO DIMENSIONS

To calculate angles or lengths, first identify a right-angled triangle containing the unknown quantity. This often involves adding lines to a diagram.

EXERCISE 7b

1. A is the point $(4, -3)$, B is the point $(4, 3)$ and C is the point $(-2, -3)$. Find AC and A$\hat{C}$B.

2. From the top of a cliff, the angle of depression of a boat at sea is $23°$. The cliff is 40 m high. Calculate the distance of the boat from the foot of the cliff.

3.

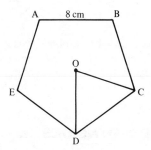

ABCDE is a regular pentagon. The point O is equidistant from all five vertices. Find

a) O$\hat{C}$D   b) the distance of O from DC   c) the area of the pentagon.

4. In a circle, centre O, radius 9 cm, a chord AB of length 12 cm is drawn. Find the distance of the chord from O.

5. Two tangents are drawn from a point P to a circle centre O. The radius of the circle is 8.2 cm and P is 12.4 cm from O. Find the angle between the tangents.

6. A rectangular frame measures 6 m by 2.5 m. The frame is kept in shape by two diagonal struts. Find
a) the length of a strut   b) the angle between the struts.

7.

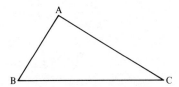

The diagram shows a vertical section through an asymmetrical roof.

The ridge A is 2.5 m above the base BC. AC is inclined at $30°$ to BC and AB is inclined at $60°$ to BC. Find the width (BC) of the base of the roof.

8. The road from village A runs due west for 5 miles to a fort B. A television mast is due south of B and 4 miles from B. Find the distance and bearing of the mast from A.

9. A is the point $(1,1)$, B is the point $(4,5)$ and C is the point $(5,3)$. Find
a) the angle between AB and a line parallel to the y-axis
b) the angle between BC and a line parallel to the y-axis.
c) Hence find $\widehat{ABC}$.

10.

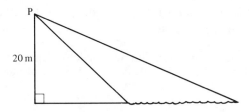

A river runs parallel to a block of flats. The roof of this block is 20 m above the water level of the river. From a point P on the roof, looking directly across the river, the angle of depression of the nearer bank is 43° and the angle of depression of the farther bank is 23°. Find the width of the river.

PROBLEMS IN THREE DIMENSIONS

To find angles or lengths in solid figures, a right-angled triangle containing the unknown quantity has to be identified. This triangle should then be drawn separately from the solid figure.

EXERCISE 7c

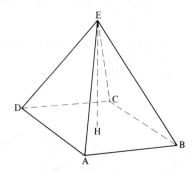

The figure is a pyramid on a square base ABCD. The edges of the base are 30 cm long and the height EH of the pyramid is 42 cm. Find   a) the length of AC   b) the angle $\widehat{EAH}$.

a)

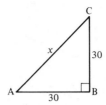

AC is a diagonal of the square base ABCD.

Using $\triangle ABC$, $\widehat{B} = 90°$, $AB = BC = 30$ cm

$$AC^2 = AB^2 + BC^2 \quad (\text{Pythag. th.})$$

$$\therefore \quad x^2 = 900 + 900$$

$$= 1800$$

$$x = 42.42$$

$$\therefore \quad AC = 42.4 \text{ cm correct to 3 s.f.}$$

b)

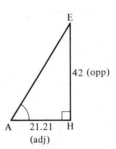

$E\widehat{A}H$ is in $\triangle AHE$, in which $\widehat{H} = 90°$ and

$AH = \frac{1}{2}AC = 21.21$ cm

$$\tan \widehat{A} = \frac{\text{opp}}{\text{adj}} = \frac{42}{21.21}$$

$$= 1.9802$$

$$\therefore \quad \widehat{A} = 63.2°$$

$$\text{i.e.} \quad E\widehat{A}H = 63.2°$$

1.

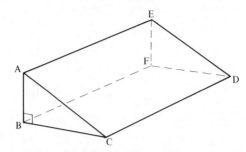

ABCDEF is a prism with a triangular cross-section. $\widehat{ABC} = 90°$, BC = 8 m, AB = 6 m and CD = 20 m.

Find   a) AC   b) BD   c) $\widehat{ACB}$ d) $\widehat{ADB}$
e) the surface area of the prism.

2.

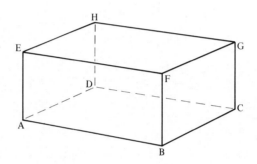

In the cuboid, AB = 12 cm, BF = 6 cm and BC = 8 cm.

Find   a) FC   b) AF   c) DB   d) $\widehat{HBD}$.
Draw △AFC accurately.

3.

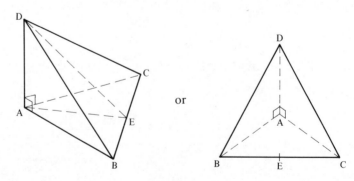

ABCD is a pyramid on an equilateral triangular base ABC of side 15 cm. AD is perpendicular to AC and AB, and is also 15 cm long. E is the midpoint of BC.

Find   a) AE   b) $\widehat{ABD}$ c) $\widehat{AED}$.

4.

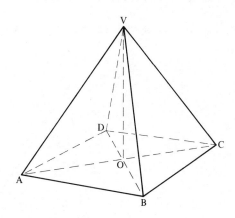

VABCD is a square-based pyramid with V vertically above O, the centre of the square ABCD.

If AB = 15 cm and VO = 20 cm find

a) AC   b) VÂO   c) AV   d) VÂB.

5.

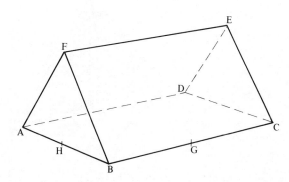

ABCDEF is a prism with cross-section an equilateral triangle of side 2 cm. BC = 5 cm, G is the midpoint of BC and H is the midpoint of AB.

Find   a) FG   b) BF̂G   c) AĜB   d) FĜH.

Sketch the solid whose vertices are F, H, G and B.

THE ANGLE BETWEEN A LINE AND A PLANE

Sometimes we are asked to find the angle between a line and a face in a solid.

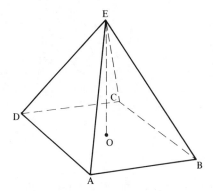

This is a square pyramid with E vertically above O, the centre of the base. Suppose that we want to find the angle between the edge EB and the base ABCD.

The angle between a line and a face is in the section of the solid that contains the line and is perpendicular to the face. In this case it is the section through E, B and D.

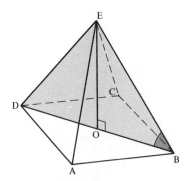

The required angle is the angle between the given line and the line where the section cuts the face. In this case the angle is EB̂D. This angle can be found using △EBO as there is a right angle at O.

Alternatively we can identify the angle between EB and the base ABCD by dropping a perpendicular from E to the base. The angle is the angle between EB and the line joining B to O, the foot of the perpendicular i.e. EB̂O.

EXERCISE 7d

You may find models of these solids helpful.

1.

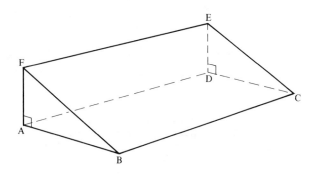

For this wedge, sketch and label a triangle containing the angle between

a) FB and the base ABCD

b) FC and the base ABCD

c) FC and the face ADEF

d) BE and the base ABCD.

2.

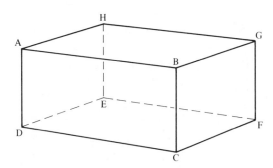

For this cuboid, sketch and label a triangle containing the angle between

a) AC and the base CDEF

b) AC and the face ADEH

c) HC and the base CDEF

d) HC and the face EFGH.

3.

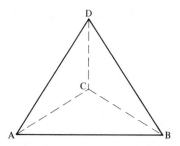

For this regular tetrahedron (each face is an equilateral triangle), sketch and label a triangle containing the angle between

a) AD and the base ABC

b) CD and the base ABC

c) AB and the face ACD.

MIXED QUESTIONS

EXERCISE 7e

1.

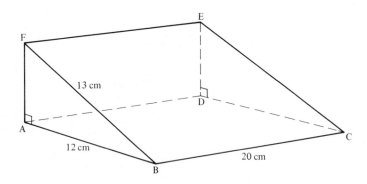

For this wedge,

a) calculate the lengths of AF and FC

b) draw the section containing the angle between the diagonal FC and the base ABCD

c) find the size of the angle that FC makes with the base ABCD.

2.

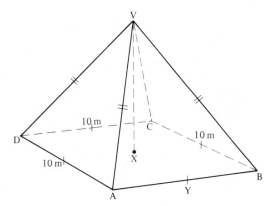

For this pyramid, whose height is 10 m,
a) calculate the length of DB
b) calculate the length of VB
c) find the angle that VB makes with the base ABCD
d) calculate the length of VY, where Y is the midpoint of AB
e) draw the section containing the angle between VY and the face ABCD
f) calculate the angle described in (e).

3.

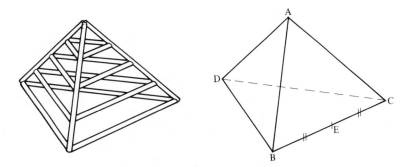

This is a sketch of a playground climbing structure. The basic shape is a pyramid with each face an equilateral triangle of side 3 m. It is possible to climb to the top in various ways.
a) Find the distance from the bottom to the top, starting at
    i) a corner   ii) the midpoint of a side.
b) Find and sketch a section that contains both the angle that AD makes with the base and the angle that AE makes with the base.
c) Using a scale of 4 cm to 1 m, draw this section accurately.
d) By taking measurements from your drawing find the angles described in (b) and the height of the structure.

4.

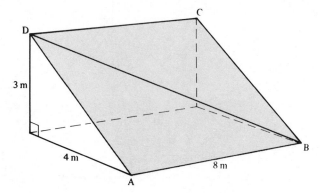

The diagram shows a part of a sea wall whose constant cross-section is a right-angled triangle.

The road is at the level of the top, DC, and the beach starts at AB.

Jim clambers from the road to the beach straight down the path DA whereas Pete goes along the path DB. Find

a) the length of each path

b) the inclination to the horizontal of each path.

5.

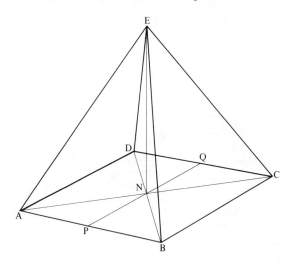

The great pyramid of Cephren at Gizeh in Egypt has a square base of side 215 m and is 225 m high. In the diagram E is the vertex of the right pyramid and ABCD is its base. The diagonals of the base intersect at N, and P and Q are the midpoints of AB and DC respectively.

a) A tunnel runs from the entrance P to the burial chamber at N. Find PN. How far is N from A ?

b) How far is it   i) from P to the top   ii) from A to the top ?

c) A climber wishes to climb from the base of the pyramid to the top. Where should he start   i) to make the shortest climb
ii) to climb in one straight line at the smallest angle to the horizontal ?

8 MATHEMATICAL STATEMENTS

The statement 'if I add 7 to a number I get 11' can be written in mathematical symbols.
If the number is x then $x + 7 = 11$.

From this particular statement we obtain an *equation* in x.

The statement 'A number is less than 6 but greater than 3' can be written as $3 < x < 6$ where x is the number. This time we have an *inequality*.

The statement 'the expenses for my car journey are made up of the number of miles I have travelled multiplied by 30 p, added to a fixed sum' can be simplified to $E = 30a + S$ provided we make it clear what each letter represents. We can call this a *formula* for E.

FORMULAE

A formula can be thought of as a set of instructions for finding the value of a particular quantity. For example, the formula $A = \pi r^2$ tells us to find the area of a circle by multiplying π by the square of the radius.

USING FORMULAE

EXERCISE 8a

1. The perimeter, P metres, of a rectangle of length l metres and width w metres is given by the formula $P = 2(l + w)$.
   a) Find P if $l = 2.4$ and $w = 1.7$.
   b) Find l if $P = 23.4$ and $w = 4.2$.
   c) Find w if $P = 0.06$ and $l = 0.024$.

2. The formula for the volume, V, of a cone of radius r and height h is $V = \frac{1}{3}\pi r^2 h$.
   a) Calculate the value of V if $r = 7$, $h = 9$ and $\pi = \frac{22}{7}$.
   b) Calculate h if $V = 14\frac{2}{3}$, $r = 2$ and $\pi = \frac{22}{7}$.

3. The equation of a straight line is $y = mx + c$ where m is the gradient and c the intercept on the y-axis.
   a) Find y if $m = 1\frac{1}{2}$, $x = 7$ and $c = -3$.
   b) Find x if $y = 9$, $m = -\frac{1}{2}$ and $c = 4$.

135

4. The formula for converting F degrees Fahrenheit to C degrees Celsius is $C = \frac{5}{9}(F - 32)$.

a) Convert $77\,°F$ to degrees Celsius.

b) Convert $15\,°C$ to degrees Fahrenheit.

5. The number, N, of diagonals of a polygon is given by the formula $N = \frac{1}{2}n(n - 3)$ where n is the number of sides of the polygon.

a) How many diagonals are there in a seven-sided polygon ?

b) How many diagonals are there in an octagon ?

c) If a polygon has 54 diagonals, how many sides has it ?

MAKING FORMULAE

If you cannot see how to make a formula, first think what you would do if you had numbers in place of the letters.

EXERCISE 8b

1. The sum of three numbers, a, b and c, is T. Give a formula for

a) T in terms of a, b and c

b) a in terms of T, b and c.

2. Six exercise books cost p pence each and 8 pencils cost q pence each. Give a formula for C if C pence is the total cost of books and pencils.

3. A ladder has N rungs. The rungs are $20\,cm$ apart and the top and bottom rungs are $20\,cm$ from the ends of the ladder. If the length of the ladder is $y\,cm$, give a formula for y in terms of N. (Draw a diagram to help.)

4. The sum of two numbers is x. One of the numbers is 6.

a) What is the second number in terms of x ?

b) The product of the two numbers is P. Give a formula for P in terms of x.

5. A furniture shop buys chairs at £C each and sells them for £S each. N chairs are sold for a total profit of £T. Give a formula for T in terms of C, S and N.

6. A room measures $x\,cm$ by $y\,cm$. The area of the floor is $A\,m^2$. Give a formula for A in terms of x and y.

7. A sequence of numbers begins 4, 7, 10, 13, ...

a) Find the eighth number in the sequence.

b) The nth term is x. Find a formula for x in terms of n. (Check your formula by using $n = 8$ and comparing the result with your answer to (a).)

LITERAL EQUATIONS AND FORMULAE

When we solve a literal equation such as $ax = bx + c$ for x we find x in terms of a, b and c. The method is the same as for equations like $5x + 2 = 3x - 1$.

1. Get rid of fractions and brackets.

2. Collect the x terms on one side.

3. Collect the non-x terms on the other side.

4. Divide both sides by the number of x's.

EXERCISE 8c

Solve $ax + b = bx + c$ for x.

$$ax + b = bx + c$$

Subtract bx from each side $ax - bx + b = c$

Take b from each side $ax - bx = c - b$

(Decide how many x's there are) $x(a - b) = c - b$

Divide both sides by $(a - b)$ $x = \dfrac{c - b}{a - b}$

Solve the following equations for x.

1. $ax + bx = c$ **4.** $ax + x = c$

2. $ax + b = c$ **5.** $x = bx - c$

3. $ax = c - bx$ **6.** $a = b - cx$

Solve the following equations for x. Get rid of fractions first, then brackets.

7. $a(x + 2) = c$ **11.** $a(x - b) + b(x + a) = c$

8. $\dfrac{x}{a} + \dfrac{x}{b} = 1$ **12.** $\dfrac{x + 1}{a} + \dfrac{x - 1}{b} = 1$

9. $\dfrac{x + a}{b} + \dfrac{x}{b} = c$ **13.** $\dfrac{x - b}{b} = \dfrac{x - a}{a}$

10. $a(x + 1) = b(x - 1)$ **14.** $a(x - 1) = b(x - 1)$

In questions 15 to 24, make the letter in the brackets the subject of the formula, i.e. solve the literal equations for the letter in the brackets.

15. $v = u + at$ $[u]$

16. $A = 2\pi r(r + h)$ $[h]$

17. $y - k = m(x - h)$ $[m]$

18. $v = u + at$ $[a]$

19. $A = \frac{1}{2}h(a + b)$ $[a]$

20. $S = 4a^2 + 4al$ $[l]$

21. $l = a + (n - 1)d$ $[d]$

22. $l = a + (n - 1)d$ $[n]$

23. $S = \dfrac{(u + v)}{2}t$ $[t]$

24. $\dfrac{1}{u} + \dfrac{1}{v} = \dfrac{1}{f}$ $[f]$

25. $F + V = E + 2$, $E = \frac{1}{2}nF$ and $V = \frac{1}{3}nF$. Find F in terms of n.

PROBLEMS

EXERCISE 8d

1.

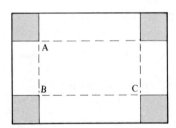

The diagram shows a rectangular piece of card measuring 40 cm by 30 cm, from which the four shaded squares each of side x cm have been cut.

The remaining piece is folded up to form an open box.
a) Give the lengths of AB and BC in terms of x.
b) If the volume of the box is V cm³, give a formula for V in terms of x.

2. Two opposite sides of a rectangular field are each of length x metres. The perimeter of the field is 600 m.
a) Find the area of the field in terms of x.
b) If the area is 21 600 m², find the dimensions of the field.

3.

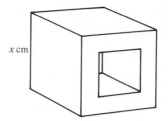

x cm

A cube of edge *x* cm has a hole of cross-sectional area $5\,\text{cm}^2$ cut through it in a direction perpendicular to one pair of faces.

a) Find in terms of *x* the remaining area of each of these faces.

b) Find in terms of *x* the volume of the remaining solid.

c) If the volume is also equal to $\frac{4}{5}x^3$, find the value of *x*.

4. The volume of a rectangular box is $288\,\text{cm}^3$. The base of the box measures *x* cm by $2x$ cm.

a) Find the height of the box in terms of *x*.

b) If the height is $4\frac{1}{2}$ cm, find the other dimensions of the box.

5. The number of diagonals of an *n*-sided polygon is $\frac{1}{2}n(n-3)$. If the number of diagonals is 35, find the number of sides of the polygon.

6. The sides of a right-angled triangle are x cm, $(x+1)$ cm and $(x+2)$ cm. Find the value of *x*.

7. A clock shows the time to be exactly 2 o'clock. A few minutes later the hour hand has moved through *x* degrees.

a) Through what angle, in terms of *x*, has the minute hand moved?

b) What angle does each hand make with the vertical line from the centre through 12?

c) If the minute hand is exactly over the hour hand, what is the value of *x*?

8. If *x* and *y* are positive integers such that $(x+y)$ is divisible by 4, show why $(5x+y)$ is also divisible by 4.

INEQUALITIES

Not all mathematical statements are as restricting as

$$x+2 = 6$$

We must also be able to put into symbols a statement such as 'A number is less than 10' and be able to handle it algebraically.

The following exercise reminds you of the meaning of inequalities and the methods for simplifying them.

Remember to use the rules for solving equations to simplify inequalities. You cannot multiply or divide by a negative number without upsetting the inequality.

EXERCISE 8e

Simplify the inequalities   a) $4 - 3x \leqslant 3$ b) $3 + x < 5 < 4 + 3x$.
In each case represent x on a number line.

a) $4 - 3x \leqslant 3$

Add $3x$ to each side $4 \leqslant 3 + 3x$

$1 \leqslant 3x$

$x \geqslant \frac{1}{3}$

(The solid ● shows that $x = \frac{1}{3}$ is included.)

b) $3 + x < 5 < 4 + 3x$

(We have two inequalities, which must be dealt with separately.)

$3 + x < 5$ and $5 < 4 + 3x$

$x < 2$ $1 < 3x$

$\frac{1}{3} < x$

so $\frac{1}{3} < x < 2$

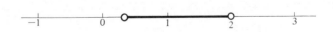

(The hollow circles ○ shows that $x = \frac{1}{3}$ and $x = 2$ are not included.)

Simplify the following inequalities and in each case represent the result on a number line.

1. $2x + 7 \geqslant 11$ **4.** $9 - 4x < 4 - x$ **7.** $18 < 2x + 1 < 24$

2. $3 - 5x > -7$ **5.** $3 < x + 2 < 7$ **8.** $2 + 3x < 8 < 15 + x$

3. $6 - 4x \leqslant 5 + x$ **6.** $-4x > 9$ **9.** $2 - x \leqslant 7 \leqslant 9 - 2x$

10. Simplify the inequality $3 + 2x < 10$ and give the largest integer which satisfies it.

11. Simplify $4 - x < 6 < 7 - 2x$ and list the integers which satisfy the inequalities.

12. Find the smallest integer which satisfies the inequality $5x - 7 > 3x + 11$.

13. Simplify the inequality $20x + 10y \geqslant 50$. If x and y are both positive, give two more inequalities concerning x and y.

14. If $x + y < 6.5$ give three pairs of integers x and y which satisfy the inequality.

15. x and y are positive integers and $2x - y > 7$. Give pairs of values of x and y which satisfy this inequality.

16. Simplify the inequalities   a) $6 - x \leqslant 4 \leqslant 8 - 2x$ b) $6 - x < 4 < 8 - 2x$
Comment on the results.

MATHEMATICAL STATEMENTS

EXERCISE 8f

Susan wishes to buy x red pens at 26 p each and y green pens at 18 p each. She cannot spend more than £1.40. Give three inequalities concerning x and/or y.

The cost of the pens is $(26x + 18y)$ p

$\therefore$ One inequality is $26x + 18y \leqslant 140$

i.e. $13x + 9y \leqslant 70$

She cannot buy a negative number of pens

$\therefore$ $x \geqslant 0$ and $y \geqslant 0$

(Since you cannot buy part of a pen, x and y are integers, so further possible inequalities are $x \leqslant 5$ and $y \leqslant 7$)

1. I think of a number and add 3. I take the first number and double it. The first result is greater than the second.

a) Write an inequality to represent this statement and simplify it.

b) Give a positive integer which satisfies the conditions.

2. Anne is more than twice as old as Mary and Mary is more than 4 years older than Sally.

a) If Anne, Mary and Sally are a, m and s years old respectively, give inequalities to represent the given conditions.

b) Suggest possible ages for the three girls if Anne is less than 16 years old.

3. A man buys x stamps at 15 p each and y stamps at 20 p each.

a) Give the total cost in pence.

b) He cannot spend more than £2. Give as many inequalities as you can think of, concerning x and/or y.

c) If the man buys seven 15 p stamps, how many 20 p stamps could he buy?

d) If the man spends £1.15, what numbers of stamps might he be buying?

4. The mathematics teacher wishes to order books for the school library. Each book is published in hardback and in paperback, the hardbacks costing £6 each and the paperbacks £2.50 each.

He decides to buy x books in hardback and y books in paperback.

a) Make a formula for the total cost $£C$ of the books.

b) If he cannot spend more than £100 on the books, give as many inequalities as you can concerning x and/or y.

c) If he spends £89 and buys 23 books altogether, give two equations in x and y, and find how many books of each sort he buys.

5. For a party, the Browns buy x cans of cola and y bottles of orange squash. Each child drinks one can of cola or two glasses of orange squash. One bottle of squash gives 12 drinks.

a) If there are 36 children of whom 9 drink cola, how many bottles of squash must the Browns buy?

b) If there are C children and all the drink is consumed, give a formula for C in terms of x and y.

c) If there are 30 children and they don't mind which drink they have, give as many relationships as you can concerning x and/or y.

d) If there are 30 children and they have a free choice of drink, give a new set of relationships concerning x and/or y.

6. A classroom is equipped with x desks and y tables. Two people can sit at a table but only one at a desk.

a) Give a formula for the number of pupils, P, that can be accommodated in this classroom.

b) If there are more than three times as many desks as there are tables give an inequality to represent this statement.

c) If a class of 22 can use the room, give an inequality concerning x and y.

d) If the room will take at the most 27 pupils and there are 16 of these pieces of furniture, find how many desks and how many tables there are.

MIXED EXERCISES

EXERCISE 8g

1. x grams of flour are mixed with y grams of lard. Find the total mass in kilograms.

2. $a = -2$, $b = 3$, $c = 4$. Find the value of

    a) $a + 2b$ b) $b^2 + c^2$ c) $3bc^2$ d) a^2

3. Simplify the inequality $7 - 2x < 9 + x$.

4. Make p the subject of the formula $A = \pi(p + 3q)$.

5. If $\dfrac{a + 2b}{a} = \dfrac{7}{3}$,

    a) find b in terms of a

    b) find the value of $\dfrac{b}{a}$.

EXERCISE 8h

1. Find the largest integer which satisfies the inequality $14 < 7 - 2x$.

2. Find a when $b = 3.6$ and $c = 2.2$, if $b^2 = c^2 + a^2$.

3. Given that $a * b$ denotes $\dfrac{a + b}{a}$, evaluate

    a) $3 * 1$ b) $1 * 3$ c) $(3 * 1) * (1 * 3)$

4. Make h the subject of the formula $A = \pi r^2 + \pi rh$.

5. A canteen has x chairs and y tables. A chair is allowed $\frac{1}{2}$ m² of floor space and a table 2 m². If there is 70 m² of floor space available, give an inequality concerning x and y.

EXERCISE 8i

1. Solve for x the equation $p(x-4)+q(x-2)=p+q$.

2. In a two-digit number, the first digit is x and the units digit is y. What is the value of the number in terms of x and y ?

3. Find the smallest even number which satisfies the inequality $5x+9>-11$.

4. a) Find the values of $x^2-x-\dfrac{1}{x}$ when $x=1$ and when $x=2$.

b) Give a value of x for which the value of $x^2-x-\dfrac{1}{x}$ is roughly zero.

5. If $y=ax^2$ and $x=\dfrac{b}{z}$, express y in terms of a, b and z.

9 STRAIGHT LINE GRAPHS AND INEQUALITIES

The gradient, or slope, of the straight line joining the points A and B is

$$\frac{\text{increase in } y \text{ value from A to B}}{\text{increase in } x \text{ value from A to B}}$$

The equation $y = mx + c$ represents a straight line, where m is the gradient and c is the y intercept.

The larger the value of m, the steeper is the slope.

If m is positive, the line slopes *up* from left to right and makes an acute angle with the positive x-axis.

If m is negative, the line slopes *down* from left to right and makes an obtuse angle with the positive x-axis.

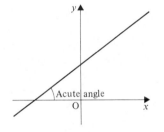

Positive m and
positive gradient

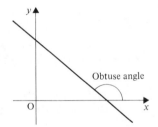

Negative m and
negative gradient

If the y intercept, c, is *positive*, the line crosses the y-axis c units *above* the origin.

If c is *negative*, the line crosses the y-axis c units *below* the origin.

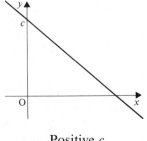

Positive c

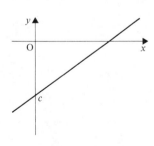

Negative c

The equation $x = a$ represents a straight line parallel to the y-axis, passing through the point $(a, 0)$.

The equation $y = b$ represents a straight line parallel to the x-axis, passing through the point $(0, b)$.

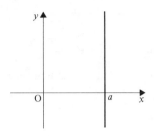

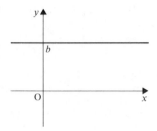

a may be positive or negative.

b may be positive or negative.

EXERCISE 9a

> Write down the gradient and y intercept for the straight line with equation $3x - 2y = 10$.
>
> Rearranging $3x - 2y = 10$ in the form $y = mx + c$ gives
> $$2y = 3x - 10$$
> $$y = \tfrac{3}{2}x - 5$$
> (Comparing with $y = mx + c$, $m = \tfrac{3}{2}$ and $c = -5$.)
>
> $\therefore$ gradient is $\tfrac{3}{2}$ and y intercept is -5.

1. Write down the gradient and y intercept for each of the following straight line equations.

a) $y = 5x + 2$

b) $2y = 10x - 7$

c) $5y = 2x + 1$

d) $x - y = 2$

e) $2y - x = 4$

f) $4x + 2y = 3$

g) $x + 2y - 4 = 0$

h) $3x - 5y - 2 = 0$.

In questions 2 to 10 several alternative answers are given. Write down the letter that corresponds to the correct answer.

2. The gradient of the straight line whose equation is $y = \frac{1}{2}x + 7$ is

    A 2   **B** $\frac{1}{2}$ **C** 7   **D** $-\frac{1}{2}$

3. The gradient of the straight line whose equation is $5y = 10x - 12$ is

    A 10   **B** 2   **C** 5   **D** $-2\frac{2}{5}$

4. The gradient of the straight line whose equation is $x + 3y - 6 = 0$ is

    A $\frac{1}{3}$ **B** -3 **C** $-\frac{1}{3}$ **D** -2

5. The gradient of the straight line whose equation is $2y = 5 - 8x$ is

    A 4   **B** -8 **C** -4 **D** $2\frac{1}{2}$

6. The y intercept for the straight line whose equation is $y = 4x - 2$ is

    A -2 **B** 4   **C** 2   **D** $\frac{1}{2}$

7. The y intercept for the straight line whose equation is $3y = 5x - 9$ is

    A 3   **B** -3 **C** -9 **D** $\frac{5}{3}$

8. The sketch of the straight line with equation $y = 2x - 4$ is

A
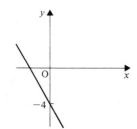

C
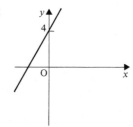

B
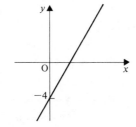

D
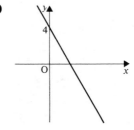

9. The sketch of the straight line with equation $y = 5 - \frac{1}{2}x$ is

A
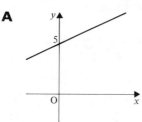

C
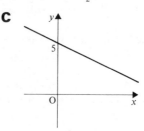

B
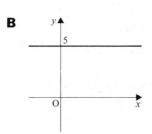

D
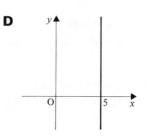

10. The sketch of the straight line whose equation is $3x + y = 4$ is

A
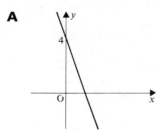

C
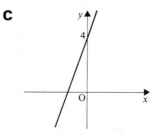

B
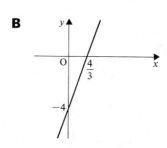

D
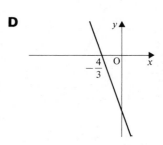

11. Sketch the straight line with the given equation.

a) $y = 4x + 2$ b) $y = -2x - 5$

c) $y = 8 + \frac{1}{2}x$ d) $y = \frac{1}{3}x - 4$

e) $2y = 2x + 5$ f) $3y = 9x - 4$

g) $2x + y = 4$ h) $3x + 4y = 8$

i) $x - 4y + 7 = 0$ j) $2x - 3y - 5 = 0$

12. Determine whether or not the following pairs of equations represent parallel lines.

a) $y = 3x - 7$, $y = 3x + 2$ b) $y = -5x + 2$, $y = 4x - 3$

c) $y = 3x + 2$, $2y = 6x - 9$ d) $x + 2y = 5$, $3x + 6y = 1$

e) $3x + y = 4$, $x + \frac{1}{3}y = 2$ f) $2x + 3y = 4$, $6y = 3 - 4x$

13. Write down the equation of the straight line that has the given gradient and y intercept.

a) gradient 2, y intercept 3   b) gradient -3, y intercept 4

c) gradient $\frac{1}{2}$, y intercept -2 d) gradient $-\frac{2}{3}$, y intercept $\frac{1}{3}$

e) gradient $\frac{2}{5}$, y intercept $-\frac{4}{5}$ f) gradient $-\frac{1}{2}$, y intercept $\frac{2}{3}$

14. Find the y coordinate of the point on the line $y = 6 - 3x$ whose x coordinate is   a) 1   b) 3   c) -4

15. Find the x coordinate of the point on the line $y = 5x + 2$ whose y coordinate is   a) 2   b) 7   c) -8

16. If the points $(2, a)$, $(-2, b)$ and $(c, 4)$ lie on the straight line with equation $y = 4 - 2x$, find the values of a, b and c.

Find the gradient of the line joining the points $(5, 4)$ and $(-3, 8)$.

$$\text{Gradient} = \frac{\text{difference in } y \text{ coordinates}}{\text{difference in } x \text{ coordinates}}$$

$$= \frac{8 - 4}{-3 - 5}$$

$$= \frac{4}{-8}$$

$$= -\frac{1}{2}$$

Note   Be careful to take the y coordinates and the x coordinates in the same order. Alternatively we could have written

$$\text{Gradient} = \frac{4 - 8}{5 - (-3)}$$

$$= \frac{-4}{8}$$

$$= -\frac{1}{2}$$

17. Find the gradient of the straight line passing through the points

a) $(2,4)$ and $(6,6)$ b) $(3,2)$ and $(2, -5)$

c) $(8,2)$ and $(2,4)$ d) $(-1,-2)$ and $(-5,-5)$

e) $(-4,-1)$ and $(5,1)$ f) $(-3,2)$ and $(4,6)$

In each question from 18 to 21, write down

a) the coordinates of A and B

b) the gradient of the straight line shown

c) the y intercept

d) the equation of the line AB in the form $y = mx + c$.

18.

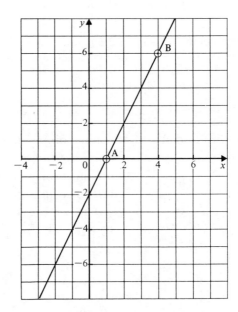

19.

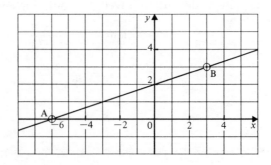

20.

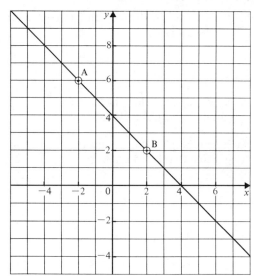

21.

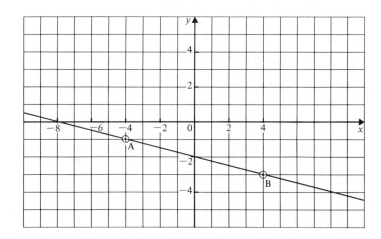

22. Plot on a graph the points A(6,2), B(8,4) and C(−2,−6), and draw the straight line that passes through them.

Use your graph to find the gradient of this line and its y intercept. Hence write down the equation of the line in the form $y = mx + c$.

23. Plot on a graph the points P(−6,4), Q(−3,0) and R(3,−8), and draw the straight line that passes through them. Use your graph to find the gradient of this line and its y intercept. Hence write down the equation of the line in the form $y = mx + c$.

For questions 24 to 26 draw x and y axes scaled from -6 to 6 using 1 cm for 1 unit on both axes.

24. On a graph, plot the point $A(2,1)$ and through this point draw the straight line that has a gradient of 2.

25. On a graph, plot the point $P(-4,2)$ and through this point draw the straight line that has a gradient of $\frac{1}{2}$.

26. On a graph, plot the point $L(3,-4)$ and through this point draw the straight line that has a gradient of $-\frac{1}{3}$.

EXERCISE 9b

Find the equation of the straight line that passes through the point $(4,-3)$ and has a gradient -1.

Let the equation of the line be $y = mx + c$
m is the gradient, therefore $m = -1$.

$\therefore$ equation of line is $y = -x + c$

$(4,-3)$ is on this line so $-3 = -4 + c$

$\therefore$ $c = 1$

$\therefore$ the equation of the line is $y = -x + 1$

---

1. Find the equation of the straight line that
a) passes through the point $(0,2)$ and has a gradient of 3
b) passes through the point $(1,3)$ and has a gradient of $\frac{1}{2}$
c) passes through the point $(4,2)$ and has a gradient of -2
d) passes through the point $(-2,1)$ and has a gradient of $\frac{2}{3}$
e) passes through the point $(-5,-2)$ and has a gradient of $-\frac{3}{4}$

2. Find the equation of the straight line passing through the following pairs of points.
a) $(2,3)$ and $(6,5)$ b) $(0,4)$ and $(2,2)$
c) $(-4,0)$ and $(2,-2)$ d) $(3,2)$ and $(1,-3)$
e) $(-2,-2)$ and $(2,8)$ f) $(-1,4)$ and $(6,2)$
g) $(2,-3)$ and $(7,-1)$ h) $(-2,3)$ and $(0,3)$

Does the point $(5, -2)$ line on the line whose equation is $y = -2x + 7$?

On the given line,

when $x = 5$, $y = -2 \times 5 + 7$

$$= -3$$

For the given point,

when $x = 5$, $y = -2$

So the point $(5, -2)$ is not on the given line.

3. In each of the following questions determine whether or not the given point lies on the given straight line.

a) $(3, 8)$, $y = 3x - 1$

b) $(-2, 3)$, $y = 2x + 7$

c) $(4, -3)$, $y = 7x + 2$

d) $(3, 0)$, $y = 2x - 5$

e) $(-1, 7)$, $y = 5 - 2x$

f) $(-4, -3)$, $x + y + 1 = 0$

g) $(-3, 2)$, $3x + 5y = 1$

4. The graph of the line with equation $4x + 3y = 12$ cuts the x-axis at A and the y-axis at B.

a) Find the coordinates of   i) A,   ii) B.

b) Find the gradient of AB.

c) Use Pythagoras' theorem to find the length of AB.

d) Find the area of triangle OAB where O is the origin.

5. The graph of the line with equation $5x - 12y + 60 = 0$ cuts the x-axis at P and the y-axis at Q. Find

a) the coordinates of P

b) the coordinates of Q

c) the gradient of PQ

d) the length of PQ

e) the area of triangle POQ where O is the origin.

6.

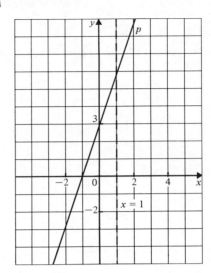

The diagram shows a line p which meets the x-axis at the point $(-1,0)$ and the y-axis at $(0,3)$. When p is reflected in the line $x = 1$, its image is the line q.

a) What is the gradient of p ?

b) Copy the diagram and draw the line q on your diagram.

c) Write down the equation of the line q.

7.

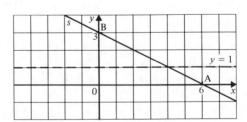

The diagram shows a line s which meets the x-axis at the point $A(6,0)$ and the y-axis at the point $B(0,3)$. When s is reflected in the line $y = 1$ its image is the line s'.

a) What is the gradient of s ?

b) Copy the diagram and draw the line s' on it.

c) Write down the gradient of s' and the coordinates of C, the point where it crosses the y-axis.

d) What is the equation of s' ?

e) Find the area of triangle ABC.

8.

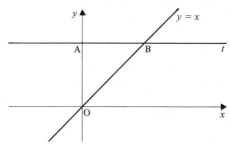

The diagram shows a horizontal line t which meets the y-axis at A, and the straight line with equation $y = x$ at the point B with coordinates $(4, 4)$. When it is reflected in the line $y = x$ its image is the line t'.

a) Write down the coordinates of A.

b) Copy the diagram and draw the line t' on it.

c) Write down the equations of the lines t and t'.

d) If the line t' crosses the x-axis at C write down the coordinates of C.

e) What is the area of triangle ABC ?

GRAPHICAL SOLUTION OF SIMULTANEOUS LINEAR EQUATIONS

If we draw two straight lines on a graph the values of x and y at the point of intersection of the two lines satisfy both the given equations, that is they are the solutions of the two simultaneous equations which are represented by the straight line graphs.

To plot a point that lies on a straight line with a given equation we can *either* substitute our chosen value of x in the given equation to find the corresponding value of y, *or* we can rearrange the equation before we make the substitution.

Consider the equation $3x + 4y = 13$

When $x = 1$
$$3 + 4y = 13$$
$$4y = 10$$
$$y = 2.5$$

Alternatively
$$4y = 13 - 3x$$
$$y = \tfrac{1}{4}(13 - 3x)$$

When $x = 1$
$$y = \tfrac{1}{4}(13 - 3)$$
$$y = 2.5$$

i.e. in both cases, if $x = 1$, $y = 2.5$

EXERCISE 9c

Using graph paper, draw axes for $-4 \leqslant x \leqslant 8$ and $-3 \leqslant y \leqslant 10$.
Take 1 cm for 1 unit on both axes.

Solve graphically the simultaneous equations

$$2x + y = 4$$
$$5x - 8y = -8$$

$2x + y = 4$	$5x - 8y = -8$
i.e. $y = -2x + 4$	$5x + 8 = 8y$
	i.e. $8y = 5x + 8$
	$y = \frac{5}{8}x + 1$

x	-2	0	2
y	8	4	0

x	-4	0	8
y	-1.5	1	6

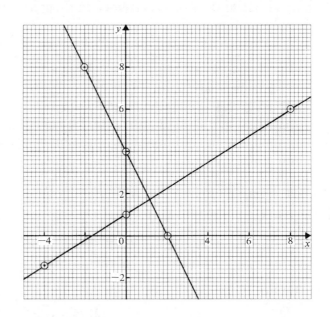

The straight lines intersect at the point $(1.1, 1.7)$.

The solutions of the given equations are $x = 1.1$ and $y = 1.7$.

Solve the following equations graphically. In each case draw axes for x and y, and use values in the ranges indicated, taking 1 cm for 1 unit.

1. $2x + 3y = 12$ $-4 \leqslant x \leqslant 6,$ $-2 \leqslant y \leqslant 8$
  $x - 2y = -2$

2. $8x + 5y = 40$ $-4 \leqslant x \leqslant 6,$ $-2 \leqslant y \leqslant 12$
  $5x - 2y = 2$

3. $6x - 5y = 30$ $-4 \leqslant x \leqslant 8,$ $-10 \leqslant y \leqslant 4$
  $2x + y = 4$

4. $8x - 3y = 24$ $-6 \leqslant x \leqslant 6,$ $-10 \leqslant y \leqslant 4$
  $2x + 5y = -10$

5. $x - 6y = -18$ $-2 \leqslant x \leqslant 8,$ $-2 \leqslant y \leqslant 10$
  $7x - 6y = -6$

6. $2x + y = -6$ $-8 \leqslant x \leqslant 4,$ $-12 \leqslant y \leqslant 2$
  $7x - 4y = -4$

7. $7x + 4y = 28$ $-2 \leqslant x \leqslant 8,$ $-10 \leqslant y \leqslant 10$
  $5x - 3y = 15$

REAL LIFE SITUATIONS

The gradient of a straight line graph tells us how the quantity on the vertical axis changes with respect to a unit increase in the quantity on the horizontal axis.

For a straight line whose equation is in the form $y = mx + c$, m gives the rate at which the value of y changes with respect to a unit increase in the value of x.

This is often abbreviated to:

Gradient is the rate of change of y w.r.t. x.

EXERCISE 9d

1. The graph given below can be used to find the cost to the householder of consuming any number of therms of gas from 0 to 240.

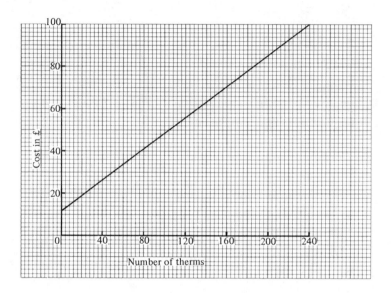

Use this graph to find

a) the cost of using 176 therms

b) the number of therms a householder could buy for £36

c) the payment due before any gas is paid for

d) the gradient of the line. Can you attach a meaning to this value ?

2. The graph opposite shows the cost of electricity to the domestic user for two different tariffs, Tariff A and Tariff B.

Use this graph to answer the following questions.

a) What is the cost of 500 units using tariff A ?

b) How many units can I buy for £55 using tariff B ?

c) For how many units is the cost the same, whichever tariff is used ?

d) Find the gradient and vertical intercept for the straight line representing each tariff. Attach a meaning to each of these values.

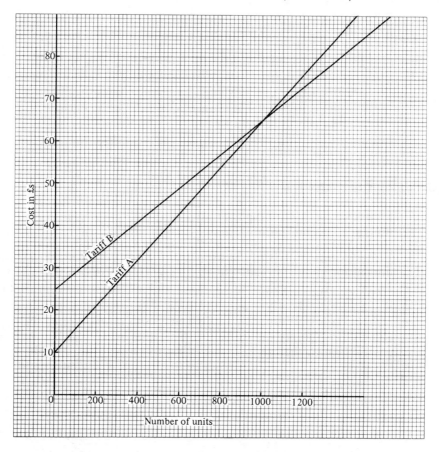

3. Details of the quarterly telephone bills for three households are given in the table.

Name	Number of metered units used (x)	Total cost in £s (y)
Arnott	500	55
Burley	800	76
Compton	1200	104

Represent this data on a graph using 1 cm for 100 units for x and 1 cm for £10 for y. Draw the straight line that passes through these points.

Use your graph to find

a) the quarterly standing charge

b) the cost for a householder using 640 metered units

c) the number of units used by a householder whose bill is £100

d) the slope of the straight line. How can you use this to find the cost of one metered unit ?

4. In the borough of Nashford, Mrs Croake lives in a house with a rateable value of £220 and pays £308 in rates, whereas Mr Loxton pays rates of £490 for his bungalow which has a rateable value of £350. Assuming that there is a linear relationship between rates and rateable value, draw a graph to represent this data taking 4 cm as £100 on both axes. Use your graph to find

a) the rates due on a house with a rateable value of £300

b) the rateable value of a house for which the rates demand is £440

c) the slope of the straight line.

Use your answer to (c) to determine the 'rate in the £' levied by the borough.

CONVERSION GRAPHS

Simple straight line graphs are frequently very useful to convert one set of units into another.

Suppose that we wish to convert from miles per gallon to kilometres per litre. If we know, for example, that 26.4 m.p.g. is equivalent to 9.3 km/l and 36.7 m.p.g. is equivalent to 13.0 km/l we can draw a graph plotting miles per gallon along the horizontal axis and kilometres per litre along the vertical axis. Plotting the two given points and joining them gives the straight line graph shown below. Note that when the line is produced it passes through the origin.

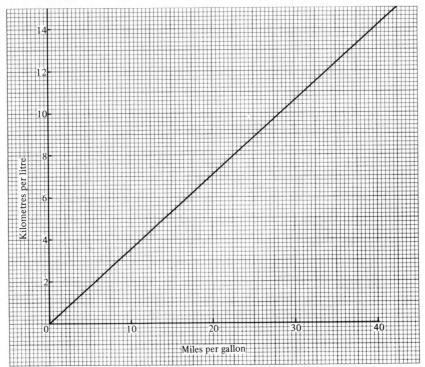

From this straight line graph we can convert, within the given ranges, from m.p.g. to km/l and vice versa.

For example   32 m.p.g. $\equiv$ 11.2 km/litre

and   8 km/litre $\equiv$ 22.8 m.p.g.

EXERCISE 9e

1. On a certain day the exchange value of £1 in dollars was $1.30. Draw a conversion graph in the range 0 to £20.

Use 8 cm to represent 10 units on both axes.

Use the graph to find

a) how many dollars could be bought for £8.50

b) how many pounds it cost to buy 18 dollars

c) the value in pounds of $5

d) the value in dollars of £14.60.

2. In the United Kingdom the area of a farm is given in acres whereas on the continent it is given in hectares. If 100 hectares is equivalent to 247 acres draw a conversion graph and use it to find

a) 193 acres in hectares

b) 47 hectares in acres

c) 46 acres in hectares

d) 88 hectares in acres.

3. A tonne, i.e. 1000 kg, is equivalent to 2200 lb. Draw a conversion graph between these two units, in the range 0 to 2000 kg.

Use your graph to find

a) 1700 kg in lb   b) 1950 lb in kg.

SCATTER DIAGRAMS

In the straight line graphs we have considered previously, the straight line drawn has passed through all the points that were plotted. In the 'real world' varying quantities are not usually related in such an exact way. In the exercise that follows, we plot points to represent experimental data and then draw the straight line that best represents these points. We draw, *by eye,* a straight line such that the sum of all the distances of the plotted points from this line is as small as possible. This may mean that the line does not go through any of the plotted points, and that some points will be above the line and some points will be below it.

EXERCISE 9f

The lengths and masses of six fish of a certain species are given in the table. Lengths are given to the nearest centimetre and masses to the nearest gram.

Length of fish x (cm)	25	36	23	32	29
Mass of fish y (g)	485	500	480	495	489

a) Draw, on graph paper, a scatter diagram to illustrate the data, and draw a line of best fit.

b) Hence estimate   i) the mass of a fish that has a length of 34 cm   ii) the length of a fish that has a mass of 488 g.

c) Find the gradient of this line and explain what meaning can be given to it.

a) (The range of values for x and for y suggests that we can have a good spread for the plotted points if x varies from 20 to 36 and y varies from 475 to 500. Suitable scales are 1 cm for 1 unit for x, and 2 cm for 5 units for y. The graph is given below.)

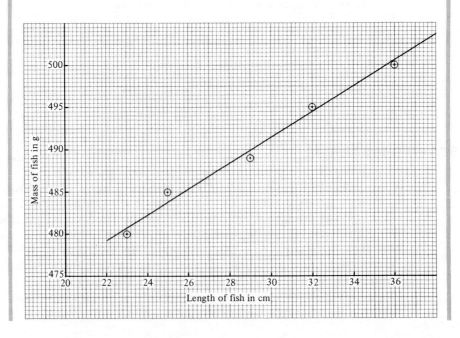

(A line of best fit is drawn by eye so that it looks as though the sum of the distances of the points from the line is as small as possible.)

b) From the graph
   i) a fish of length 34 cm would be expected to have a mass of 497 g
   ii) a fish that has a mass of 488 g would be expected to be about 28 cm long.

c) Gradient $= \dfrac{15}{10} = 1.5$. This means that for each centimetre increase in length the mass of the fish increases by 1.5 g.

---

1. The heights and weights of twelve boys in a class are given in the following table:

Boy	A	B	C	D	E	F	G	H	I	J	K	L
Height (cm)	153	171	164	167	157	173	162	161	174	154	178	164
Weight (kg)	50	68	60	72	52.5	75	57	63	80	55	79	66

On graph paper, construct a scatter diagram for this data. Use 4 cm to represent 10 units on both axes.

a) Draw a line of best fit.

b) Use your line to estimate
   i) the weight of a new pupil who is 171 cm tall
   ii) the height of another pupil who weighs 70 kg.

2. Ten pupils received the following marks in two tests.

Geography	54	60	49	74	61	35	56	42	80	70
Physics	73	71	84	53	64	100	76	87	42	53

a) If the maximum score was obtained in each subject, find
   i) the highest score possible in geography
   ii) the highest score possible in physics
   iii) the score in physics of the pupil who scored maximum marks in geography.

b) Plot points to represent the marks given in the table. Use 2 cm to represent 10 marks on both axes.
   i) Does it look as though ability in physics is linked to ability in geography?
   ii) Draw a line of best fit.
   iii) Peter scored 50 in geography but missed the physics test. Use your graph to estimate what he would have scored had he sat the physics test. Comment on the answer.

3. The table shows the prices of some school textbooks and the number of pages in them.

Price (£)	2.60	4	2	3.30	3.40	4.20	2	4	3
Number of pages	280	385	270	350	400	455	220	410	300

On graph paper, construct a scatter diagram for this information. Use 2 cm to represent 50 pages and 2 cm to represent 40 p. You do not need to begin each scale at 0.

a) Draw a line of best fit to represent the points you have plotted.

b) Use your line to estimate
  i) the cost of a book with 250 pages
  ii) the number of pages I can expect if I buy a book costing £2.75

4. The table shows the relation between the selling prices and ages of six second-hand bicycles, as noted from a local newspaper.

Age x (years)	3	8	2	6	1	4	10
Price y (£)	45	25	55	25	60	40	15

a) Plot a scatter diagram for this data on graph paper. One point is obviously out of place. Ignoring this point, draw a line of best fit.

b) Use this line to estimate
  i) what you would expect to pay for a six year old bicycle.
  ii) how old a bicycle would be that you were prepared to buy for £30.

c) Can you explain why one point was out of place ?

5. The table shows the maximum daily temperature at Severnsea during the first week of July last summer, together with the total amount of ice-cream sold on each day.

Day	1	2	3	4	5	6	7
Temperature x (°C)	18	20	15	25.2	17	22	24.2
Ice-cream sold y (gallons)	36	37.5	32.5	42	34.5	39	41.5

Plot a scatter diagram for this data on graph paper. Use 1 cm to represent 1 °C and 4 cm to represent 5 gallons. You do not need to begin each scale at 0. Draw a line of best fit. Use your line to estimate

a) the amount of ice-cream sold when the maximum temperature is 23 °C

b) the maximum temperature recorded when 35 gallons of ice-cream were sold.

INEQUALITIES AND REGIONS

These topics were dealt with thoroughly in Chapter 21 of Book 3A. The most important points to remember are:

a continuous line is used for a boundary which is included but a broken line is used for a boundary which is not included. When we have the choice we will shade the region that we do *not* want.

e.g. we represent $x \geqslant 3$ by the following diagram.

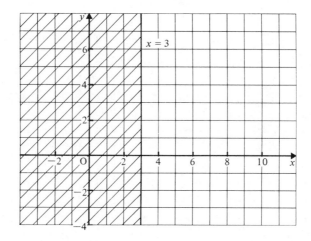

The inequality $x > -1$ tells us that x may not take the value -1. In this case we use a broken line for the boundary, e.g.

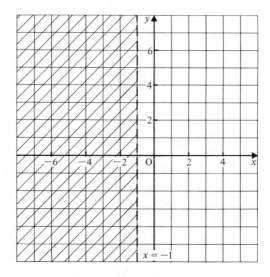

EXERCISE 9g

Draw a diagram to represent each of the following inequalities.

1. $x \geqslant 4$ **2.** $y \leqslant 2$ **3.** $y > -2$ **4.** $0 < x$

Draw a diagram to represent each of the following pairs of inequalities.

5. $2 \leqslant x \leqslant 6$ **7.** $-3 \leqslant y < -1$

6. $3 < y < 5$ **8.** $-2 < x \leqslant 3$

Give the inequalities that define each of the *unshaded* regions:

9.

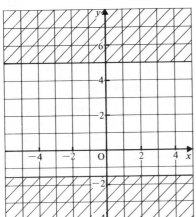

10.

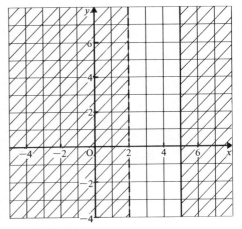

Give the inequalities that define each of the *shaded* regions:

11.

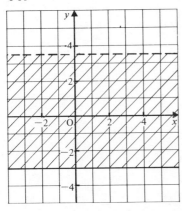

12.

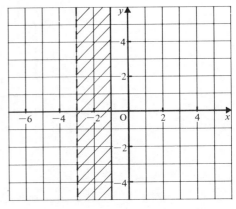

Draw a diagram to represent the region defined by the set of inequalities $-4 \leqslant x \leqslant 2$, $-3 \leqslant y < \frac{3}{2}$.

The boundary lines are

$x = -4$ and $x = 2$

$y = -3$ and $y = \frac{3}{2}$ (not included)

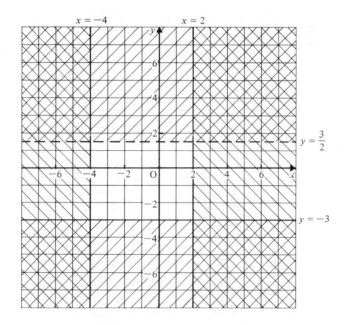

The unshaded area represents the inequalities.

Draw a diagram to represent the region described by each of the following sets of inequalities. In each case, draw axes for values of x and y from -5 to 5.

13. $-3 < x < 0,$ $-4 < y < 2$ **14.** $x \geqslant 2, -1 \leqslant y \leqslant 3$

Give the set of inequalities that describes each of the following unshaded regions.

15.

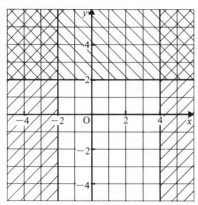

16.

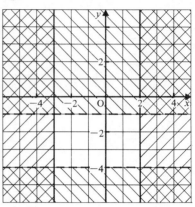

Leave unshaded the region defined by the inequality $x + y > 5$.

The boundary line (not included) is $x + y = 5$

x	0	3	5
y	5	2	0

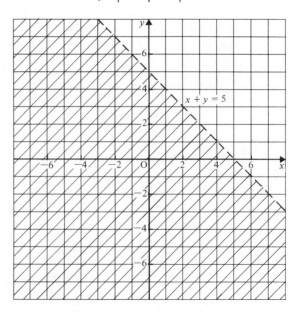

In questions 17 to 20 leave unshaded the region defined by the inequality.

17. $x + y < 3$

19. $2x - 3y \leqslant 6$

18. $x - y > 4$

20. $4x + 3y \leqslant 12$

Leave unshaded the regions defined by the set of inequalities $y > x$, $y < 4x$, $x + y \leqslant 5$.

First boundary line (not included) $y = x$

x	-3	0	3
y	-3	0	3

Second boundary line (not included) $y = 4x$

x	-2	0	2
y	-8	0	8

Third boundary line (included) $x + y = 5$

x	-2	0	5
y	7	5	0

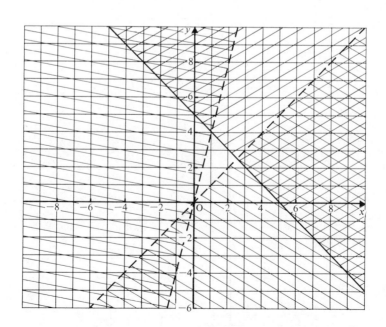

In questions 21 to 23 leave unshaded the region defined by the following set of inequalities.

21. $y \geqslant 0, \quad x \geqslant 0, \quad x+y \leqslant 2$

22. $x > -1, \quad -1 < y < 4, \quad x+y < 6$

23. $x+y \leqslant 4, \quad 3x+y \geqslant 3, \quad y \geqslant 0, \quad x \leqslant 3$

In questions 24 to 26 shade the region defined by the set of inequalities.

24. $x < 4, \quad y \leqslant 3, \quad x+y > 1$

25. $y > \frac{1}{2}x, \quad 0 < x < 3, \quad x+y < 5, \quad y < 3$

26. $5x + 2y \leqslant -10, \quad y \geqslant -2, \quad x \geqslant -5$

In questions 27 and 28 give the inequalities that define the unshaded region.

27.

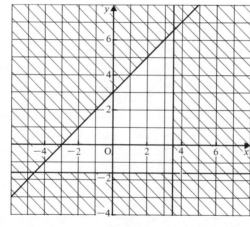

28.

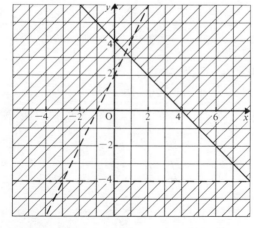

29.

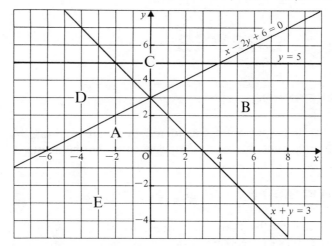

Use inequalities to describe the regions.

a) A   b) B   c) C   d) A + B

(The *y*-axis is not a boundary line but the *x*-axis is.)

10 CURVES

CURVE SKETCHING

Curved graphs were considered in detail in Book 4A, where it is shown that the general shape of a curve can often be deduced from its equation.

QUADRATIC GRAPHS

All the graphs that have equations of the type $y = ax^2 + bx + c$ $(a \neq 0)$ are curves of the same basic shape. They are called parabolas.

If a is positive the vertex is at the bottom and there is no highest value of y.

On the other hand when a is negative the vertex is at the top and there is no lowest point.

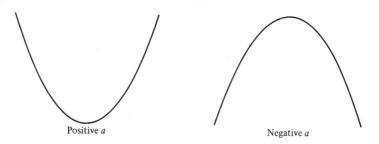

Positive a Negative a

CUBIC GRAPHS

All the graphs that have equations of the type $y = ax^3 + bx^2 + cx + d$ $(a \neq 0)$ are curves of the same general shape. They are called cubic curves.

If a is a positive they look like

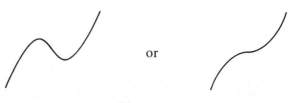

or

while if *a* is negative they look like

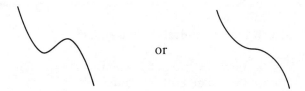

or

EQUATIONS OF THE FORM $y = \dfrac{a}{x}$

Equations of the form $y = \dfrac{a}{x}$ give distinctive 'two part' curves. There is no value for *y* when $x = 0$.

If *a* is positive it looks like this

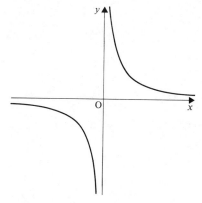

while if *a* is negative it looks like this

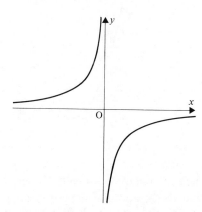

(These curves are called hyperbolas.)

EXERCISE 10a

Sketch the curve given by the equation $y = (x - 3)(x + 5)$.

$y = 0$ when $x = 3$ and when $x = -5$. The graph therefore crosses the x-axis at these values of x.

$y = x^2 + 2x - 15$.

The x^2 term is positive, so the vertex is at the bottom and there is no highest value of y.

When $x = 0$, $y = -15$.

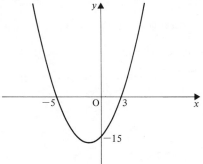

1. Draw a sketch of the curve given by the equation
   a) $y = (x - 1)(x - 6)$ b) $y = (x + 4)(x + 7)$
   c) $y = x^2 - 2x - 3$ d) $y = x^2 + 8x + 12$
   e) $y = x(4 - x)$ f) $y = x^2 + 5$
   g) $y = 4 - x^2$ h) $y = (4 - x)(5 + x)$

2.

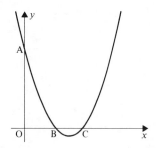

The sketch shows the graph of $y = x^2 - 5x + 6$.
   a) Find the coordinates of A, B and C.
   b) Find the equation of the straight line joining A and C.
   c) Copy the sketch and add the graph of $y = x$. Let this graph intersect the given graph at D and E. What equation will have the x-coordinates of D and E as roots?

3.

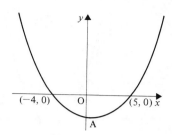

The equation of the curve shown in the sketch is $y = x^2 + px + q$.
Find   a) p and q b) the coordinates of A.

4.

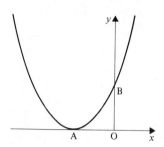

The sketch shows the graph of $y = x^2 + 6x + 9$.
a) Find the coordinates of A and B.
b) Without plotting points sketch, on the same axes, the graph of $y = x^2 - 6x + 9$.

5. Sketch, on the same axes, the graphs of $y = x^2 - 1$, $y = x^2$ and $y = x^2 + 2$, clearly distinguishing between them.

6. On the same axes sketch the graphs of $y = x^2$ and $y = -x^2$. Describe a transformation that maps the first curve to the second.

7. Sketch, on the same axes, the graphs of $y = x^2 - 4$ and $y = -x^2 + 4$. Describe a transformation that maps the first curve to the second.

8.

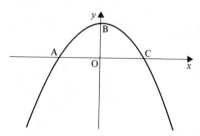

The sketch shows the graph of $y = 9 - x^2$.
a) Find the coordinates of A, B and C.
b) Without plotting points, sketch on the same axes the graph of $y = x^2 - 9$.

9.

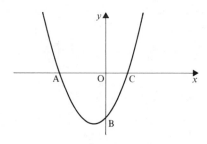

The sketch shows the graph of $y = x^2 + 2x - 8$.

a) Find the coordinates of A, B and C.

b) Without plotting points, sketch on the same axes the graph of $y = 8 - 2x - x^2$.

10. Sketch the graph of $y = 2x - x^2$, showing clearly where it crosses the axes. On the same axes sketch the graphs of

a) $y = 4x - 2x^2$ b) $y = x^2 - 2x$

Sketch the cubic graph given by the equation $y = x^3 - 4x$.

The x^3 term is positive and therefore the shape of the curve is either

   or

$$y = x^3 - 4x$$
$$= x(x^2 - 4)$$
$$= x(x + 2)(x - 2)$$

The curve crosses the x-axis when $y = 0$, i.e. when $x = 0$, $x = -2$ and $x = 2$.

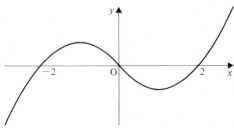

11. Draw a sketch of the curve given by the equation
   a) $y = (x-1)(x-3)(x-5)$
   b) $y = (x+1)(x-2)(x+7)$
   c) $y = 9x - x^3$
   d) $y = 3x(4-x^2)$
   e) $y = x^3 - 2x^2 + x$

In each case give a clear indication of where the graph crosses or touches the axes.

12.

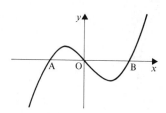

The sketch shows the graph of $y = x^3 - x^2 - 12x$. Find the coordinates of A and B.

13.

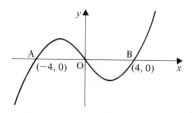

The sketch shows the graph of $y = (x-3)(x^2-4)$. Find the coordinates of A, B, C and D.

14.

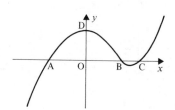

The equation of the cubic curve shown in the sketch is $y = x^3 + ax^2 + bx$. If the curve passes through the points $A(-4,0)$ and $B(4,0)$ find the values a and b.

15. Sketch the graph of $y = \dfrac{10}{x}$ for values of x between 1 and 10.

16. Sketch the graph of $y = -\dfrac{12}{x}$ showing clearly the two parts of the curve.

In questions 17 to 19 several possible answers are given. Write down the letter that corresponds to the correct answer.

17. The graph of $y = x^2 + 6x + 8$ could be

A

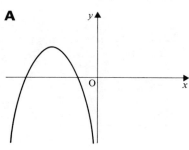

C

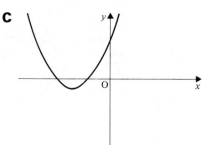

B

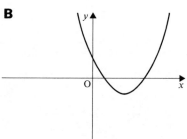

D

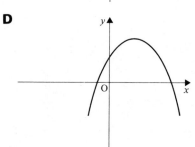

18. The graph of $y = x(x^2 - 9)$ could be

A

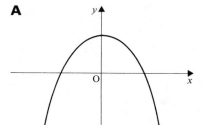

C

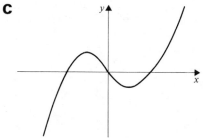

B

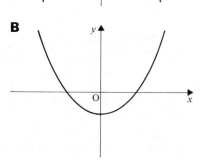

D

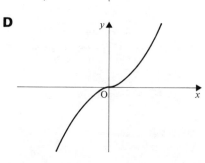

19. The graph of $y = -\dfrac{15}{x}$ could be

A

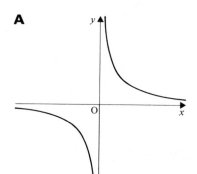

C

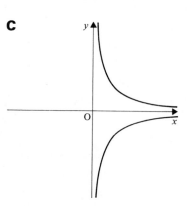

B

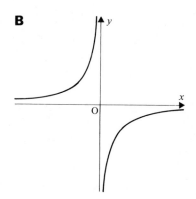

D

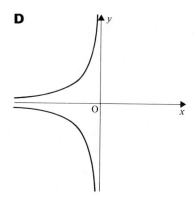

20.

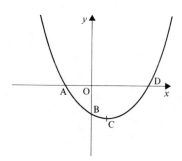

The sketch shows the graph of $y = x^2 - 4x - 32$.

The coordinates of D are

A $(32,0)$ **B** $(4,0)$ **C** $(8,0)$ **D** $(6,0)$

21.

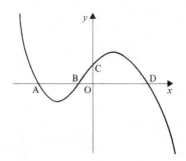

The sketch shows the graph of $y = (x+5)(x+1)(4-x)$.

The coordinates of C are

A $(0,5)$ **B** $(0,1)$ **C** $(0,4)$ **D** $(0,20)$

22. The graph of $y = x(x+2)(x-2)$ could be

A

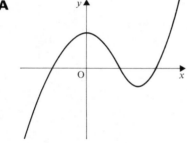

C

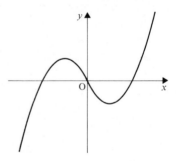

B

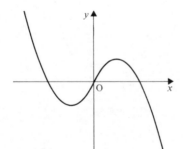

D

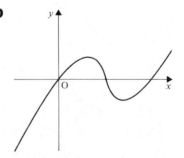

There is further curve sketching practice in Chapter 17, Exercise 17d, which can be worked independently of the rest of that chapter.

DRAWING GRAPHS

Sketches frequently tell us all we need to know about the curve for a particular equation, but sometimes the information we seek can be found only if an accurate graph is drawn.

When curved graphs are being drawn the following advice should be kept clearly in mind.

1. Do not take too few points. About eight or ten are usually required.

2. To decide where to draw the *y*-axis look at the range of *x*-values, and vice versa.

3. In some questions you will be given most of the *y*-values but you may have to calculate a few more for yourself. If so, always plot first those points that you were given and, from these, get an idea of the shape of the curve. Then you can plot the points you calculated and see if they fit on to the curve you have in mind. If they do not, go back and check your calculations. Always have a clear idea of the shape of the resulting curve before you begin to draw it.

4. When you draw a smooth curve to pass through the points, always turn the page into a position where your wrist is on the inside of the curve.

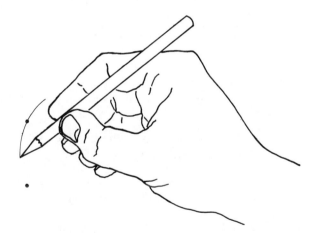

It is interesting to note that there are graphics programs available for microcomputers that will plot curves and give the coordinates of points of intersection as accurately as anyone could want. There is even a pocket calculator on the market which has this facility. If you have access to either of these, you may like to use them to answer some of the following questions. If you explore these graphics capabilities further, you will find that you can do many of the remaining questions in this chapter far more accurately than it is possible to do by drawing.

EXERCISE 10b

Draw the graph of $y = 5 + 3x - x^2$ for whole number values of x from -2 to 5.

Use your graph to find the highest value of $5 + 3x - x^2$, and the corresponding value of x.

Draw, on the same axes, the graph of $y = \frac{1}{3}x - 1$.

a) Write down the values of x at the points of intersection of the two graphs.

b) Use your graph to find the range of values of x for which $5 + 3x - x^2$ is greater than $\frac{1}{3}x - 1$.

c) Find, in its simplest form, the equation whose roots are the values of x at the points of intersection.

d) What other straight line graph should be drawn to solve the equation $x^2 - 2x - 4 = 0$?

x	-2	-1	0	1	2	3	4	5
5	5	5	5	5	5	5	5	5
$3x$	-6	-3	0	3	6	9	12	15
$-x^2$	-4	-1	0	-1	-4	-9	-16	-25
$y = 5 + 3x - x^2$	-5	1	5	7	7	5	1	-5

The symmetry of the y-values suggests that the value of y corresponding to $x = 1.5$ gives the highest point of the graph.

a) From the graph, the highest value of $5 + 3x - x^2$ is 7.25 and it occurs when x is 1.5.

b) (The graph of $y = \frac{1}{3}x - 1$ is a straight line, so we take only three values of x and find the corresponding values of y.)

x	-2	0	3
$y = \frac{1}{3}x - 1$	$-1\frac{2}{3}$	-1	0

The graphs intersect when $x = -1.46$ and 4.12.
From $x = -1.46$ to $x = 4.12$ the curve $y = 5 + 3x - x^2$ is above the straight line $y = \frac{1}{3}x - 1$.

Therefore $5 + 3x - x^2 > \frac{1}{3}x - 1$ for $-1.46 < x < 4.12$.

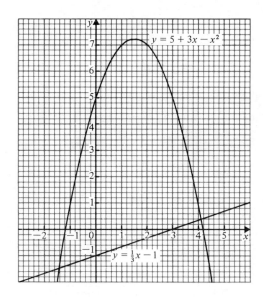

c) At $x = -1.46$ and $x = 4.12$, the values of y for the curve and for the straight line are equal,

i.e.
$$5 + 3x - x^2 = \tfrac{1}{3}x - 1$$
$$15 + 9x - 3x^2 = x - 3$$
$$3x^2 - 8x - 18 = 0$$

Therefore $3x^2 - 8x - 18 = 0$ is the equation whose roots are the values of x at the points of intersection of the two graphs.

d) (To solve the equation $x^2 - 2x - 4 = 0$ we convert it so that one side becomes $5 + 3x - x^2$.)
$$x^2 - 2x - 4 = 0$$
$$0 = 4 + 2x - x^2$$
$$1 + x = 5 + 3x - x^2$$

(Therefore we draw the line with equation $y = x + 1$ on the same axes as the curve $y = 5 + 3x - x^2$.)

The points of intersection of the curve whose equation is $y = 5 + 3x - x^2$ and the straight line that has equation $y = x + 1$ will give the roots of the equation $x^2 - 2x - 4 = 0$.

1. Draw the graph of $y = x^2 - 5x + 3$ for values of x from 0 to 5 at half-unit intervals. Take 4 cm as 1 unit on the x-axis and 2 cm as 1 unit in the y-axis. Use your graph to find

a) the value of y when x is 3.2

b) the values of x when the value of $x^2 - 5x + 3$ is -2

c) the lowest value of $x^2 - 5x + 3$ and the value of x for which it occurs

d) the values of x when $x^2 - 5x + 3 = 0$.

2. Use the graph of $y = 4 - 2x - x^2$, which is given below, to solve the equations

a) $4 - 2x - x^2 = 0$ b) $x^2 + 2x - 4 = 0$

c) $2 - 2x - x^2 = 0$ d) $13 - 2x - x^2 = 0$

e) $x^2 + 2x - 10 = 0$

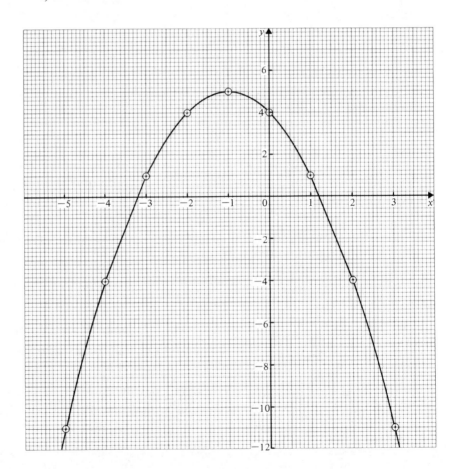

3. Complete the following table which gives values of $3x^2 - 3x - 2$ for values of x from -2 to 3.

x	-2	$-1\frac{1}{2}$	-1	$-\frac{1}{2}$	0	$\frac{1}{2}$	1	$1\frac{1}{2}$	2	$2\frac{1}{2}$	3
$3x^2$	12	$6\frac{3}{4}$		$\frac{3}{4}$		$\frac{3}{4}$	3	$6\frac{3}{4}$		$18\frac{3}{4}$	27
$-3x$	6	$4\frac{1}{2}$		$\frac{3}{2}$		$-\frac{3}{2}$	-3	$-4\frac{1}{2}$		$-7\frac{1}{2}$	-9
-2	-2	-2		-2		-2	-2	-2		-2	-2
$3x^2 - 3x - 2$	16	$9\frac{1}{4}$		$\frac{1}{4}$			-2	$\frac{1}{4}$		$9\frac{1}{4}$	16

Hence draw the graph of $y = 3x^2 - 3x - 2$ for values of x from -2 to 3. Take 3 cm as 1 unit for x and 1 cm as 1 unit for y.

Use your graph to solve the equations

a) $3x^2 - 3x - 2 = 0$

b) $3x^2 - 3x - 5 = 0$

c) $x^2 - x - 3 = 0$

4. Draw the graph of $y = x^2 - 4x - 4$ for whole number values of x from -1 to 5. Take 2 cm as 1 unit on both axes.

Use your graph to find

a) the value of y when $x = 3.3$

b) the values of x when $y = -2.4$

Draw, on the same axes, the graph of $y = x - 6$.

c) Write down the values of x at the points of intersection of the two graphs, and find, in its simplest form, the equation whose roots are these x values.

d) Use your graph to find the range of values of x for which $x^2 - 4x - 4$ is less than $x - 6$.

5. Draw the graph of $y = 5 - 2x - x^2$ for values of x from -4 to 3, calculating the values of x at unit intervals, but adding other values if they are needed. Take 2 cm as 1 unit on the x-axis and 1 cm as 1 unit on the y-axis.

Use your graph

a) to find the values of x when the graph crosses the x-axis, and the equation that has these x values as roots

b) to solve the equation $x^2 + 2x - 1 = 0$.

Copy and complete the following table, which gives values of $x(x+2)(x-4)$ for values of x from -3 to 4.

x	-3	-2	-1	0	1	2	3	4
$x(x+2)(x-4)$	-21	0		0	-9	-16		

Hence draw the graph of $y = x(x+2)(x-4)$ for values of x from -3 to 4. Take $2\,$cm as 1 unit for x and $2\,$cm as 5 units for y.

Draw, on the same axes, the graph of $y = 2x - 10$.

Write down the values of x at the points of intersection of the two graphs and find, in its simplest form the equation for which these values are the roots.

x	-1	3	4
x	-1	3	4
$x+2$	1	5	6
$x-4$	-5	-1	0
$x(x+2)(x-4)$	5	-15	0

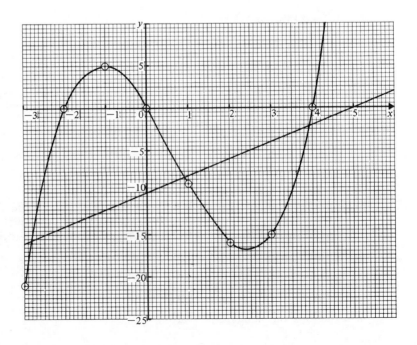

(The graph of $y = 2x - 10$ is a straight line, so we take only three values of x and the corresponding values of y.)

x	-2	0	3
$y = 2x - 10$	-14	-10	-4

The graphs intersect when x is -2.8, -0.9 and 3.9.

When x has these values, the values of y for the curve and for the straight line are equal.

i.e. $$x(x+2)(x-4) = 2x - 10$$

i.e. $$x^3 - 2x^2 - 8x = 2x - 10$$

i.e. $$x^3 - 2x^2 - 10x + 10 = 0$$

Therefore the roots of the equation $x^3 - 2x^2 - 10x + 10 = 0$ are the values of x at the points of intersection of the two graphs.

6. The following table gives values of $4(x^3 - x^2 - 6x)$ for values of x from -3 to 4.

x	-3	$-2\frac{1}{2}$	-2	$-1\frac{1}{2}$	-1	$-\frac{1}{2}$	0	$\frac{1}{2}$
$4(x^3 - x^2 - 6x)$	-72	$-27\frac{1}{2}$	0	$13\frac{1}{2}$		$10\frac{1}{2}$		$-12\frac{1}{2}$

x	1	$1\frac{1}{2}$	2	$2\frac{1}{2}$	3	$3\frac{1}{2}$	4
$4(x^3 - x^2 - 6x)$	-24	$-31\frac{1}{2}$	-32	$-22\frac{1}{2}$	0	$38\frac{1}{2}$	

a) Write down the three missing values.

b) Draw the graph of $y = 4(x^3 - x^2 - 6x)$ for values of x from -3 to 4. Take 2 cm as 1 unit on the x-axis and 1 cm as 10 units on the y-axis. Draw on the same axes the graph of $y = 5(4 - x)$.

c) Write down the value(s) of x at the point(s) of intersection of the two graphs.

d) Use your graph to find the range of values of x for which $4(x^3 - x^2 - 6x)$ is greater than $5(4 - x)$.

e) Find, in its simplest form, the equation whose roots are the values of x at the points of intersection of the two graphs.

7. Draw the graph of $y = 3x - x^3$ for values of x from -2 to 2 at half-unit intervals, but adding other values if you think they are needed. Use your graph to solve the equations

a) $x^3 - 3x = 0$

b) $x^3 - 3x + 1 = 0$

c) $x^3 - 3x - 1 = 0$

Can you use your graph to find the square roots of 3 ? If you can, write down their values.

8. Copy and complete the following table, which gives values of $\frac{1}{5}x^3$ for values of x from -4 to 4.

x	-4	-3.5	-3	-2.5	-2	-1.5	-1	0
$\frac{1}{5}x^3$	-12.8	-8.6	-5.4	-3.2	-1.6	-0.68	-0.2	0

x	1	1.5	2	2.5	3	3.5	4
$\frac{1}{5}x^3$		0.68		3.2		8.6	12.8

Hence draw the graph of $y = \frac{1}{5}x^3$ for values of x from -4 to 4. Take 2 cm as 1 unit on the x-axis and 1 cm as 1 unit on the y-axis.

Use your graph to solve the equations

a) $\frac{1}{5}x^3 = -8$ b) $x^3 = 35$ c) $x^3 + 20 = 0$.

9. Draw, on the same axes, the graphs of $y = x^3$ and $y = 9x - 4$ for values of x from -4 to 4. Take 2 cm as 1 unit on the x-axis and 2 cm as 10 units on the y-axis.

a) Write down the values of x at the points where the two graphs intersect.

b) Write down and simplify the equation whose roots are these x values.

c) Without drawing any more graphs determine the number of roots of the equation $x^3 + 3x - 2 = 0$.

10. Draw the graph of $y = \dfrac{24}{x}$ for values of x from 1 to 12. Take 1 cm as 1 unit on both axes. Draw on the same axes the graph of $x + y = 13$.

a) Write down the values of x at the points where the graphs intersect.

b) What equation has these x values as its roots ?

c) Write down the range of values of x for which $13 - x \geqslant \dfrac{24}{x}$.

d) Why do we not continue the graph of $y = \dfrac{24}{x}$ towards the y-axis ?

11. Draw the graph of $y = \dfrac{12}{x+1}$ taking 0, 1, 2, 3, 5, 7, 9, 11 and 13 as the values for x in the table. Take 1 cm as 1 unit on the x-axis and 2 cm as 1 unit on the y-axis.

a) Use your graph to solve the equation $\dfrac{12}{x+1} = 5.2$.

On the same axes draw the graph of $x + y = 10$.

b) Write down the values of x at the points where these graphs intersect.

c) What equation has these values as its roots ?

12. Copy and complete the table which gives values of $\dfrac{1}{x^2}$ for values of x from 0.5 to 4.

x	0.5	0.6	0.75	1	1.5	2	2.5	3	3.5	4
$\dfrac{1}{x^2}$		2.78	1.78		0.44	0.25	0.16	0.11	0.08	0.06

a) Draw the graph of $y = \dfrac{1}{x^2}$ for values of x from 0.5 to 4, taking 4 cm as 1 unit on both axes.

13. Sketch the graph of $y = 1 + \dfrac{1}{x^2}$ for values of x from 0.5 to 4.

14. Sketch the graph of $y = 1 - \dfrac{1}{x^2}$ for values of x from 0.5 to 4.

15. Sketch the graph of $y = \dfrac{1}{x^2} - 1$ for values of x from 0.5 to 4.

16. Copy and complete the table which gives values of $1 + \dfrac{2}{x^2}$ for values of x from 0.75 to 4.

x	0.75	1	1.5	2	2.5	3	3.5	4
$1 + \dfrac{2}{x^2}$	4.56	3	1.89		1.32	1.22		1.125

a) Draw the graph of $y = 1 + \dfrac{2}{x^2}$ for values of x from 0.75 to 4, taking 4 cm as 1 unit on both axes.

b) Using the same axes and scales, draw the graph of the straight line whose equation is $y = \frac{1}{2}(x+3)$.

c) Write down, and simplify, the equation satisfied by the value of x at the point of intersection of the curve and the straight line. Read off from your graph an approximate solution of this equation.

17. a) Draw the graph of $y = x + \dfrac{1}{x}$ for values of x from $\frac{1}{2}$ to 4, taking the x values at half-unit intervals. Take 4 cm as 1 unit on both axes.

  b) Draw, on the same axes, the graph of $x + y = 4$.

  c) Write down the value of x at the point where the graphs intersect.

  d) Write down, and simplify, the equation which is satisfied by this value of x.

PROBLEM SOLVING USING GRAPHS

The types of graph studied in the two previous exercises can help us to solve certain real life problems. The next exercise considers some typical problems.

EXERCISE 10c

1. A subsidiary of a large company produces electric switchboards. When it produces x thousand switchboards it makes a profit of y thousand pounds, where $y = 7x - (x^2 + 2)$.

Corresponding values of x and y are given in the following table.

x	0	1	2	3	4	5	6
y	-2	4	8	10	10	8	4

Draw a graph to represent this data using 4 cm as the unit for x and 1 cm as the unit for y. You may find it an advantage to plot additional points.

Use your graph to find

a) the maximum profit the company can make and the number of switchboards it must produce to give this profit

b) the minimum number of switchboards the company must produce in order at least to break even.

The parent company decides that the subsidiary must make a minimum profit of £6000 to remain in production. Within what range must the number of switchboards lie in order to achieve this ?

2. After t seconds, the height of a stone above its point of projection is s metres, where $s = 35t - 5t^2$. Draw a graph to represent this data for values of t from 0 to 7. Take 2 cm as 1 unit for t and 2 cm as 5 units for s.

Use your graph to find

a) the height of the stone after 2.6 s

b) at what times the stone is 40 m above the ground

c) the maximum height the stone reaches and the time it takes to reach this height.

d) What is the height of the stone after 7 s ?

3. A farmer has a roll of wire 140 metres long, and wishes to make a rectangular enclosure using a straight wall as one side.

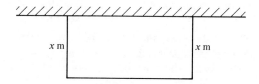

If the rectangle is x metres wide, find its length in terms of x, and hence show that the area of the enclosure, A m², is given by $A = 2x(70 - x)$.

Complete the following table to find the values for A corresponding to the given values for x.

x	0	10	20	30	40	50	60	70
$A = 2x(70 - x)$	0	1200		2400		2000	1200	

Draw the graph of $A = 2x(70 - x)$ for values of x from 0 to 70. Take 1 cm to represent 5 units for x and 1 cm to represent 100 units for A.

Use your graph to find

a) the width of the enclosure when its area is 1500 m²

b) the maximum area that the farmer can enclose and the corresponding value of x.

c) the area enclosed when the width of the enclosure is 56 m

d) the area enclosed when the length of the enclosure is 56 m.

4.

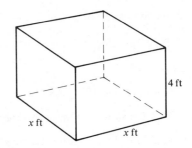

The diagram shows a four-foot-deep water tank with a square base of side x feet.

a) Show that the capacity of the tank, C cubic feet, is given by the formula $C = 4x^2$.

b) Complete the table below which shows the value of C for various values of x.

x	0.5	1	1.3	1.7	2	2.3	2.7	3	3.3	3.6	4
C	1	4	6.76	11.56			29.2	36	43.6	51.8	64

c) Draw the graph of $C = 4x^2$ for values of x from 0 to 4. Take 4 cm as 1 unit on the x-axis and choose your own scale for the C-axis.

d) What value for x will give a capacity of 40 ft³.

e) A householder needs a new tank with a capacity between 30 ft³ and 40 ft³. To fit the tank into the available space the side of the base must be less than 3 feet. What range of values for x will satisfy these conditions ?

5. A piece of wire, 16 cm long, is cut into two pieces. One piece is $8x$ cm long and is bent to form a rectangle measuring $3x$ cm by x cm. The other piece is bent to form a square.

a) Find, in terms of x,
  i) the length of wire used to make the square
  ii) the length of a side of this square
  iii) the area of the rectangle
  iv) the area of the square.

b) Show that the combined area of the rectangle and the square is A cm² where $A = 7x^2 - 16x + 16$.

Corresponding values for x and A are given below.

x	0	0.5	1	1.5	2	2.5	3	3.5
A	16	9.75	7	7.75	12	19.75	31	45.75

c) Draw the graph of $A = 7x^2 - 16x + 16$ for values of x in the range 0 to 3.5. Take 4 cm as the unit on the x-axis and 4 cm as 10 units on the A-axis.

d) Use your graph to find the value of x for which the combined area of the rectangle and the square is smallest.

6. A plastics firm is asked to make an open rectangular container with a square base. The pieces making up the faces of the container are cut from a rectangular strip of plastic 20 cm wide.

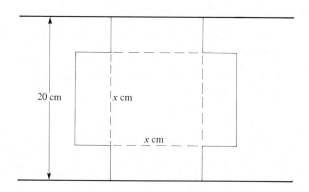

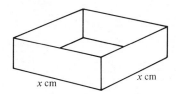

a) If the side of the base is x cm write down, in terms of x, an expression for the depth of the container.

b) Show that the capacity of the container, C cm^3, is given by the formula
$$C = x^2 \left(10 - \frac{x}{2} \right).$$

c) Construct a table which shows the value of C for various values of x from 0 to 20.

d) Draw the graph of $C = x^2 \left(10 - \frac{x}{2} \right)$ for values of x from 0 to 20. Take 1 cm as 1 unit on the x-axis and choose your own scale for the C-axis.

e) What is the capacity when x is 3.75 ?

f) What values of x will give a capacity of 400 cm^3 ?

g) What is the largest capacity possible for this container ? For what value of x does this occur ?

7. The graph below shows the price of Brited plc shares at the close of business each week day over a twelve week period.

a) During which periods is the price of the share rising ?

b) During which periods is the price of the share falling ?

c) After week 3 when would the best time have been   i) to buy   ii) to sell ?

d) What do you think is likely to happen to the price of the share during week 13 ?

e) "The price of the share doubled in the first six weeks." Is this statement true or false ?

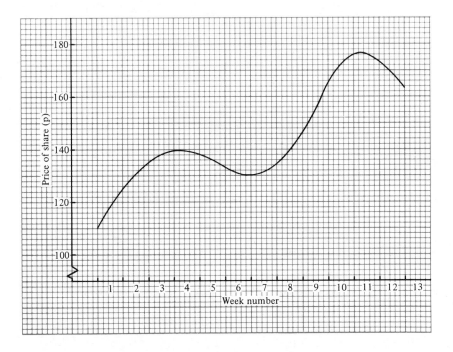

8. As soon as Peter was admitted to hospital he was linked up to a machine that recorded his temperature every five minutes. The graph opposite shows his temperature for the first ten hours after being admitted.

a) At what time was Peter admitted.

b) For how long did his temperature continue to rise ?

c) What was the highest temperature recorded ?

d) How much did it fall before it began to rise again ?

e) Peter's normal temperature is 37 °C. At what time should his temperature become normal if it continues to fall at the same rate ?

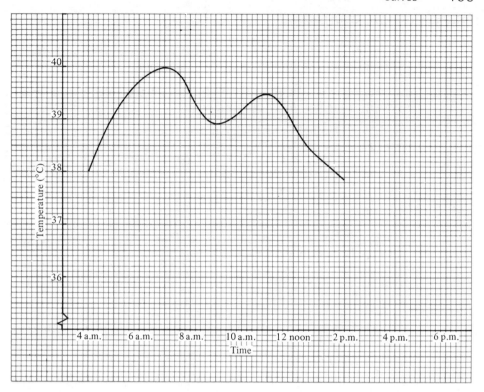

DISTANCE–TIME GRAPHS

When an object is travelling at a constant speed, the distance–time graph is a straight line. The speed is found from the gradient of this line.

i.e. $$\text{Speed} = \frac{\text{distance travelled}}{\text{time taken}}$$

When the velocity of an object is constantly changing, the distance–time graph is a curved line.

The *average velocity* from time t_1 to time t_2 may be found from this graph by finding the gradient of the chord joining the points on the curve that correspond to times t_1 and t_2.

The *velocity* at the instant when the time is T is given by the gradient of the tangent to the distance–time curve at the point where the time is T. To find the velocity from a graph, the tangent has to be drawn by eye. The gradient can then be found from the graph.

VELOCITY-TIME GRAPHS

The acceleration at time T is represented by the gradient of the tangent to the velocity–time graph at the point where the time is T.

The *distance travelled* in an interval of time is represented by the area under the velocity–time graph for that interval.

Methods for finding areas under curves are fully discussed in Book 4A, Chapter 17.

EXERCISE 10d

1. The table shows the distance, d metres, of a ball from its starting point, t seconds after being thrown into the air.

t	0	0.5	1	1.5	2	2.5	3	3.5	4
d	0	8.75	15	18.75	20	18.75	15	8.75	0

Draw the graph of d against t using 4 cm as 1 unit on the t-axis and 1 cm as 1 unit on the d-axis.

From your graph find
a) when the ball returns to the starting point
b) the average velocity of the ball from $t = 1$ to $t = 2$
c) the average velocity of the ball from $t = 1.5$ to $t = 2$
d) the velocity of the ball when $t = 1$
e) the velocity of the ball when $t = 3$
f) the velocity of the ball when $t = 2.5$

2. A ball was thrown vertically upwards and, after t seconds, its height, h metres, above the ground was given by $h = 10 + 15t - 5t^2$

a) Copy and complete the following table

t	0	0.5	1	1.5	2	2.5	3	3.5	4
10	10	10	10	10	10	10	10	10	10
$15t$		7.5	15	22.5		37.5		52.5	60
$-5t^2$		-1.25	-5	-11.25		-31.25		-61.25	-80
h		16.25	20	21.25		16.25		1.25	-10

b) Use scales of 4 cm to represent 1 second and 4 cm to represent 5 m and draw the distance–time graph.

c) From your graph find the velocity of the ball
   i) at the start   ii) when $t = 2.5$

d) What is the greatest distance of the ball from the ground ?

e) How long does the ball take before it hits the ground ?

3. A particle travels in a straight line and its speed, v metres per second, after t seconds, is given by $v = \frac{1}{10}t(11-t)$. Draw the graph of $v = \frac{1}{10}t(11-t)$ for whole number values of t from 0 to 11. Take 2 cm as 1 unit for t and 4 cm as 1 unit for v.

Use your graph to find

a) the maximum value of v

b) the acceleration of the particle, in metres per second per second, when $t = 3$

c) the time at which the acceleration is zero

d) the total distance covered by the particle in 11 seconds.

4.

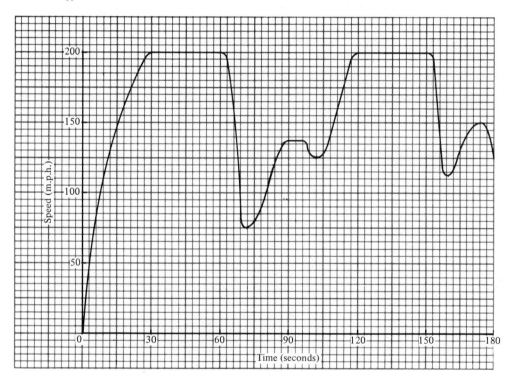

The sketch shows the velocity-time graph for a racing car during the first three minutes of a race.

a) During which period does the acceleration have its greatest positive value ?

b) What is the maximum speed of the car ?

c) For how long is the car driven at this maximum speed ?

d) What is the lowest speed to which the car slows ?

e) How far does the car travel in the first minute ?

5. A rocket is fired and its velocity, v km/minute, t minutes after firing is given by

$$v = t^3 - \tfrac{1}{2}t^2$$

Copy and complete the following table.

t	0	1	1.5	2	2.5	3	3.5	4
v	0	0.5	2.25		12.5		36.75	56

Use scales of 4 cm $\equiv$ 1 minute and 4 cm $\equiv$ 10 km/min to draw the velocity–time graph. From your graph find

a) the acceleration 3 minutes after firing

b) the velocity after $2\tfrac{3}{4}$ minutes

c) the time when the velocity is 30 km/min

d) the distance covered in the first 3 minutes

e) the distance covered in the third minute.

INTERPRETATION OF RATES OF CHANGE

The graph shows the distance travelled by a hill-climb cyclist from the time he leaves the bottom of a hill until he reaches the top.

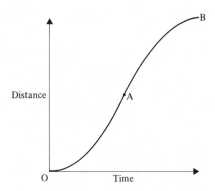

The gradient of any tangent to this curve gives the rate of change of distance with respect to time, i.e. it gives the speed of the cyclist at that instant.

From O to A the gradient of the tangent is increasing, i.e. the speed is increasing. From A to B the gradient of the tangent is decreasing, until at B it has become very small. This indicates that the speed is gradually decreasing as the cyclist travels towards the top of the hill. By the time he reaches the top his speed is quite slow.

In general terms, if we draw the graph of a variable y plotted against a variable x, and the graph is a smooth curve, the gradient of the tangent at any point on the curve gives the rate of change of y with respect to x.

EXERCISE 10e

1. Steve is a 400 metre runner. The distance–time graph for one of his races is given below.

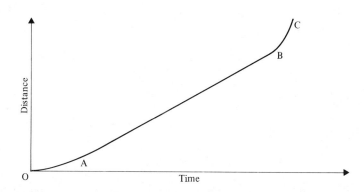

Explain what is happening between

a) O and A   b) A and B   c) B and C.

2. In another 400 m race Steve's speeds were recorded at different times. The velocity–time graph for this race is given below.

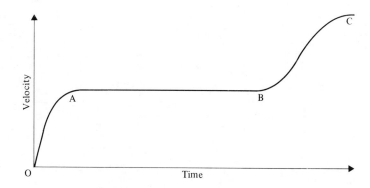

Explain what is happening between

a) O and A   b) A and B   c) B and C.

3. 'The rate of inflation is slowing down'. Which ONE of these graphs best represents this statement ?

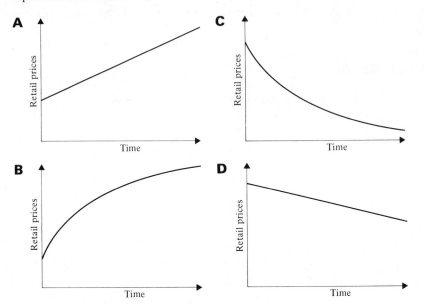

4. 'Building Society interest rates rose several times during the year'. Which ONE of the following graphs best represents this statement ?

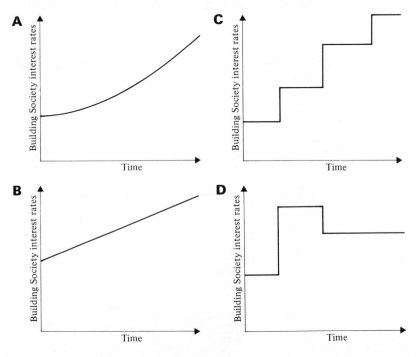

5.

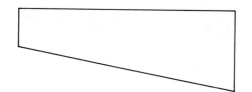

The diagram shows the cross-section of a swimming pool. Water enters the pool at a constant rate. Which ONE of the following graphs best represents how the depth of water in the pool is changing ?

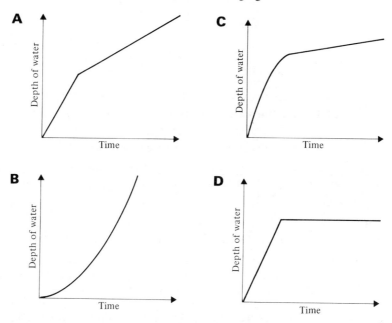

6. The population of Blackborough doubles every hundred years. In the year AD 1000 it was 8. Draw a sketch to show how it has changed from the year AD 1000 to the present day.

7. Margaret O'Neil lived to the ripe old age of 90. She was 50 cm long when she was born, 150 cm tall at the age of 30 but lost 5 cm in height before she was finally laid to rest. She grew fastest in her teens. Draw a sketch to show Margaret's height throughout her lifetime.

8. A child blows up a spherical balloon so that its volume increases by 5 cm^3 every second. Draw a sketch to show how the volume of the sphere changes with time.

9. The radius of a circle is increasing at a constant rate of 2 centimetres each second. The graph shows how the area of the circle A, changes with time (t).

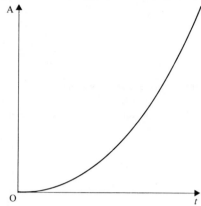

State whether each of the following statements is true or false.

a) The area increases with time.

b) The area increases at constant rate.

c) The area increases at an ever increasing rate.

10. A car accelerates from rest to a maximum speed of 100 m.p.h. in 32 seconds. What would you expect the velocity–time graph to look like. (Any suitable sketch will do.)

11. A sprinter covers 100 metres at a constant speed.
Sketch   a) the distance–time graph,   b) the velocity–time graph.

12. A husband and wife team have an idea for a business venture. They wish to manufacture, and market, an easy-to-use binder. As they become successful they begin to employ staff. The number of staff they employ doubles every three months until, in two years, it is approximately 500. The workforce then remains fairly stable for three years, but in the sixth year demand for the binders declines. A small number of employees leave who are not replaced, and by the seventh year further problems cause the workforce to be reduced by half. The problems continue and become so acute that the business is forced to cease trading even though it still employs 200 people.

Draw a sketch graph to show how the size of the workforce varies throughout the life of the business.

13. Rabbits colonise a hitherto rabbit-free common. To begin with, the numbers are small, but soon they treble in number every year until a population of approximately two thousand is reached after five years. This number remains stable for three years when, unfortunately, the viral disease of myxomatosis is introduced into the colony. The size of the colony falls, slowly at first, but then with increasing rapidity, until the common is once again rabbit-free after ten years.

Draw a sketch graph to show how the rabbit population on the common changes with time.

11 SYMMETRY, SOLIDS AND NETS

LINE SYMMETRY

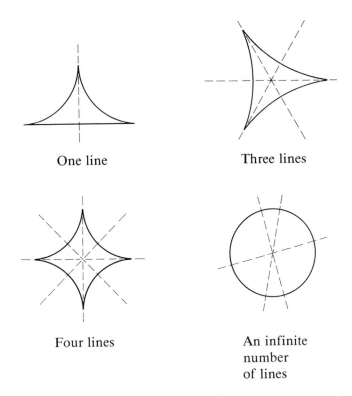

One line

Three lines

Four lines

An infinite
number
of lines

A figure has line symmetry if we can create the whole by reflecting half the figure in the line of symmetry. Notice that the diagonal of a rectangle is *not* a line of symmetry because reflecting half a rectangle in a diagonal produces a kite.

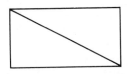

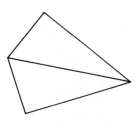

ROTATIONAL SYMMETRY

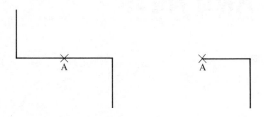

This figure has no line of symmetry but by rotating half of it about A through 180° we produce the whole of it. It has *rotational symmetry* of order 2.

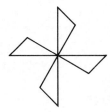

This figure can be created by rotating one triangle through 90° about the centre. It takes four such rotations to complete the figure so it has rotational symmetry of order 4. (360° ÷ 90° = 4.)

EXERCISE 11a

1. Draw diagrams of the following figures and mark any lines of symmetry.

a) A square

b) A rhombus

c) A parallelogram

d) An equilateral triangle

e) A rectangle

f) An isosceles triangle.

2. Draw a diagram of each of the figures in question 1 and mark with a cross the centre of rotational symmetry if any. State the order of rotational symmetry.

3.

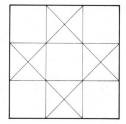

Copy the figure above on squared paper.

a) Mark, in colour, any lines of symmetry.

b) If the figure has a centre of rotational symmetry, mark it with a cross and give the order of rotational symmetry.

4. Repeat question 3 for each of the diagrams below.

a)
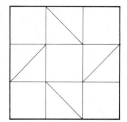

b)
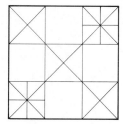

5. Make up figures like those in questions 3 and 4 and investigate their symmetry.

6. Investigate the connection between the rotational and line symmetry of a figure. Use the information from questions 1 to 5 and draw other figures to test your conclusions.

7. Sketch a regular octagon.

a) How many lines of symmetry has it ?

b) What is the order of rotational symmetry ?

c) What is the size of the angle between two adjacent lines of symmetry ?

d) Construct a regular octagon inscribed in a circle of radius 5 cm.

8. Repeat question 7(a), (b) and (c) for a regular *n*-sided polygon.

TESSELLATIONS

The word "tessellation" comes from *tessera* which was the name for a small tile or block used in making mosaics. We use the word in mathematics to mean a continuous regular pattern made of shapes fitting together.

The important points to look for are that the pattern can be *continued* and that there are *no holes* in it.

The most obvious one is made with squares but rectangles, equilateral triangles and rhombuses make equally simple tiling patterns.

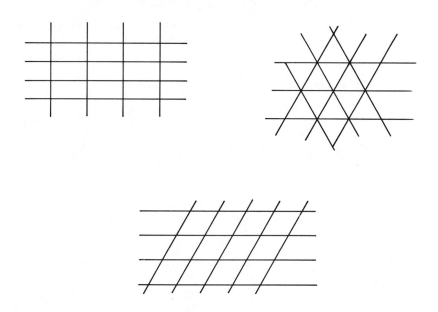

Using rectangles, more complicated patterns can be made, such as those in herringbone wood block flooring or a brick wall.

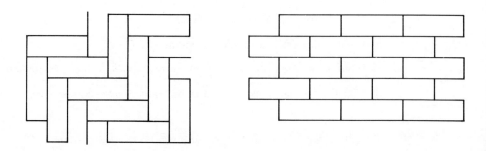

EXERCISE 11b

Draw tessellations of the given figure.

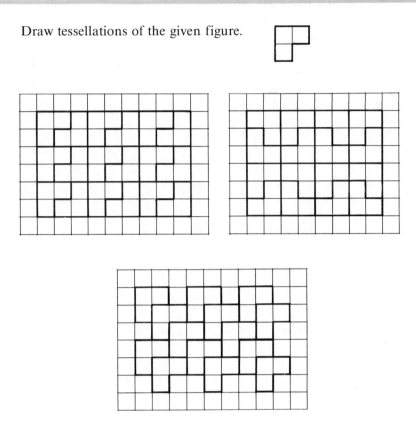

(There are a great many more possibilities but notice that some patterns, such as the one given below, are *not* tessellations as it is not possible to continue the pattern.)

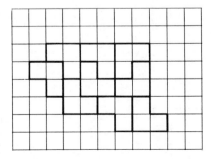

Squared paper is useful in this exercise.

1. Draw tessellations of the given figures.

a) 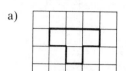   b) 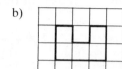   c)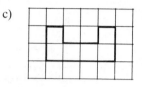

2. Draw tessellations of the given figures. They may not be turned over,

i.e. 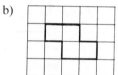   is not the same figure as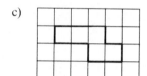

a) 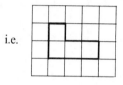   b)   c)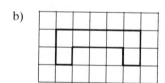

3. Is it possible to tessellate with each of the following shapes? Where possible draw the tessellation.

a) 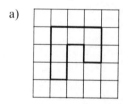   b)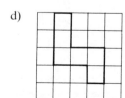

c) 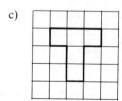   d)

4. a) Sketch six regular *n*-sided polygons for $n = 3, 4, 5, 6, 7$ and 8 (e.g. for $n = 4$, the polygon is a square). Calculate the size of an interior angle of each polygon.

   b) Find which of the polygons will tessellate. How can you tell from the size of the interior angle whether the polygon will tessellate or not? Are there any other regular polygons that will tessellate?

5. Draw tessellations of the following figures. Tracing paper can be used to complete the drawings.

   a)   b)   c)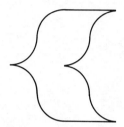

6. Make up other shapes that will tessellate. Show enough of each repeating pattern to make sure that it will continue.

7. a)

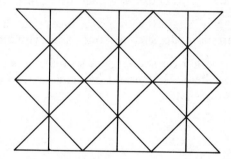

   b)

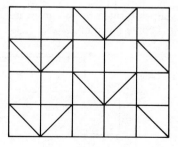

The two diagrams above each show a tessellation of a square and a triangle used together. (The squares and triangle are not of the same relative size in (a) and (b).)

Draw other tessellations using a square and a triangle.

NETS

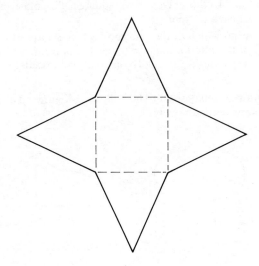

The diagram above shows the *net* of a square pyramid; if it were drawn on paper and cut out it could be folded up to form the shape.

Most nets give shapes with *plane* faces, but a few objects with curved surfaces can be made, such as cones and cylinders.

Flaps must be left on some edges so that the shape can be stuck together. If in doubt where they should be, leave a flap on every edge and cut off the surplus ones later.

EXERCISE 11c

1. Name the shapes formed by the given nets below. In each case, state which lettered points come together.

a)

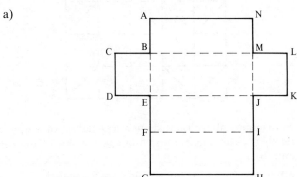

b)

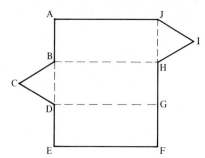

c)

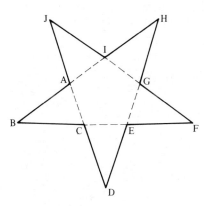

d)

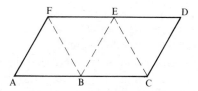

2. For each of the shapes whose nets are given in question 1, answer the following questions.

a) How many edges has the solid ?

b) How many vertices has the solid ?

c) What is the least number of flaps needed to stick all the edges together ?

d) Draw a net with a different arrangement of the faces. It should not be a reflection or rotation of the given net.

e) Does your alternative net need more or fewer flaps ? Investigate the number of flaps on other possible nets and comment on your result.

3. a)   b)

c)   d)

Which of the diagrams above are nets of a cube ? Sketch other possible nets (no reflections or rotations).

4.

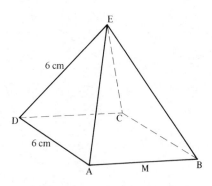

In the square pyramid above, E is above the centre of the base, AD = ED = 6 cm and M is the midpoint of AB.

a) Sketch a net of the pyramid.

b) Draw the net accurately and find the length of EM.

c) F is the centre of the base. Draw △EFM accurately and hence find the height of the pyramid.

d) Calculate the volume of the pyramid.

5.

The diagram shows an open cone of radius 5 cm and slant height 10 cm. Sketch its net.

6.

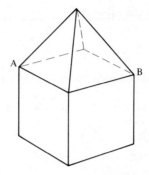

The solid above consists of a cube surmounted by a square pyramid. All edges measure 5 cm.

b) Sketch a net of the solid.

b) Draw an accurate diagram to find the shortest distance, over the surface, from A to B. (It may be necessary to use a diagram different from that in (a).)

7. Each of the following nets has a face missing.

a) Sketch the net and add the missing face in one of its possible positions.

b) In how many different positions could the missing face be drawn ?

i)

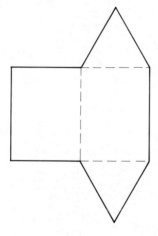

ii)

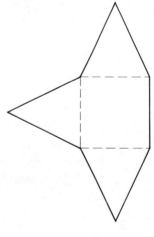

iii)

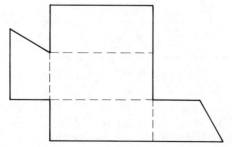

EDGES, FACES AND VERTICES ───────────────────

EXERCISE 11d

1. a) Copy and complete the following table.

Solid	Number of faces (F)	Number of edges (E)	Number of vertices (V)
Cube	6	12	
Tetrahedron	4		
Square pyramid			
Triangular prism			

b) What is the connection between the numbers in each row of the completed table ? Give a formula connecting F, E and V.

c) Add to your table any solids that you considered in previous exercises. Does the formula still hold ?

d) A solid has 8 faces and 6 vertices. How many edges would you expect it to have ?

e) A solid is said to have 5 faces, 5 edges and 8 vertices. Is this possible ?

2.

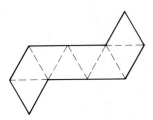

The diagram above shows the net of an octahedron.

a) How many faces, edges and vertices has it ?

b) Does the information in (a) fit the formula obtained in question 1 ?

3.

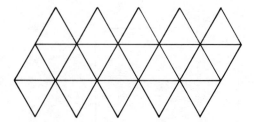

The diagram above shows the net of an icosahedron.
Repeat questions 2(a) and (b) for the icosahedron.

4.

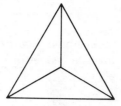

a) The diagram is a plane figure. How many lines (L) are used to draw it and how many points (P) are there where lines meet ?

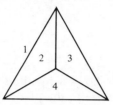

Counting the space outside the triangle as one region, the number of regions (R) in the figure is 4.

b) Copy and complete the following table.

Figure	Number of regions (R)	Number of lines (L)	Number of points (P)
	4		
	2		
			
		8	

c) Draw other figures and add the information to the table.

d) Give a formula for the connection between R, L and P.

e) What connection is there between this result and that in question 1 ? Can you explain it ?

PLANES OF SYMMETRY

Just as some two dimensional figures have lines of symmetry so a three dimensional solid can have *planes* of symmetry.

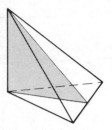

The diagram above shows a pyramid with an isosceles triangle for a base. The shaded triangle lies in the plane of symmetry.

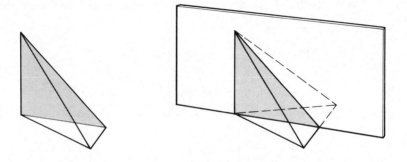

If half the pyramid were to be reflected in a mirror, as in the diagram, we should see the complete solid. The mirror is the plane of symmetry.

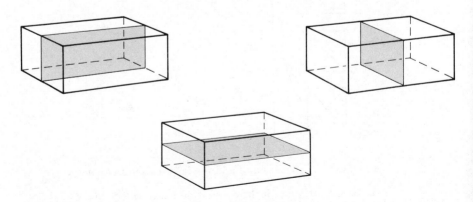

A solid may have more than one plane of symmetry. A cuboid, for instance, has three, while a sphere has an infinite number.

EXERCISE 11e

1. How many planes of symmetry has each of the following solids ?

a)
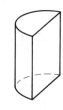

b)
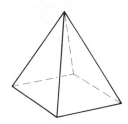

c)
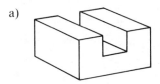

d)
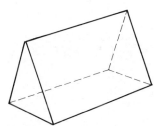

2. How many planes of symmetry has each of the following solids ?
a) A cube
b) A circular cone
c) A hexagonal prism
d) A cylinder
e) A hemisphere
f) A regular tetrahedron

3. Some solids have only approximate symmetry. A human head for example, or a leaf, is only roughly symmetrical, with one plane of symmetry. A book is roughly symmetrical with two planes of symmetry (ignoring the printing).

How many approximate planes of symmetry do each of the following solids have ?
a) Aeroplane
b) Cup
c) Plate
d) Fountain pen
e) Chair
f) Bicycle
g) Clock
h) Hexagonal pencil
i) Name other solids with only one approximate plane of symmetry.
j) Name other solids with more than one approximate plane of symmetry, giving the number of planes in each case.

PLANS AND ELEVATIONS

If we look down on a *solid* we can see its *plan*. For instance, if we look down on this metal waste-paper bin

we see that its plan consists of two concentric circles; we can see the rim and the inside edge of the base.

a)   b)

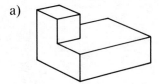   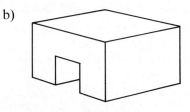

The plans of the two solids above are given below.

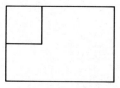   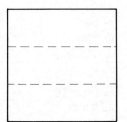

Edges in sight are drawn with plain lines, but hidden edges are drawn with broken lines.

We could also draw a side view or *elevation* of the wastepaper bin. This is not a picture with perspective, but a diagram, so we draw a trapezium with horizontal lines top and bottom.

A view from the left of solid (a) has a broken line because there is a hidden edge.

A view from the front (front elevation) of the same solid has all the edges either in plain view or behind edges which are in plain view so there are no broken lines.

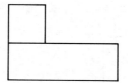

EXERCISE 11f

1. Sketch the plans of the following solids.

a)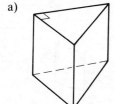

b)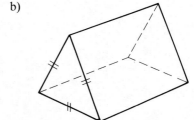

c)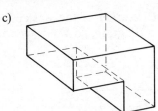

d)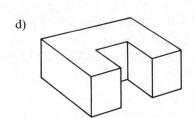

e)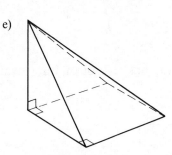

f)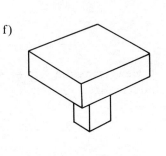

2. Name the solids whose plans are given below. In several cases there are two or more possible answers; give as many as you can, and sketch the solids if you cannot give them a name.

a)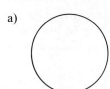

b)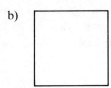

c)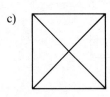

d)

3. Sketch the two elevations of each of the solids, looked at in the directions indicated by the arrows.

a)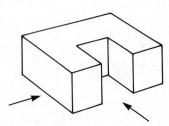

b)

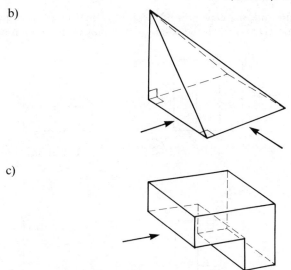

c)

4. The following solids are built up of cubes with edges of length two centimetres. For each solid draw accurately a plan and elevations in the directions indicated by the arrows.

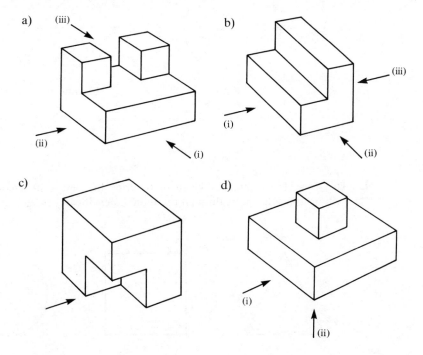

5. The plan and one elevation of a solid (in the direction of the arrow on the plan) are given. Sketch the solid. (There may be more than one.)

a)

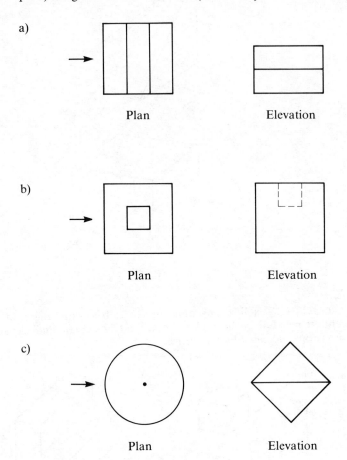

6. This is a classic puzzle. Sketch or describe the one solid which will pass, with no space to spare, through the three holes drawn below.

SECTIONS

If we have a solid made of modelling clay we can slice it into two parts.

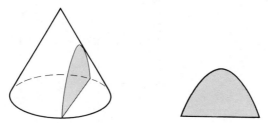

The cut surface so produced is called a *section*. We have already encountered the *cross-section* of a prism.

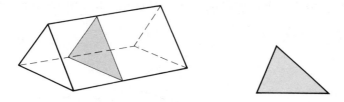

If we cut a prism anywhere with a slice parallel to its ends we produce the same section, in this case a triangle. The prism has a *uniform* cross-section.

If we cut a square pyramid with a slice parallel to its base, the resulting section is a square but its size depends on the distance of the cut from the base.

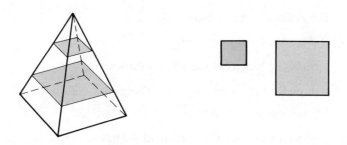

EXERCISE 11g

1.

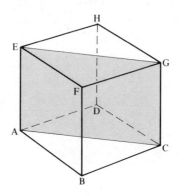

The diagram shows a cube of side 4 cm.

a) Sketch the shaded section and calculate its area.

b) Is the shaded plane a plane of symmetry ?

c) Sketch the section EBCH. How does it compare with the shaded section ?

2.

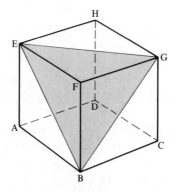

The diagram shows a cube of side 6 cm.

a) Sketch the shaded section.

b) Find the lengths of the sides of the section.

c) Find the area of the section. (If necessary draw the section accurately.)

d) Find the area of △EFB.

e) Find the volume of the pyramid EFBG.

3.

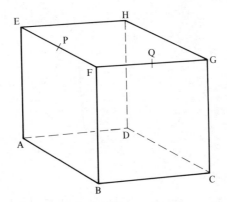

The diagram shows a cube of side 5 cm. ABCD is horizontal. P and Q are midpoints of EF and FG. The cube is cut into two parts by a vertical slice through PQ.

a) Sketch the resulting section.

b) Find the area of the section.

c) Find the volumes of the two solids into which the cube has been cut.

4.

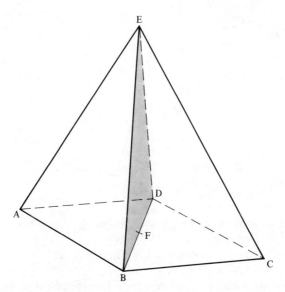

The diagram shows a square pyramid. Its base is horizontal and E is vertically above F, the centre of the base. AB = 8 cm and EF = 10 cm.

a) Sketch section EBD. What type of triangle is it? Find BD and the area of the section.

b) Sketch the vertical section through EF, parallel to BC. Find its area.

5.

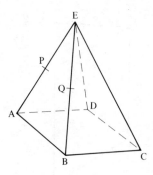

The diagram shows a square pyramid. Its base is horizontal and E is vertically above the centre of the base. AB $= 6$ cm   and   AE $= 10$ cm. P and Q are midpoints of EA and EB.

a) Find the length of PQ.

b) Sketch the horizontal section through PQ. What shape is it ? Find its area.

c) Sketch the vertical section through PQ. What shape is it ?

d) Give the ratio of the volumes of the two solids into which section (b) cuts the pyramid.

6.

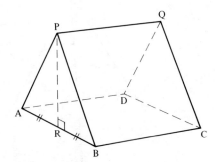

The diagram shows a triangular prism. AB $= 6$ cm, AP $=$ PB $= 5$ cm and PQ $= 8$ cm. ABCD is horizontal.

a) Draw the section given by a vertical cut through PQ. Find the area of the section.

b) Draw the section given by a horizontal cut through the midpoint of AP and find its area.

c) Find the surface area of the prism.

d) Find the volume of the prism.

e) Find the volume of the smaller of the two solids into which the cut in (b) divides the prism.

f) Which of the cuts described in (a) and (b) is in a plane of symmetry ?

MIXED QUESTIONS

EXERCISE 11h

1.

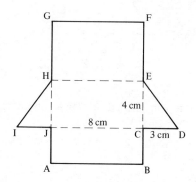

The net above is formed of three rectangles and two right-angled triangles.

a) Name the solid formed by the net and sketch it.

b) How many vertices and edges has the solid ?

c) Find the missing measurements.

d) Find the surface area of the solid.

e) Find the volume of the solid.

f) Draw an alternative net which is not a reflection or rotation of the given net.

g) Calculate the length AE on the given net.

h) Calculate the length AE in the solid.

i) A fly crawls on the surface of the solid from G to C by the shortest route. What is the distance crawled ?

2.

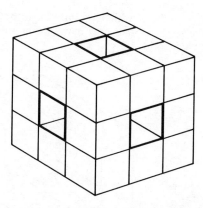

The cube above has three holes cut through it each from one face to the opposite face. How many small cubes make up the solid ?

3.

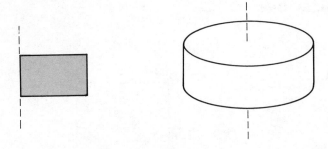

If we rotate the shaded rectangle about the broken line we generate the cylinder on the right.

Sketch or describe the solids generated by rotating the figures below about the broken line.

a)   b)   c)   d)   e)

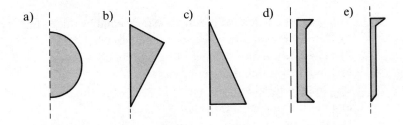

f) Sketch or describe other solids which can be generated by this method and in each case sketch the shape used to generate the solid.

4. The given solids, of which we only have the front views, are made of cubes stuck face to face. State how many different numbers of cubes can be used to make each solid.

a)

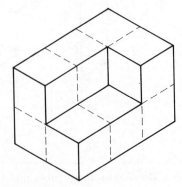

b)

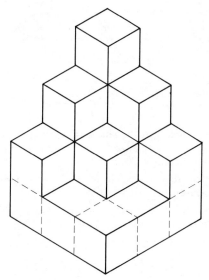

5.

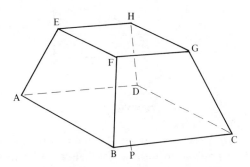

The diagram shows the *frustum* of a pyramid (i.e. a pyramid with its top sliced off). ABCD and EFGH are both horizontal squares.

EA = FB = GC = HD = 6 cm,   AB = 10 cm   and   EF = 6 cm.

a) How many planes of symmetry are there ?

b) Sketch the face FBCG. What shape is it ?

c) P is the point on BC such that $\widehat{FPB} = 90°$. Calculate the length of FP.

d) Find the surface area of the frustum.

e) Calculate FH and BD.

f) Sketch the section FHDB and draw it accurately.

g) Sketch the vertical section through FE.

6.

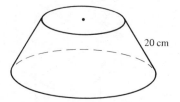

20 cm

A lampshade is in the shape of a frustum of a cone. The radius of the circle forming the lower edge is 16 cm and the radius of the upper circle is 8 cm. The slant height is 20 cm.

a) Sketch the net used to make the lampshade, marking in as many measurements as possible.

b) Describe, with a diagram, how you would draw the net accurately. Make sure you have all necessary lengths and angles.

c) Find the area of the net.

7.

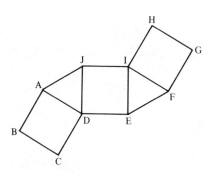

The figure above is formed of squares and equilateral triangles.

a) Is it possible to form a tessellation with this figure ? Give reasons, with a diagram if necessary, for your answer.

b) If the diagram represents the net of a solid, with which letter or letters will B join ?

8.

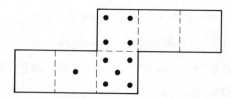

The diagram represents the net of a die. The number of dots on pairs of opposite faces add up to seven.

a) Copy and complete the net.

b) Draw a different net for the same die.

12 TRANSFORMATION AND MATRICES

COMMON TRANSFORMATIONS

These are dealt with fairly thoroughly in Books 2 and 4A, so here we give only a summary and a revision exercise.

Reflection

A reflection is defined by the mirror line.

Rotation

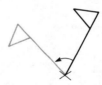

A rotation is defined by the centre of rotation and the angle of rotation.

The convention in mathematics is that angles are measured anticlockwise (except for bearings), so a positive angle indicates an anticlockwise turn and a negative one a clockwise turn. In work at this level, however, it is better to specify the direction in order to avoid misunderstandings.

Translation

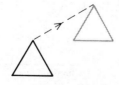

A translation is defined by a description of the displacement, usually in the form of a vector.

Enlargement

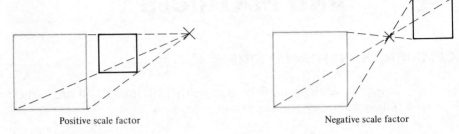

Positive scale factor   Negative scale factor

An enlargement is defined by the centre of enlargement and the scale factor.

OTHER TRANSFORMATIONS

Although these are not common they do sometimes occur.

One-way stretch

Invariant line

The object is stretched in one direction only. A one-way stretch is defined by the direction of the stretch, the scale factor and the invariant line.

Shear

This is dealt with in Book 3A.

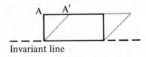

Invariant line

A shear is defined by the invariant line and the displacement of a point not on the invariant line.

EXERCISE 12a

1. Draw x and y axes for values from -7 to 5 on the x-axis and from -2 to 4 on the y-axis.

 A is the quadrilateral with vertices $(1,2)$, $(2,2)$, $(2,4)$ and $(1,3)$.

 a) Reflect A in the line $x = -1$ and label this image B.

 b) Rotate A through $180°$ about the point $(-2,1)$ and label the image C.

 c) Translate A using the vector $\begin{pmatrix} -4 \\ -2 \end{pmatrix}$ and label the image D.

 d) Give the single transformation which maps D to C.

 e) Explain why there is no single transformation which will map B to C.

2.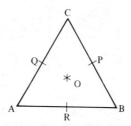

 $\triangle ABC$ is equilateral. P, Q and R are the midpoints of the sides. O is the centre of the triangle.

 a) $\triangle ABC$ is rotated through $120°$ anticlockwise about O. Which points are the images of A, B and C respectively ?

 b) Describe the transformation that will map A to B, B to A and Q to P.

 c) Describe the transformation that will map R to Q, C to B but *not* B to C. To which point will B be mapped ?

 d) $\triangle ABC$ is enlarged using A as the centre, and a scale factor of 2. Give the images of Q and R.

3.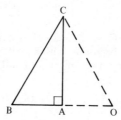

 $\triangle OBC$ is equilateral. A is the midpoint of OB. OB $= 6$ cm.

 a) Draw the diagram accurately and enlarge $\triangle ABC$ by a scale factor of 2, using O as the centre of enlargement. (Remember to draw a rough sketch of the complete diagram first.) Label the image A'B'C'.

 b) What type of quadrilateral is A'B'C'C ? Give your reason.

 c) What type of triangle is C'CB ?

4.

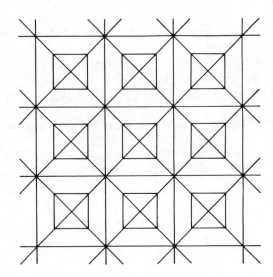

The diagram shows part of a continuous pattern. Describe the various transformations which will map the pattern to itself.

5.

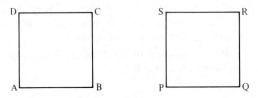

ABCD and PQRS are equal squares and BP = AB. Describe fully the various transformations which will map the first square to the second, in each case giving a diagram and indicating to which points A, B, C and D are mapped.

6. Draw x and y axes for values from -6 to 9 on each axis, using $1\,\text{cm}$ to 1 unit. Draw $\triangle ABC$ with $A(3,5)$, $B(6,5)$ and $C(3,9)$ and $\triangle A'B'C'$ with $A'(-4,0)$, $B'(-4,-3)$ and $C'(0,0)$.

a) $\triangle ABC$ is mapped to $\triangle A'B'C'$ by a rotation. By constructing the perpendicular bisectors of BB' and CC', find the centre of rotation.

b) Give the angle of rotation.

7.

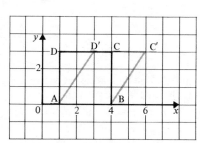

a) Square ABCD is mapped to rectangle ABLM by a one-way stretch parallel to the *y*-axis. Give the invariant line and the scale factor.

b) ABCD is mapped to rectangle PQRS by a one-way stretch with invariant line the *y*-axis. Give the direction of stretch and the scale factor.

c) Draw a diagram showing ABCD and the image A′B′C′D′ of ABCD under a one-way stretch parallel to the *x*-axis, invariant line the *y*-axis and scale factor 2.

d) Repeat (c) with invariant line $x = 1$.

8.

a) Square ABCD is mapped to ABC′D′ by a shear. Describe the transformation.

b) Describe the transformation that maps ABC′D′ to ABCD.

c) Draw a diagram showing ABCD and its image $A_1B_1C_1D_1$ under a shear, with invariant line the *y*-axis, such that A is mapped to $(1,1)$. (Produce BA and CD to the *y*-axis to extend the object and make it easier to see what the image is.)

IDENTITY TRANSFORMATION

An identity transformation is a transformation for which the image is the same as the object, e.g. a rotation of $360°$ or a translation defined by the vector $\begin{pmatrix} 0 \\ 0 \end{pmatrix}$.

COMPOUND TRANSFORMATIONS

We can perform two or more successive transformations on an object, producing a final image.

It is possible that the final image could be produced by a single, different, transformation of the object, so this single transformation is equivalent to the combined transformations.

INVERSE TRANSFORMATIONS

The inverse of a transformation is such that it reverses that transformation.

The inverse of a rotation of $90°$ clockwise about O is a rotation of $90°$ anticlockwise about O.

Some transformations are *self-inverses,* e.g. a reflection in the *x*-axis. Two successive reflections in the *x*-axis give a final image which is the same as the original object.

EXERCISE 12b

1. Draw *x* and *y* axes for values of *x* from -6 to 10 and for values of *y* from 0 to 5. A is the quadrilateral with vertices $(3,2)$, $(5,2)$, $(6,5)$ and $(3,5)$.

a) Reflect A in the *y*-axis and label the image B.

b) Reflect B in the line $x = 2$ and label the image C.

c) What single transformation will map A to C ?

d) Reflect A in the line $x = 2$ to give image D, and reflect D in the *y*-axis to give image E.

e) What single transformation will map A to E ?

f) Comment on the results of these reflections. Experiment with successive reflections in, say, the lines $x = -1$ and $x = 2$ or the line $x = -3$ and the *y*-axis.

2. Draw *x* and *y* axes for values from -6 to 6 on each axis. A is the quadrilateral described in question 1.

a) Reflect A in the line $y = x$ and label the image B.

b) Reflect B in the line $y = -x$ and label the image C.

c) What single transformation will map A to C ?

d) Reflect B in the *y*-axis and label the image D.

e) What single transformation will map A to D ?

f) Comment on the results of these reflections. Experiment with successive reflections in pairs of non-parallel lines such as $y = x$ and $y = -x$ or $y = -1$ and $y = x$.

3.   X is a reflection in the *x*-axis.
Y is a reflection in the *y*-axis.
R is a rotation of 180° about the origin.
I is an identity transformation.

The quadrilateral P has vertices (1,1), (3,1), (4,3) and (1,3).
Draw *x* and *y* axes for values from −5 to 5 on each axis.

a) Find the image of P under the transformation X and label it X(P).

b) Find the images Y(P) and R(P) where Y(P) is the image of P under the transformation Y etc.

c) Find the image of X(P) under the transformation *Y* and label it YX(P). Is it true that YX = R ?

d) Find XY(P), XR(P) and RY(P). (In XR(P), R is nearer to P and so is used first, followed by X.)

e) Find R²(P) (i.e. RR(P)).

f) Is it true that R² = I ?

g) Investigate other combinations of X, Y and R.

4.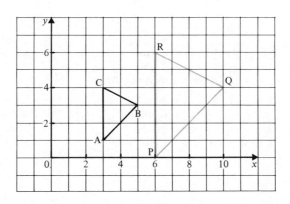

a) △PQR is the image of △ABC under an enlargement. Give the centre of enlargement and the scale factor.

b) Describe the inverse of the transformation in (a).

5.   Draw *x* and *y* axes for values from −10 to 10 on each axis. (3,1), (5,2) and (2,4) are vertices of a triangle A. (−6,1), (−10,−1) and (−4,−5) are vertices of a second triangle B.

a) Describe the transformation T that maps A to B.

b) Describe the inverse transformation T⁻¹.

c) R is a reflection in the *y*-axis. Draw R(A) and TR(A).

d) What single transformation will map B to TR(A) ?

MATRICES

Matrices are rectangular arrays of numbers and can be used in many different ways. They can represent vectors, information about, say, prices or routes, they can give transformations and they can be used to solve simultaneous equations. The rest of this chapter deals with some of these uses.

We start with a summary of the operations possible when using matrices.

THE SIZE OF A MATRIX

The size, or order, of a matrix is given in terms of the number of rows and columns. The number of rows is given first, then the number of columns,

e.g. $\begin{pmatrix} 1 & 4 & 6 \\ 3 & 2 & 3 \end{pmatrix}$ is a 2×3 matrix,

$(1 \quad 2 \quad 4 \quad 0)$ is a 1×4 matrix and is a *row matrix*.

$\begin{pmatrix} 3 \\ -4 \end{pmatrix}$ is a 2×1 matrix and is a *column matrix*. A matrix of this type is sometimes called a column vector.

$\begin{pmatrix} 4 & 1 \\ 3 & 2 \end{pmatrix}$ is a 2×2 square matrix. One of the common uses of this type of matrix is to represent a transformation.

EXERCISE 12c

In each question state the size of the given matrix.

1. $\begin{pmatrix} 3 & 2 \\ -1 & 4 \\ 3 & 5 \end{pmatrix}$

4. $\begin{pmatrix} 5 \\ 2 \\ -1 \\ 4 \end{pmatrix}$

2. $\begin{pmatrix} 5 & 2 & 4 \\ 4 & 2 & -1 \\ 3 & 3 & 5 \\ -4 & -2 & 1 \end{pmatrix}$

5. $(3 \quad 2 \quad -6)$

3. $\begin{pmatrix} -5 & 3 \\ 2 & 4 \end{pmatrix}$

6. $\begin{pmatrix} 5 & 4 \\ -3 & 2 \end{pmatrix}$

ADDITION AND SUBTRACTION

If matrices are of the same order they can be added or subtracted.

$$\begin{pmatrix} 6 & 2 & -3 \\ 1 & -4 & 2 \end{pmatrix} + \begin{pmatrix} -1 & 3 & 1 \\ 2 & 0 & 4 \end{pmatrix} = \begin{pmatrix} 5 & 5 & -2 \\ 3 & -4 & 6 \end{pmatrix}$$

$$\begin{pmatrix} 6 & 2 & -3 \\ 1 & -4 & 2 \end{pmatrix} - \begin{pmatrix} -1 & 3 & 1 \\ 2 & 0 & 4 \end{pmatrix} = \begin{pmatrix} 7 & -1 & -4 \\ -1 & -4 & -2 \end{pmatrix}$$

$$3\begin{pmatrix} 6 & 2 & 1 \\ 1 & -4 & 4 \end{pmatrix} = \begin{pmatrix} 18 & 6 & 3 \\ 3 & -12 & 12 \end{pmatrix}$$

ZERO MATRIX

Every entry in a zero matrix is 0, e.g. $\begin{pmatrix} 0 & 0 & 0 \\ 0 & 0 & 0 \end{pmatrix}$.

EXERCISE 12d

In questions 1 to 9, $A = \begin{pmatrix} 2 & -3 \\ 4 & 1 \end{pmatrix}$, $B = \begin{pmatrix} 1 \\ 3 \end{pmatrix}$, $C = \begin{pmatrix} -2 \\ 4 \end{pmatrix}$

and $D = \begin{pmatrix} -1 & 0 \\ 3 & 2 \end{pmatrix}$.

Find, where possible,

1. $A + B$

2. $A + D$

3. $A - D$

4. $D + A$

5. $B + C$

6. $2A$

7. $2D + A$

8. $3D - 4A$

9. $A + B + C$

10. If $\begin{pmatrix} x & y \\ z & 3 \end{pmatrix} + \begin{pmatrix} 4 & -2 \\ z & 1 \end{pmatrix} = \begin{pmatrix} 6 & -3 \\ -8 & 4 \end{pmatrix}$, find x, y and z.

11. If $\begin{pmatrix} p & p & 3 \\ 1 & p & q \end{pmatrix} - \begin{pmatrix} q & -p & 1 \\ 4 & r & r \end{pmatrix} = \begin{pmatrix} 2 & 10 & 2 \\ -3 & 8 & 5 \end{pmatrix}$, find the value of
the letters.

MULTIPLICATION

Some matrices can be multiplied, but only when they are *compatible* for multiplication: the number of columns in the first matrix must be the same as the number of rows in the second matrix because each entry in a row of the first matrix is multiplied by the corresponding entry in a column of the second matrix.

e.g.
$$\begin{pmatrix} 4 & 2 \\ 1 & 6 \\ 3 & 4 \end{pmatrix} \times \begin{pmatrix} 4 \\ 1 \end{pmatrix} = \begin{pmatrix} 4 \times 4 + 2 \times 1 \\ 1 \times 4 + 6 \times 1 \\ 3 \times 4 + 4 \times 1 \end{pmatrix} = \begin{pmatrix} 18 \\ 10 \\ 16 \end{pmatrix}$$

$$3 \times \boxed{2} \qquad 2 \times 1 \qquad\qquad\qquad 3 \times 1$$

When the sizes of the two matrices are written down, the two middle numbers must be the same. The outer two numbers give the size of the product matrix.

$$\begin{pmatrix} 3 & 1 \\ 4 & 5 \end{pmatrix} \times (1 \quad 4 \quad -1) \quad \text{cannot be found.}$$

$$2 \times 2 \qquad\qquad 1 \times 3$$

Except in a very few cases the order of multiplication does matter. In general, **AB ≠ BA**.

UNIT MATRIX

$\begin{pmatrix} 1 & 0 \\ 0 & 1 \end{pmatrix}$ is an example of a unit matrix.

Multiplying a matrix by $\begin{pmatrix} 1 & 0 \\ 0 & 1 \end{pmatrix}$ does not change the matrix.

e.g.
$$\begin{pmatrix} 1 & 0 \\ 0 & 1 \end{pmatrix} \times \begin{pmatrix} 6 & 2 & 1 \\ 4 & 3 & 2 \end{pmatrix} = \begin{pmatrix} 6 & 2 & 1 \\ 4 & 3 & 2 \end{pmatrix}$$

and
$$\begin{pmatrix} 2 & 4 \\ 6 & 8 \end{pmatrix} \times \begin{pmatrix} 1 & 0 \\ 0 & 1 \end{pmatrix} = \begin{pmatrix} 2 & 4 \\ 6 & 8 \end{pmatrix}$$

EXERCISE 12e

1. $A = \begin{pmatrix} 3 \\ 2 \end{pmatrix}$ $B = (4 \ 2)$ $C = \begin{pmatrix} 4 & -1 \\ 0 & 3 \end{pmatrix}$ $D = \begin{pmatrix} 1 & 2 & 1 \\ 4 & 3 & 2 \end{pmatrix}$

Find, where possible,

a) **AB**　　　　　　　b) **BA**　　　　　　　c) **AC**
d) **CA**　　　　　　　e) **DA**　　　　　　　f) **AD**
g) $\mathbf{A}^2$　　　　　　　h) $\mathbf{C}^2$　　　　　　　i) **BD**

2. If $\begin{pmatrix} a & 1 \\ 2 & b \end{pmatrix}\begin{pmatrix} 2 & 1 \\ 3 & 1 \end{pmatrix} = \begin{pmatrix} 5 & 2 \\ 10 & 4 \end{pmatrix}$, find a and b.

3. If $\begin{pmatrix} x+1 & 2 \\ 3 & y \end{pmatrix}\begin{pmatrix} 1 \\ 4 \end{pmatrix} = \begin{pmatrix} 11 \\ 7 \end{pmatrix}$, find x and y.

4. If $\begin{pmatrix} p & 2 \\ 3 & -1 \end{pmatrix}\begin{pmatrix} p & -1 \\ p & r \end{pmatrix} = \begin{pmatrix} q & -2 \\ 4 & -3 \end{pmatrix}$, find p, q and r.

5. If $3A = \begin{pmatrix} 1 & 0 \\ 2 & 3 \end{pmatrix}\begin{pmatrix} 3 & 3 \\ -1 & 0 \end{pmatrix}$, find **A**.

MIXED QUESTIONS

EXERCISE 12f

$P = (4 \quad 1 \ -2),$ $Q = \begin{pmatrix} 2 & 1 \\ 0 & 3 \end{pmatrix},$ $R = \begin{pmatrix} -4 & 2 \\ -2 & 3 \end{pmatrix},$ $S = \begin{pmatrix} 1 \\ 2 \end{pmatrix},$

$T = (-2 \quad 3).$

Find, where possible,

1. P + Q　　　　　　　**5.** QR　　　　　　　**9.** TS

2. PQ　　　　　　　**6.** Q − R　　　　　　**10.** T + S

3. QP　　　　　　　**7.** RS　　　　　　　**11.** SP

4. Q + 2R　　　　　　**8.** SR　　　　　　　**12.** TP

In questions 13 to 15, several alternative answers are given. Write down the letter that corresponds to the correct answer.

13. If $M = \begin{pmatrix} 3 & 1 \\ -4 & 5 \end{pmatrix}$ and $N = \begin{pmatrix} 1 & 4 \\ 3 & -1 \end{pmatrix}$ the value of $2M - N$ is

A $\begin{pmatrix} 7 & 6 \\ -5 & 9 \end{pmatrix}$ **B** $\begin{pmatrix} 5 & -6 \\ -14 & 12 \end{pmatrix}$ **C** $\begin{pmatrix} 12 & 22 \\ 22 & -42 \end{pmatrix}$ **D** $\begin{pmatrix} 5 & -2 \\ -11 & 11 \end{pmatrix}$

14. If $H = \begin{pmatrix} 1 & 4 \\ 3 & -1 \end{pmatrix}$ and $K = \begin{pmatrix} 3 & -2 \\ 1 & 4 \end{pmatrix}$ then HK is

A $\begin{pmatrix} -13 & 14 \\ 13 & 0 \end{pmatrix}$ **B** $\begin{pmatrix} 4 & 2 \\ 4 & 3 \end{pmatrix}$ **C** $\begin{pmatrix} 7 & 14 \\ 8 & -10 \end{pmatrix}$ **D** $\begin{pmatrix} 7 & -18 \\ 8 & -10 \end{pmatrix}$

15. If $E = \begin{pmatrix} 1 & 3 \\ 0 & 4 \end{pmatrix}$ and $F = \begin{pmatrix} -1 & 2 \\ 2 & 1 \end{pmatrix}$ then $(E + F)^2$ is

A $\begin{pmatrix} 10 & 25 \\ 10 & 35 \end{pmatrix}$ **B** $\begin{pmatrix} 6 & 13 \\ 0 & 21 \end{pmatrix}$ **C** $\begin{pmatrix} 11 & 30 \\ 12 & 35 \end{pmatrix}$ **D** $\begin{pmatrix} 2 & 5 \\ -10 & 7 \end{pmatrix}$

INVERSE MATRICES

The inverse of a matrix A is denoted by A^{-1} and $AA^{-1} = I$, where I is the unit matrix.

e.g. $\begin{pmatrix} 4 & 3 \\ 1 & 2 \end{pmatrix} \begin{pmatrix} \frac{2}{5} & -\frac{3}{5} \\ -\frac{1}{5} & \frac{4}{5} \end{pmatrix} = \begin{pmatrix} 1 & 0 \\ 0 & 1 \end{pmatrix}$

so $\begin{pmatrix} \frac{2}{5} & -\frac{3}{5} \\ -\frac{1}{5} & \frac{4}{5} \end{pmatrix}$ is the inverse of $\begin{pmatrix} 4 & 3 \\ 1 & 2 \end{pmatrix}$ and vice versa.

To find the inverse of A where $A = \begin{pmatrix} 4 & 3 \\ 1 & 2 \end{pmatrix}$

we start by interchanging the numbers in the leading diagonal $\begin{pmatrix} 2 & \\ & 4 \end{pmatrix}$

and changing the sign of the numbers in the other diagonal $\begin{pmatrix} & -3 \\ -1 & \end{pmatrix}$

This gives $\begin{pmatrix} 2 & -3 \\ -1 & 4 \end{pmatrix}$

Then we divide by the determinant of A.

THE DETERMINANT OF A MATRIX

The determinant of a matrix $\mathbf{A}$ can be denoted by $|\mathbf{A}|$

One way to find the determinant is to use a formula, as follows:

if $\mathbf{A} = \begin{pmatrix} a & b \\ c & d \end{pmatrix}$ then $|\mathbf{A}| = ad - bc$.

e.g. for the matrix $\begin{pmatrix} 4 & 3 \\ 1 & 2 \end{pmatrix}$ the determinant is $4 \times 2 - 3 \times 1$ i.e. 5.

An alternative method for finding a determinant is given in the next worked example.

If $|\mathbf{A}| = 0$ there is no inverse matrix because we cannot divide by zero.

EXERCISE 12g

Find the inverse of $\begin{pmatrix} 3 & 1 \\ 4 & 2 \end{pmatrix}$.

Try $\begin{pmatrix} 2 & -1 \\ -4 & 3 \end{pmatrix}$

$$\begin{pmatrix} 3 & 1 \\ 4 & 2 \end{pmatrix}\begin{pmatrix} 2 & -1 \\ -4 & 3 \end{pmatrix} = \begin{pmatrix} 2 & 0 \\ 0 & 2 \end{pmatrix}$$

∴ the determinant is 2

∴ the inverse of $\begin{pmatrix} 3 & 1 \\ 4 & 2 \end{pmatrix}$ is $\frac{1}{2}\begin{pmatrix} 2 & -1 \\ -4 & 3 \end{pmatrix}$, i.e. $\begin{pmatrix} 1 & -\frac{1}{2} \\ -2 & \frac{3}{2} \end{pmatrix}$

1. Find the determinants of

a) $\begin{pmatrix} 4 & 3 \\ 2 & 2 \end{pmatrix}$ b) $\begin{pmatrix} 3 & 4 \\ 2 & 2 \end{pmatrix}$ c) $\begin{pmatrix} 4 & 3 \\ 1 & 1 \end{pmatrix}$ d) $\begin{pmatrix} 5 & -6 \\ -3 & 4 \end{pmatrix}$

2. Find the inverses of

a) $\begin{pmatrix} 2 & 1 \\ 1 & 3 \end{pmatrix}$ b) $\begin{pmatrix} 4 & 5 \\ 3 & 4 \end{pmatrix}$ c) $\begin{pmatrix} 2 & 2 \\ 5 & 4 \end{pmatrix}$ d) $\begin{pmatrix} 1 & -1 \\ 1 & 1 \end{pmatrix}$

3. If $A = \begin{pmatrix} 4 & 3 \\ 3 & 3 \end{pmatrix}$, $B = \begin{pmatrix} 2 & -6 \\ -1 & 4 \end{pmatrix}$ and $C = \begin{pmatrix} -2 & 3 \\ 0 & 1 \end{pmatrix}$ find

a) A^{-1} b) $|B|$ c) $|2C|$ d) $|AB|$

e) $|A+C|$ f) C^{-1} g) $(AC)^{-1}$ h) $A^{-1}C^{-1}$

4. $A = \begin{pmatrix} -5 & 7 \\ 3 & -4 \end{pmatrix}$.

If **B** is the inverse of **A**, find **B**.

5. $P = \begin{pmatrix} 6 & 4 \\ 4 & 3 \end{pmatrix}$ and $Q = \begin{pmatrix} 1 & 1 \\ 1 & 1 \end{pmatrix}$. Find, where possible,

a) **P**   b) **Q**   c) P^{-1} d) Q^{-1} e) $(PQ)^{-1}$

MIXED QUESTIONS

EXERCISE 12h

1. $A = \begin{pmatrix} 6 & 2 \\ 5 & 2 \end{pmatrix}$ and $B = \begin{pmatrix} 2 & -3 \\ -1 & 2 \end{pmatrix}$

Give as a single matrix

a) **AB**   b) **BA**   c) A^{-1} d) B^{-1}

e) A^2 f) B^3 g) $2A$ h) $3B$

2. $A = \begin{pmatrix} 0 & 1 \\ 1 & 0 \end{pmatrix}$. Find

a) $2A$ b) A^{-1} c) A^2 d) A^3

e) A^4 f) A^5 g) $|A|$ h) $|A^2|$

i) A^n if n is even   j) A^n if n is odd

3. If $4\begin{pmatrix} 6 & -3 \\ 1 & 0 \end{pmatrix} + \begin{pmatrix} 2 & -1 \\ 3 & 4 \end{pmatrix}\begin{pmatrix} 1 & -2 \\ 3 & 0 \end{pmatrix} = \begin{pmatrix} a & b \\ c & d \end{pmatrix}$ find a, b, c and d.

4. $A = \begin{pmatrix} 2 & 0 \\ 3 & 1 \end{pmatrix}$ and $B = \begin{pmatrix} 1 & 0 \\ 2 & k \end{pmatrix}$.

a) Find **AB** and **BA**.

b) If $AB = BA$ find k.

c) If $k = 0$ why does **B** have no inverse ?

5. If $\begin{pmatrix} 1 & 2 \\ 3 & 8 \end{pmatrix}P = \begin{pmatrix} 4 & 0 \\ 0 & 4 \end{pmatrix}$ find **P**.

THE USE OF MATRICES FOR TRANSFORMATIONS

EXERCISE 12i

> A(1,1), B(3,1), C(3,2) and D(1,3) are the vertices of a quadrilateral ABCD. Find the image of ABCD under the transformation defined by the matrix $\begin{pmatrix} 1 & 2 \\ 1 & 1 \end{pmatrix}$.
>
> $$\begin{pmatrix} 1 & 2 \\ 1 & 1 \end{pmatrix}\overset{A\ \ B\ \ C\ \ D}{\begin{pmatrix} 1 & 3 & 3 & 1 \\ 1 & 1 & 2 & 3 \end{pmatrix}} = \overset{A'\ \ B'\ \ C'\ \ D'}{\begin{pmatrix} 3 & 5 & 7 & 7 \\ 2 & 4 & 5 & 4 \end{pmatrix}}$$
>
>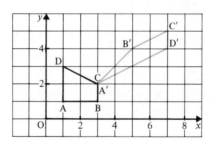
>
> (There is no simple way to describe this transformation.)

1. P(0,1), Q(3,1), R(3,3) and S(0,2) are the vertices of a quadrilateral PQRS. Find the image of PQRS under the transformation given by each of the following matrices and, where possible, describe the transformation.

a) $\begin{pmatrix} 1 & 0 \\ 0 & -1 \end{pmatrix}$ b) $\begin{pmatrix} 3 & 0 \\ 0 & 3 \end{pmatrix}$ c) $\begin{pmatrix} \frac{1}{2} & 0 \\ 0 & \frac{1}{2} \end{pmatrix}$ d) $\begin{pmatrix} 3 & 0 \\ 0 & 2 \end{pmatrix}$ e) $\begin{pmatrix} 0 & -1 \\ 1 & 0 \end{pmatrix}$

2. Choose your own object and identify the eight transformations given by the matrices.

a) $\begin{pmatrix} -1 & 0 \\ 0 & -1 \end{pmatrix}$ c) $\begin{pmatrix} 1 & 0 \\ 0 & -1 \end{pmatrix}$ e) $\begin{pmatrix} 0 & -1 \\ -1 & 0 \end{pmatrix}$ g) $\begin{pmatrix} 1 & 0 \\ 0 & 1 \end{pmatrix}$

b) $\begin{pmatrix} -1 & 0 \\ 0 & 1 \end{pmatrix}$ d) $\begin{pmatrix} 0 & 1 \\ 1 & 0 \end{pmatrix}$ f) $\begin{pmatrix} 0 & -1 \\ 1 & 0 \end{pmatrix}$ h) $\begin{pmatrix} 0 & 1 \\ -1 & 0 \end{pmatrix}$

3. Choose your own object and identify the transformations given by the matrices.

a) $\begin{pmatrix} 2 & 0 \\ 0 & 2 \end{pmatrix}$ b) $\begin{pmatrix} -2 & 0 \\ 0 & -2 \end{pmatrix}$ c) $\begin{pmatrix} \frac{1}{4} & 0 \\ 0 & \frac{1}{4} \end{pmatrix}$ d) $\begin{pmatrix} -1 & 0 \\ 0 & -1 \end{pmatrix}$

(There are two ways of describing (d).)

$\triangle ABC$ has vertices $A(4,0)$, $B(0,3)$ and $C(1,1)$.

$\triangle A'B'C'$ has vertices $A'(8,12)$, $B'(3,0)$ and $C'(3,3)$.

Find the matrix of the transformation that maps $\triangle ABC$ to $A'B'C'$.

Let the matrix be $\begin{pmatrix} a & b \\ c & d \end{pmatrix}$

then
$$\begin{matrix} & A & B & C & & A' & B' & C' \\ \begin{pmatrix} a & b \\ c & d \end{pmatrix} & \begin{pmatrix} 4 & 0 & 1 \\ 0 & 3 & 1 \end{pmatrix} & & = & \begin{pmatrix} 8 & 3 & 3 \\ 12 & 0 & 3 \end{pmatrix} \end{matrix}$$

Multiplying out, $4a = 8$, $3b = 3$,   (and $a + b = 3$)

$4c = 12$, $3d = 0$ (and $c + d = 3$)

so $a = 2$, $b = 1$, $c = 3$ and $d = 0$.

The matrix is $\begin{pmatrix} 2 & 1 \\ 3 & 0 \end{pmatrix}$.

$\left(\text{Check that C is mapped to C': } \begin{pmatrix} 2 & 1 \\ 3 & 0 \end{pmatrix}\begin{pmatrix} 1 \\ 1 \end{pmatrix} = \begin{pmatrix} 3 \\ 3 \end{pmatrix}. \right)$

---

4. ABCD is a rectangle with vertices $A(2,0)$, $B(5,0)$, $C(5,0$), and $D(2,2)$. A'B'C'D' is the rectangle such that A' is $(4,0)$, $B'(10,0)$, $C'(10,6)$ and $D'(4,6)$.

ABCD is mapped to A'B'C'D' by the transformation with matrix $\begin{pmatrix} a & b \\ c & d \end{pmatrix}$.

Find this matrix.

5. The vertices of $\triangle PQR$ are $P(1,4)$, $Q(2,5)$ and $R(0,5)$ and of $\triangle P'Q'R'$ are $P'(0,2)$, $Q'(12,-2)$ and $R'(-20,10)$. If $\triangle PQR$ is mapped to $\triangle P'Q'R'$ find the matrix of the transformation.

INVERSE TRANSFORMATIONS

EXERCISE 12j

1. $A(1,1)$, $B(3,1)$, $C(3,3)$ and $D(1,3)$ are vertices of the square ABCD.

P is the matrix $\begin{pmatrix} 2 & 3 \\ 1 & 2 \end{pmatrix}$.

a) Find the image A'B'C'D' of ABCD under the transformation using the matrix **P**.

b) Find $\mathbf{P}^{-1}$.

c) Find the image of A'B'C'D' under the transformation using the matrix $\mathbf{P}^{-1}$.

d) Comment on and explain the result.

2. A unit square has vertices $(0,0)$, $(1,0)$, $(1,1)$ and $(0,1)$. **M** is the matrix $\begin{pmatrix} 3 & 0 \\ 0 & 3 \end{pmatrix}$.

a) Find the image of the unit square under the transformation given by the matrix **M**.

b) Describe the transformation.

c) Find $\mathbf{M}^{-1}$ and describe the transformation given by $\mathbf{M}^{-1}$.

3. a) By choosing your own object, find the transformations defined by the matrices $\mathbf{A} = \begin{pmatrix} 0 & 1 \\ -1 & 0 \end{pmatrix}$ and $\mathbf{B} = \begin{pmatrix} 0 & 1 \\ 1 & 0 \end{pmatrix}$.

b) Find the matrices defining the inverses of the transformations given in (a).

4. The matrix $\mathbf{P} = \begin{pmatrix} -2 & 4 \\ -2 & 3 \end{pmatrix}$ represents the transformation T

where $\mathrm{T}: \begin{pmatrix} x \\ y \end{pmatrix} \rightarrow \begin{pmatrix} -2 & 4 \\ -2 & 3 \end{pmatrix} \begin{pmatrix} x \\ y \end{pmatrix}$

(i.e., as before, **P** defines the transformation.)

a) Find the inverse of **P**.

b) Find the coordinates of the point to which $(3,2)$ is mapped by the transformation T.

c) Find the coordinates of the point which is mapped to $(10,7)$ by T.

COMPOUND TRANSFORMATIONS

EXERCISE 12k

1. A square ABCD has vertices at $A(2,0)$, $B(4,0)$, $C(4,2)$ and $D(2,0\ )$.

$$P = \begin{pmatrix} 1 & 0 \\ 0 & -1 \end{pmatrix} \quad \text{and} \quad Q = \begin{pmatrix} 0 & 1 \\ -1 & 0 \end{pmatrix}.$$

a) Find the images of ABCD under the two transformations given by P and Q and identify the transformations. Label the two images $A_1B_1C_1D_1$ and $A_2B_2C_2D_2$.

b) Find the image of $A_1B_1C_1D_1$ under the transformation given by Q. Label the image $A_3B_3C_3D_3$.

c) Describe the transformation that maps ABCD to $A_3B_3C_3D_3$.

d) Find the matrices R and S where $R = QP$ and $S = PQ$.

e) Find the image of ABCD under the transformation given by R.

f) Find the image of ABCD under the transformation given by S.

g) Comment on the results of (b), (e) and (f). Explain the significance of the order in which P and Q occur.

2. The vertices of a triangle P are $(2,1)$, $(4,1)$ and $(4,4)$.

$$M = \begin{pmatrix} -1 & 0 \\ 0 & -1 \end{pmatrix} \quad \text{and} \quad N = \begin{pmatrix} -1 & 0 \\ 0 & 1 \end{pmatrix} \quad \text{define transformations M and}$$
N respectively.

a) Find $M(P)$ and $NM(P)$.

b) Describe the transformations M and NM.

c) Find the matrix $L = NM$.

d) The matrix L defines a transformation L. Find $L(P)$ and describe the transformation L.

e) Is L the same transformation as NM ?

3.

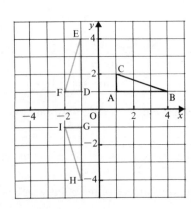

Find the matrices **P**, **Q** and **R** if

a) **P** is the matrix of the transformation which maps △ABC to △ADEF

b) **Q** is the matrix of the transformation which maps △DEF to △GHI

c) **R** is the matrix of the transformation which maps △ABC to △GHI.

d) Give an equation linking **P**, **Q** and **R**.

4. If $Q = \begin{pmatrix} 0 & -1 \\ 1 & 0 \end{pmatrix}$ find $Q\begin{pmatrix} x \\ y \end{pmatrix}$

a) Draw a diagram and describe the transformation given by **Q**.

b) Hence, or otherwise, write Q^8 as a single matrix.

INFORMATION MATRICES

Matrices can be used as stores of information. This was discussed thoroughly in Book 4A and a revision exercise is given here.

EXERCISE 12I

Simon and Martin go by bus to work and the following tables give information about their journeys during one week.

	Number of tickets				Cost in pence
	Single	Return			
Simon	2	4		Single	40
Martin	6	3		Return	70

a) Form two matrices **A** and **B** from the information in the two tables.

b) Find **AB** and state what information this matrix gives.

a) $A = \begin{pmatrix} 2 & 4 \\ 6 & 3 \end{pmatrix}$ $B = \begin{pmatrix} 40 \\ 70 \end{pmatrix}$

b) $AB = \begin{matrix} \text{Simon} \\ \text{Martin} \end{matrix} \begin{pmatrix} 2 & 4 \\ 6 & 3 \end{pmatrix} \begin{matrix} \text{cost} \\ \begin{pmatrix} 40 \\ 70 \end{pmatrix} \end{matrix} = \begin{matrix} \text{Simon} \\ \text{Martin} \end{matrix} \begin{matrix} \text{cost} \\ \begin{pmatrix} 360 \\ 450 \end{pmatrix} \end{matrix}$

The bus journeys cost Simon £3.60 and Martin £4.50.

1. Mrs Jarman buys butter and margarine in two successive weeks.

	Amount bought in kg	
	Butter	Margarine
First week	$\frac{1}{2}$	1
Second week	1	$1\frac{1}{2}$

	Cost in p per kg
Butter	120
Margarine	106

Calculate $\begin{pmatrix} \frac{1}{2} & 1 \\ 1 & 1\frac{1}{2} \end{pmatrix} \begin{pmatrix} 120 \\ 106 \end{pmatrix}$ and label the rows and columns of the resulting matrix to show what the numbers represent.

2. A television engineer visits three areas of Bristol and makes the number of calls given in the table.

	Cotham	Redland	Bishopston
First week	5	6	0
Second week	4	1	7
Third week	3	5	3

a) Cotham, Redland and Bishopston are respectively 2, 3 and 4 km from his headquarters. Write this information as a column matrix **A**.

b) Use the information in the table to give a matrix **B**.

c) Find **AB** or **BA**, whichever is possible, and state what information is given by the resulting matrix.

d) If $\mathbf{C} = (1 \quad 1 \quad 1)$ and $\mathbf{D} = \begin{pmatrix} 1 \\ 1 \\ 1 \end{pmatrix}$, what information is given by

i) the matrix **CD**   ii) the matrix **BD** ?

3. A toy shop sells three types of jigsaw and details of the prices and sales during the two weeks are given below.

	Sales		
	A	B	C
First week	1	3	5
Second week	0	2	7

	Cost in pounds
A	3
B	1
C	2

a) $\mathbf{P} = \begin{pmatrix} 1 & 3 & 5 \\ 0 & 2 & 7 \end{pmatrix}$ and $\mathbf{Q} = \begin{pmatrix} 3 \\ 1 \\ 2 \end{pmatrix}$

Find **PQ** and state what information is given by the product.

b) Type A has 2000 pieces, B has 500 and C has 1000. Using this information, form a column matrix **R** and find **PR**. What information is given by the product **PR** ?

c) $S = \begin{pmatrix} 1 \\ 1 \\ 1 \end{pmatrix}$ and $T = (1 \quad 1)$

Find the following products and in each case state what information is given by the product.

i) **PS**   ii) **TP**   iii) **TPR**   iv) **TPQ**

4. During one week a milkman delivers milk to four families as follows:

Occleshaw	:	5 gold top, 1 silver top
Pottinger	:	7 gold top, 10 silver top
Reynolds	:	12 silver top
Lowe	:	6 gold top

a) Copy and complete the matrix **M**:

$$\begin{array}{c} \\ G \\ S \end{array} \begin{pmatrix} \overset{\displaystyle O \qquad P \qquad R \qquad L}{} \\ \end{pmatrix}$$

b) $N = (1 \quad 1)$. Find **NM** and state what information **NM** gives.

c) Gold top costs 24 p and silver top costs 21 p.

$$H = \begin{pmatrix} 24 \\ 21 \end{pmatrix} \quad \text{and} \quad K = (24 \quad 21)$$

Which of these two matrices will combine with **M** to give each family's milk bill ?

d) The total number of bottles of gold top and the total number of bottles of silver top delivered by the milkman each week are given either by **ML** or by **LM** where **L** is another matrix. Find **L** and state which product should be used.

e) Give the combination of matrices which will give
   i) the total money collected by the milkman for the week
   ii) the total number of bottles delivered in the week.

ROUTE MATRICES

We can use matrices to represent information about routes between points.
Consider the following map showing roads that link A, B and C.

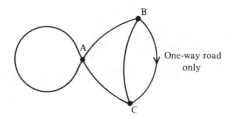

The matrix below shows the number of one-stage routes from one point
to another, i.e. routes which do not go through any other point. There are
two routes from A to A because the loop road can be travelled in either
direction

$$
\begin{array}{c}
\text{To}\\
\begin{array}{ccc} \text{A} & \text{B} & \text{C} \end{array}\\
\text{From}\begin{array}{c}\text{A}\\\text{B}\\\text{C}\end{array}
\begin{pmatrix}
2 & 1 & 1\\
1 & 0 & 2\\
1 & 1 & 0
\end{pmatrix}
\end{array}
$$

EXERCISE 12m

In each question from 1 to 4 give a matrix to show the number of one-stage
routes from one point to another.

1.

3.
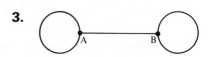

2.
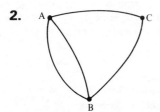

4.
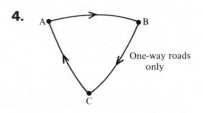

In each question from 5 to 8 draw maps using the information given by the matrices. (There may be several possible maps in each case.)

5.
$$\begin{array}{c} \text{To} \\ \begin{array}{cc} \text{A} & \text{B} \end{array} \\ \text{From} \quad \begin{array}{c} \text{A} \\ \text{B} \end{array} \left(\begin{array}{cc} 0 & 1 \\ 2 & 0 \end{array} \right) \end{array}$$

7.
$$\begin{array}{c} \text{To} \\ \begin{array}{ccc} \text{A} & \text{B} & \text{C} \end{array} \\ \text{From} \quad \begin{array}{c} \text{A} \\ \text{B} \\ \text{C} \end{array} \left(\begin{array}{ccc} 0 & 1 & 1 \\ 1 & 0 & 1 \\ 1 & 1 & 0 \end{array} \right) \end{array}$$

6.
$$\begin{array}{c} \text{To} \\ \begin{array}{ccc} \text{A} & \text{B} & \text{C} \end{array} \\ \text{From} \quad \begin{array}{c} \text{A} \\ \text{B} \\ \text{C} \end{array} \left(\begin{array}{ccc} 1 & 1 & 1 \\ 1 & 1 & 1 \\ 1 & 1 & 1 \end{array} \right) \end{array}$$

8.
$$\begin{array}{c} \text{To} \\ \begin{array}{cccc} \text{A} & \text{B} & \text{C} & \text{D} \end{array} \\ \text{From} \quad \begin{array}{c} \text{A} \\ \text{B} \\ \text{C} \\ \text{D} \end{array} \left(\begin{array}{cccc} 0 & 1 & 2 & 1 \\ 1 & 0 & 1 & 0 \\ 2 & 1 & 0 & 1 \\ 1 & 0 & 1 & 0 \end{array} \right) \end{array}$$

9.

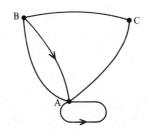

a) Give the route matrix **R** for this map.

b) Find **R**^{2}.

c) How many two-stage routes are there from A to C i.e. that involve passing through one other point on the way ? (AAC is one example.)

d) Give the two-stage routes from A to B, A to A, B to A, etc. Compare these with the matrix **R**^{2}. What information does **R**^{2} contain ?

e) What information would **R**^{3} give you ?

10.

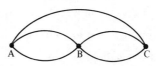

a) Give the route matrix **R** for this map.

b) Find **R**^{2} and **R**^{3}.

c) How many two-stage routes are there from A to C ?

d) How many three-stage routes are there from A to C ?

13 VECTORS

DEFINITION OF A VECTOR

A vector quantity has magnitude (i.e. size) and direction and can be represented by a directed line segment.

Velocity is a vector (e.g. 16 m.p.h. west) but mass (e.g. 4 kg) is not. We can represent 16 m.p.h. west by a line.

The length of the line represents the magnitude, and the direction of the line represents the direction, of the vector.

EXERCISE 13a

1. Which of the following quantities are vectors ?

a) The length of a piece of string

b) The force needed to move a lift up its shaft

c) A move from the door to your chair

d) The speed of a galloping horse

e) The distance between Bristol and Edinburgh

2. Represent each of the following vector quantities by a suitable directed line.

a) A force of 6 newtons acting vertically downwards

b) A velocity of 3 m/s on a bearing of 035 °

c) An acceleration of 2 m/s² northwest

d) A displacement of 7 km due south

REPRESENTATION OF A VECTOR

If we use squared paper then the line representing the vector can be described by the displacements parallel to the *x* and *y* axes.

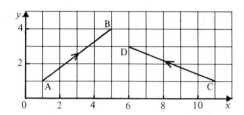

The vector can be named either by using the end letters or by using a small letter in heavy type. In writing we cannot use heavy type so the letter is underlined, e.g. $\underline{a}$.

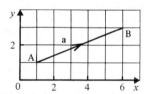

$$\overrightarrow{AB} = \mathbf{a} = \underline{a} = \begin{pmatrix} 5 \\ 2 \end{pmatrix}$$

DISPLACEMENT

We have already used vectors to describe translations. A translation is a *displacement,* or shift of position from one point to another.

EXERCISE 13b

1.

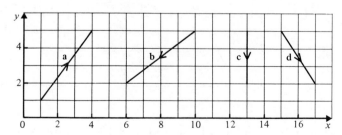

Give the vectors **a, b, c** and **d** in the form $\begin{pmatrix} x \\ y \end{pmatrix}$.

2. Draw lines to represent the following vectors. Remember to include the arrow.

a) $\begin{pmatrix} 1 \\ 4 \end{pmatrix}$ b) $\begin{pmatrix} -6 \\ -4 \end{pmatrix}$ c) $\begin{pmatrix} -1 \\ 5 \end{pmatrix}$ d) $\begin{pmatrix} 3 \\ -4 \end{pmatrix}$ e) $\begin{pmatrix} 4 \\ 0 \end{pmatrix}$

3. A, B and C are the points $(2,1)$, $(5,7)$ and $(3,-2)$. Draw a diagram and give the vectors $\overrightarrow{AB}$, $\overrightarrow{BA}$ and $\overrightarrow{AC}$.

4. A is the point $(-2,1)$. $\overrightarrow{AB} = \begin{pmatrix} 4 \\ -3 \end{pmatrix}$ and $\overrightarrow{AC} = \begin{pmatrix} 5 \\ 2 \end{pmatrix}$.

a) Find the coordinates of the points B and C.

b) Give the vector $\overrightarrow{BC}$.

5. B is the point $(2,1)$, $\overrightarrow{AB} = \begin{pmatrix} 6 \\ -4 \end{pmatrix}$. Give the coordinates of A.

6. A, B and C are the points $(-2,1)$, $(2,0)$ and $(1,3)$.

a) $\triangle ABC$ is translated using the vector $\begin{pmatrix} 4 \\ 1 \end{pmatrix}$. Label the image $A_1B_1C_1$.

b) $\triangle ABC$ is translated using the vector $\begin{pmatrix} -3 \\ 2 \end{pmatrix}$. Label the image $A_2B_2C_2$.

c) Give the vector defining the translation which maps
   i) $\triangle A_1B_1C_1$ to $\triangle A_2B_2C_2$ ii) $\triangle A_2B_2C_2$ to $\triangle A_1B_1C_1$

EQUAL AND PARALLEL VECTORS

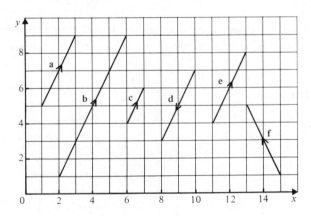

From the diagram, $a = \begin{pmatrix} 2 \\ 4 \end{pmatrix}$ and $e = \begin{pmatrix} 2 \\ 4 \end{pmatrix}$ so we can say that $a = e$.

The lines representing a and e are parallel and equal in length.

Now $\mathbf{b} = \begin{pmatrix} 4 \\ 8 \end{pmatrix}$ and $\mathbf{c} = \begin{pmatrix} 1 \\ 2 \end{pmatrix}$ so $\mathbf{b} = 2\mathbf{a}$ and $\mathbf{c} = \frac{1}{2}\mathbf{a}$.

a, **b** and **c** are parallel but **b** is twice the size of **a** and **c** is half the size of **a**.

$\mathbf{d} = \begin{pmatrix} -2 \\ -4 \end{pmatrix}$ so $\mathbf{d} = -\mathbf{a}$.

d and **a** are parallel and the same size but they are in opposite directions.

$\mathbf{f} = \begin{pmatrix} -2 \\ 4 \end{pmatrix}$.

f is the same size as **a** but is not parallel to **a** so **a** ≠ **f**.

EXERCISE 13c

1.

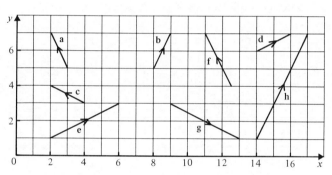

a) Give the vectors in the diagram above in the form $\begin{pmatrix} x \\ y \end{pmatrix}$.

b) Find the relationships between as many pairs of vectors as possible, giving them in the form **p** = k**q**.

2.

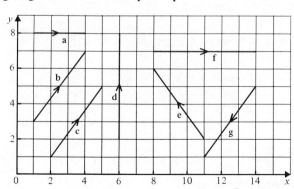

Repeat question 1 for the vectors in the diagram above.

$\overrightarrow{PQ} = \mathbf{p} = \begin{pmatrix} 10 \\ 5 \end{pmatrix}$ and R is a point on PQ such that

PR : RQ = 2 : 3. Give $\overrightarrow{PR}$ and $\overrightarrow{RQ}$ in terms of **p**.

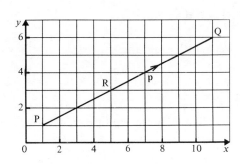

(Divide both 10 and 5 in the ratio 2 : 3.)

10 splits into 4 and 6, and 5 into 2 and 3

so $\overrightarrow{PR} = \begin{pmatrix} 4 \\ 2 \end{pmatrix}$ and $\overrightarrow{RQ} = \begin{pmatrix} 6 \\ 3 \end{pmatrix}$

$\overrightarrow{PR} = \frac{2}{5}\mathbf{p}$ and $\overrightarrow{RQ} = \frac{3}{5}\mathbf{p}$

3. $\overrightarrow{AB} = \mathbf{a} = \begin{pmatrix} 12 \\ 9 \end{pmatrix}$.

   a) Draw a diagram showing $\overrightarrow{AB}$ and mark the point C on AB such that
      AC : CB = 1 : 2.

   b) Give $\overrightarrow{AC}$ and $\overrightarrow{CB}$ in terms of **a**.

   c) Give $\overrightarrow{BA}$, $\overrightarrow{CA}$ and $\overrightarrow{BC}$ in terms of **a**.

4. $\overrightarrow{PQ} = \mathbf{p} = \begin{pmatrix} 8 \\ 12 \end{pmatrix}$ and R is a point on PQ such that $\overrightarrow{PR} = \frac{3}{4}\mathbf{p}$. Give the
   following ratios.

   a) PR : RQ   b) PR : PQ   c) RQ : PQ

5. $\overrightarrow{AB} = \mathbf{a} = \begin{pmatrix} -4 \\ 6 \end{pmatrix}$ and C is a point on AB produced such that $\overrightarrow{AC} = \frac{3}{2}\overrightarrow{AB}$.
Give the following ratios.
a) AB : BC   b) AC : AB   c) AC : BC

6. In a quadrilateral ABCD, A and B are the points $(-1, 0)$ and $(3, 1)$,
$\overrightarrow{BC} = \begin{pmatrix} -3 \\ 3 \end{pmatrix}$, BC is parallel to AD and $AD = \frac{4}{3}BC$. Find $\overrightarrow{AD}$ and give
the coordinates of D.

7. $\overrightarrow{AB} = \mathbf{a}$. C is on AB produced such that $\dfrac{AB}{AC} = \dfrac{3}{5}$. Give $\overrightarrow{AC}$ in terms
of **a**.

THE MAGNITUDE OF A VECTOR

When a vector is represented by a line segment, the *magnitude* of the vector
is the *length* of the line, e.g. if $\mathbf{a} = \begin{pmatrix} 4 \\ 2 \end{pmatrix}$ then the magnitude of **a**, which is
written as $|\mathbf{a}|$, is equal to the length of the line representing **a**.

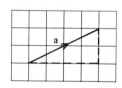

$$|\mathbf{a}| = \sqrt{4^2 + 2^2} \quad (\text{Pythag. th.})$$
$$= \sqrt{20}$$
$$= 4.47 \quad \text{correct to 2 s.f.}$$

EXERCISE 13d

1. Find the magnitude of each of the vectors in Exercise 13c, question 1.

2. Find the magnitude of each of the vectors in Exercise 13c, question 2.

ADDITION OF VECTORS

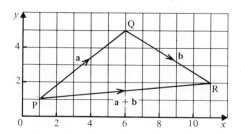

The displacement from P to Q followed by the displacement from Q to R is equivalent to the displacement from P to R, so we write

$$\overrightarrow{PQ} + \overrightarrow{QR} = \overrightarrow{PR}$$

or

$$\begin{pmatrix} 5 \\ 4 \end{pmatrix} + \begin{pmatrix} 5 \\ -3 \end{pmatrix} = \begin{pmatrix} 10 \\ 1 \end{pmatrix}$$

The vector $\overrightarrow{PR}$ is equivalent to $\overrightarrow{PQ}$ together with $\overrightarrow{QR}$ and is called the resultant of $\overrightarrow{PQ}$ and $\overrightarrow{QR}$.

To add two vectors we add the corresponding numbers. Notice that a vector in the form $\begin{pmatrix} x \\ y \end{pmatrix}$ is a 2×1 matrix. Vector addition is similar to matrix addition.

Note that in vector addition, + means 'together with'.

SUBTRACTION OF VECTORS

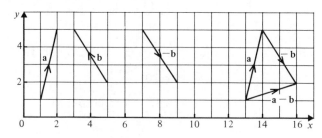

We may think of $a - b$ as $a + (-b)$.

$$a + (-b) = \begin{pmatrix} 1 \\ 4 \end{pmatrix} + \begin{pmatrix} 2 \\ -3 \end{pmatrix} = \begin{pmatrix} 3 \\ 1 \end{pmatrix}$$

or

$$a - b = \begin{pmatrix} 1 \\ 4 \end{pmatrix} - \begin{pmatrix} -2 \\ 3 \end{pmatrix} = \begin{pmatrix} 3 \\ 1 \end{pmatrix}$$

EXERCISE 13e

In this exercise $\mathbf{a} = \begin{pmatrix} 6 \\ 2 \end{pmatrix}$, $\mathbf{b} = \begin{pmatrix} -3 \\ 4 \end{pmatrix}$ and $\mathbf{c} = \begin{pmatrix} 4 \\ -5 \end{pmatrix}$.

Draw diagrams to represent the following vectors and calculate the resultant vectors. Check that your calculations agree with your diagrams.

1. a	**4.** a + b	**7.** b + c	**10.** c − a
2. b	**5.** b + a	**8.** c + a	**11.** c + b
3. c	**6.** a − b	**9.** a − c	**12.** b − a

13. Is vector addition *commutative*, i.e. does the order of adding **a** and **b** make no difference ?

14. Is vector subtraction commutative ?

15. Draw diagrams to represent
a) **a + b + c**   b) **b + a + c**
c) Are the two resultant vectors the same ?
d) In how many different orders could you add **a**, **b** and **c** ?
Would the resultant vectors be different from those in (a) and (b) ?

16. Draw diagrams to represent
a) 2a   b) 2b   c) 2a + 2b   d) a + b
e) What can you say about the lines representing **a + b** and **2a + 2b** ?

VECTORS AND GEOMETRY

Because a vector can be represented by a line segment, we can use the sides of triangles and polygons to represent vectors, and other lines in the figure can represent combinations of these vectors.

Consider, for example, a triangle XYZ.

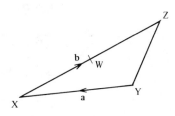

By comparing this with previous diagrams, we can see that the displacement from Y to Z, i.e. $\overrightarrow{YZ}$, is equivalent to **a** together with **b**.

i.e. $\overrightarrow{YZ} = \mathbf{a} + \mathbf{b}$

If W is the midpoint of XZ, then $XW = \frac{1}{2}XZ$

so $\overrightarrow{XW} = \frac{1}{2}\mathbf{b}$

Remember also that if one line represents the vector $\mathbf{c}$, say, and another line represents $2\mathbf{c}$ then the lines are parallel and the second line is twice the length of the first.

EXERCISE 13f

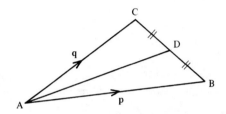

$\overrightarrow{AB} = \mathbf{p}$, $\overrightarrow{AC} = \mathbf{q}$ and D is the midpoint of BC.

Give $\overrightarrow{BC}$ and $\overrightarrow{AD}$ in terms of $\mathbf{p}$ and $\mathbf{q}$.

(Give an alternative route from B to C first.)

$$\overrightarrow{BC} = \overrightarrow{BA} + \overrightarrow{AC}$$
$$= -\mathbf{p} + \mathbf{q}$$
$$= \mathbf{q} - \mathbf{p}$$

$$\overrightarrow{AD} = \overrightarrow{AB} + \overrightarrow{BD}$$
$$= \overrightarrow{AB} + \frac{1}{2}\overrightarrow{BC}$$
$$= \mathbf{p} + \frac{1}{2}(\mathbf{q} - \mathbf{p})$$
$$= \frac{1}{2}\mathbf{p} + \frac{1}{2}\mathbf{q}$$
$$= \frac{1}{2}(\mathbf{p} + \mathbf{q})$$

1.

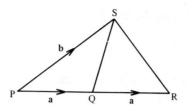

a) Is PQR a straight line ?

b) Give in terms of **a** and **b**

   i) $\overrightarrow{QS}$ ii) $\overrightarrow{SR}$ iii) $\overrightarrow{RS}$

2.

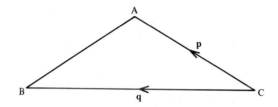

If D is the midpoint of BC, give in terms of **p** and **q**

a) $\overrightarrow{AB}$ b) $\overrightarrow{CD}$ c) $\overrightarrow{DB}$ d) $\overrightarrow{AD}$

3.

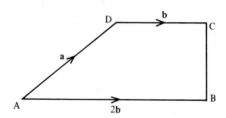

a) What type of quadrilateral is ABCD ?

b) Give in terms of **a** and **b**

   i) $\overrightarrow{BC}$ ii) $\overrightarrow{BD}$ iii) $\overrightarrow{AC}$

4. $\overrightarrow{AB} = \begin{pmatrix} 5 \\ 3 \end{pmatrix}$ and $\overrightarrow{AC} = \begin{pmatrix} -2 \\ 5 \end{pmatrix}$.

Draw a diagram and find $\overrightarrow{BC}$ and $\overrightarrow{CB}$.

5. $p = \begin{pmatrix} 3 \\ 1 \end{pmatrix}$, $q = \begin{pmatrix} -2 \\ 3 \end{pmatrix}$ and $r = \begin{pmatrix} 0 \\ 11 \end{pmatrix}$

If $r = hp + kq$, find h and k.

GEOMETRY USING VECTORS

The following fact is useful

If $a = kb$ (i.e. if a is a multiple of b) then a and b are parallel and $|a| = k|b|$.

EXERCISE 13g

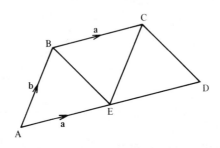

$\overrightarrow{AB} = b$, $\overrightarrow{AE} = a$, $\overrightarrow{BC} = a$ and $\overrightarrow{CD} = a - b$.

a) Find $\overrightarrow{BE}$ and $\overrightarrow{ED}$ in terms of a and b.

b) What type of quadrilateral is BEDC ?

c) Do A, E and D lie in a straight line ?

a) $\overrightarrow{BE} = \overrightarrow{BA} + \overrightarrow{AE}$ $\overrightarrow{ED} = \overrightarrow{EB} + \overrightarrow{BC} + \overrightarrow{CD}$

     $= -b + a$ $= -(a - b) + a + (a - b)$

     $= a - b$ $= a$

b) $\overrightarrow{BE} = \overrightarrow{CD}$

     ∴ CD is parallel to BE and   CD = BE.

     ∴ BEDC is a parallelogram.

c) Both $\overrightarrow{AE}$ and $\overrightarrow{ED}$ represent a.

     ∴ ED is parallel to AE.

     Also the point E is on both lines.

     ∴ A, E and D lie in a straight line.

1.

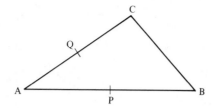

$\overrightarrow{AB} = \mathbf{b}$ and $\overrightarrow{AC} = \mathbf{c}$. P and Q are the midpoints of AB and AC respectively. Give in terms of **b** and **c**

a) $\overrightarrow{AP}$ b) $\overrightarrow{AQ}$ c) $\overrightarrow{BC}$ d) $\overrightarrow{PQ}$

e) Show that PQ is parallel to BC.

f) What can you say about the lengths of PQ and BC ?

2.

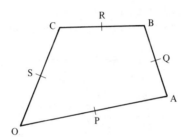

$\overrightarrow{OA} = \mathbf{a}$, $\overrightarrow{OB} = \mathbf{b}$ and $\overrightarrow{OC} = \mathbf{c}$. P, Q, R and S are the midpoints of OA, AB, BC and OC respectively. Give in terms of **a**, **b** and **c**

a) $\overrightarrow{OP}$ b) $\overrightarrow{AB}$ c) $\overrightarrow{AQ}$ d) $\overrightarrow{PQ}$ e) $\overrightarrow{SR}$

f) Show that PQ is parallel to SR.

g) What type of quadrilateral is PQRS ?

3.

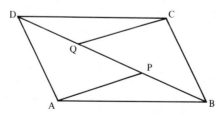

ABCD is a parallelogram. $\overrightarrow{AB} = \mathbf{a}$ and $\overrightarrow{AD} = \mathbf{b}$. P and Q are points on BD such that BP = PQ = QD. Give in terms of **a** and **b**

a) $\overrightarrow{BD}$ b) $\overrightarrow{BP}$ c) $\overrightarrow{BQ}$ d) $\overrightarrow{AP}$ e) $\overrightarrow{QC}$

f) Show that APCQ is a parallelogram.

4.

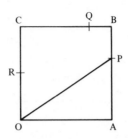

OABC is a square. $\overrightarrow{OA} = \mathbf{a}$ and $\overrightarrow{OC} = \mathbf{b}$.

P is the point on AB such that AP : PB = 2 : 1.

Q is the point on BC such that BQ : QC = 1 : 3.

R is the midpoint of OC.

Find in terms of **a** and **b**

a) $\overrightarrow{AB}$ b) $\overrightarrow{AP}$ c) $\overrightarrow{OP}$ d) $\overrightarrow{OR}$ e) $\overrightarrow{CQ}$ f) $\overrightarrow{RQ}$

g) Show that RQ is parallel to OP.

h) How do the lengths of RQ and OP compare ?

5.

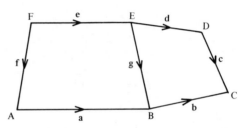

The diagram shows a rough sketch of two quadrilaterals.

a) If $\mathbf{a} = 2\mathbf{b}$ what can you say about A, B and C ?

b) If $\mathbf{a} = \mathbf{b} = \mathbf{e} = \mathbf{d}$ what type of figure is ABCDEF ?

c) If $\mathbf{g} = 2\mathbf{c}$ what type of figure is BCDE ?

d) If $\mathbf{d} + \mathbf{c} = \mathbf{e} + \mathbf{g}$ name four points that are vertices of a parallelogram.

6.

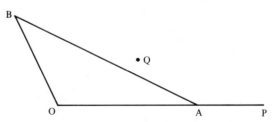

$\overrightarrow{OA} = 4\mathbf{a}$, $\overrightarrow{OB} = 2\mathbf{b}$, $\mathbf{AP} = \frac{1}{2}\mathbf{OA}$ and $\overrightarrow{OQ} = 3\mathbf{a} + \mathbf{b}$.

Give in terms of **a** and **b**

a) $\overrightarrow{BP}$ b) $\overrightarrow{BQ}$

c) Show that B, Q and P lie in a straight line.

d) Find BQ : BP.

7. In $\triangle PQR$, $\overrightarrow{PQ} = \mathbf{q}$ and $\overrightarrow{PR} = \mathbf{r}$. S is the point such that $\overrightarrow{PS} = 3\mathbf{r}$ and T is the mid-point of QS.

a) Find, in terms of **q** and **r**, the vectors

   (i) $\overrightarrow{QS}$ (ii) $\overrightarrow{QT}$ (iii) $\overrightarrow{PT}$

b) U is the point such that $PU = 2\overrightarrow{PT}$. Find $\overrightarrow{QU}$.

c) What type of quadrilateral is
   (i) QUSP   (ii) QURP ?

POSITION VECTORS

In general, vectors are not fixed in position. They are *free* vectors.

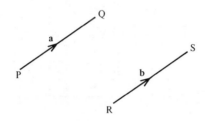

If PQ is parallel to RS and PQ = RS then we say

$$\overrightarrow{PQ} = \overrightarrow{RS} \quad \text{and} \quad \mathbf{a} = \mathbf{b}$$

If, however, we wish to fix the position of one point relative to another, we can use a *position vector*.

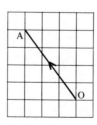

Relative to O the position vector of A is $\begin{pmatrix} -3 \\ 4 \end{pmatrix}$.

Position vectors have already been used to describe the positions of points. We used them when matrix transformations were discussed.

EXERCISE 13h

The position vectors, relative to O, of A, B and C are

$\begin{pmatrix} 7 \\ 1 \end{pmatrix}$, $\begin{pmatrix} -3 \\ 5 \end{pmatrix}$ and $\begin{pmatrix} -1 \\ -1 \end{pmatrix}$ respectively. Mark points A, B and C

on a diagram and find the position vectors of the midpoints of AB and BC.

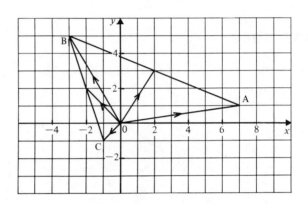

The midpoint of AB is $(2,3)$.

Its position vector is $\begin{pmatrix} 2 \\ 3 \end{pmatrix}$.

The midpoint of BC is $(-2,2)$.

Its position vector is $\begin{pmatrix} -2 \\ 2 \end{pmatrix}$.

1. The position vectors of P and Q relative to O are $\begin{pmatrix} 6 \\ 2 \end{pmatrix}$ and $\begin{pmatrix} 1 \\ 8 \end{pmatrix}$.

Find the midpoint of PQ.

2. Relative to O the position vectors of A and B are $\begin{pmatrix} 3 \\ 5 \end{pmatrix}$ and $\begin{pmatrix} -6 \\ 8 \end{pmatrix}$.

Find the position vector of the midpoint of

a) AB   b) OA   c) OB

3. Relative to O the position vectors of L and M are $\begin{pmatrix} 4 \\ 3 \end{pmatrix}$ and $\begin{pmatrix} 6 \\ 9 \end{pmatrix}$.

Find the position vector of
a) the midpoint of LM
b) the midpoint of OL
c) the point N on OM such that ON : NM = 1 : 2.

The position vectors, relative to O, of A and B are $\begin{pmatrix} 5 \\ 1 \end{pmatrix}$ and

$\begin{pmatrix} -3 \\ 3 \end{pmatrix}$. If $\overrightarrow{OA} = \mathbf{a}$, find the point D such that $\overrightarrow{BD} = 2\mathbf{a}$.

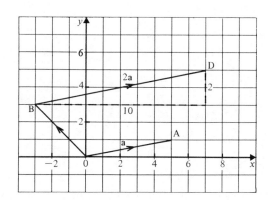

$$2\mathbf{a} = \begin{pmatrix} 10 \\ 2 \end{pmatrix}$$

Therefore, from the diagram, D is the point $(7,5)$.

4. The position vectors relative to O of A and B are **a** and **b**.
$\mathbf{a} = \begin{pmatrix} 2 \\ 4 \end{pmatrix}$ and $\mathbf{b} = \begin{pmatrix} -3 \\ 1 \end{pmatrix}$.

a) Find the point C such that $\overrightarrow{AC} = 3\mathbf{b}$.

b) Find the point D such that $\overrightarrow{BD} = \frac{1}{2}\mathbf{a}$.

c) Find the point E such that $\overrightarrow{AE} = \mathbf{b} - \mathbf{a}$.

5.

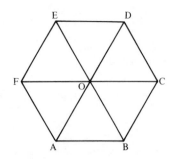

ABCDEF is a regular hexagon. Relative to O the position vectors of A and B are **a** and **b**.

a) Give the position vectors of C, D, E and F in terms of **a** and **b**.

b) Give in terms of **a** and **b** the vectors

   i) $\overrightarrow{AB}$ ii) $\overrightarrow{BC}$ iii) $\overrightarrow{CD}$ iv) $\overrightarrow{DE}$ v) $\overrightarrow{EF}$ vi) $\overrightarrow{FA}$

6.

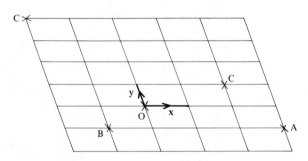

Relative to O the position vector of any point in this diagram is of the form $h\mathbf{x} + k\mathbf{y}$. For each of the vectors $\overrightarrow{OA}$, $\overrightarrow{OB}$ and $\overrightarrow{OC}$ give the values of h and k.

7.

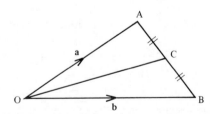

a and **b** are the position vectors of A and B relative to O.
C is the midpoint of AB. Give in terms of **a** and **b**

a) $\overrightarrow{BA}$ b) $\overrightarrow{AB}$ c) $\overrightarrow{BC}$ d) $\overrightarrow{AC}$

e) $\overrightarrow{OC}$ can be given as $\overrightarrow{OB} + \overrightarrow{BC}$ or $\overrightarrow{OA} + \overrightarrow{AC}$. Use each of these two versions to find $\overrightarrow{OC}$ in terms of **a** and **b**. Are your two answers the same?

8.

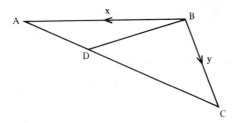

$\overrightarrow{BA} = \mathbf{x}$ and $\overrightarrow{BC} = \mathbf{y}$. D is the point on AC such that AD : DC = 1 : 2.

Give in terms of $\mathbf{x}$ and $\mathbf{y}$

a) $\overrightarrow{AC}$ b) $\overrightarrow{AD}$ c) $\overrightarrow{CA}$ d) $\overrightarrow{CD}$

e) $\overrightarrow{BD}$ can be given as $\overrightarrow{BA} + \overrightarrow{AD}$ or $\overrightarrow{BC} + \overrightarrow{CD}$.

Find $\overrightarrow{BD}$ in terms of $\mathbf{x}$ and $\mathbf{y}$ by the two different ways.

9.

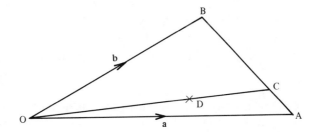

Relative to O the position vectors of A and B are $\mathbf{a}$ and $\mathbf{b}$. C is on AB such that AC : CB = 1 : 3. D is on OC such that OD : DC = 2 : 1.

Give the following vectors in terms of $\mathbf{a}$ and $\mathbf{b}$.

a) $\overrightarrow{AB}$ b) $\overrightarrow{BA}$ c) $\overrightarrow{AC}$ d) $\overrightarrow{BC}$ e) $\overrightarrow{OC}$ f) $\overrightarrow{OD}$ g) $\overrightarrow{DC}$

10.

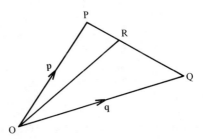

Relative to O the position vectors of P and Q are $\mathbf{p}$ and $\mathbf{q}$. R is the point on PQ such that $\overrightarrow{PR} = k\overrightarrow{PQ}$.

Give in terms of $\mathbf{p}$ and $\mathbf{q}$

a) $\overrightarrow{PQ}$ b) $\overrightarrow{PR}$ c) $\overrightarrow{RQ}$ d) $\overrightarrow{OR}$

HARDER PROBLEMS

Given two non-parallel vectors **a** and **b** there is only one way to express a third vector in terms of **a** and **b**,

i.e.   if $3\mathbf{a} + 2\mathbf{b} = h\mathbf{a} + k\mathbf{b}$

                        then $h = 3$ and $k = 2$

Sometimes the expressions need rearranging first but all the questions in the next exercise use this idea.

EXERCISE 13i

a and **b** are non-parallel vectors
a) If $3\mathbf{a} - \mathbf{b} = h\mathbf{a} + k\mathbf{b}$ find h and k.
b) If $p\mathbf{a} + \mathbf{b} = 2\mathbf{a} + q(\mathbf{a} + \mathbf{b})$ find p and q.

(There is only one way to express a given vector in terms of **a** and **b**. Therefore the coefficients of **a** are equal and the coefficients of **b** are equal.)

a) $3\mathbf{a} - \mathbf{b} = h\mathbf{a} + k\mathbf{b}$

   ∴ $h = 3$ and $k = -1$

b) $p\mathbf{a} + \mathbf{b} = 2\mathbf{a} + q(\mathbf{a} + \mathbf{b})$

   i.e. $p\mathbf{a} + \mathbf{b} = (2 + q)\mathbf{a} + q\mathbf{b}$

   ∴ $p = 2 + q$ and $1 = q$

   So $p = 3$ and $q = 1$.

In each question from 1 to 9, **x** and **y** are non-parallel vectors.
Find h and k.

1. $5\mathbf{x} + h\mathbf{y} = k\mathbf{x} - 4\mathbf{y}$

2. $h\mathbf{x} - 6\mathbf{y} = 7\mathbf{x} + k\mathbf{y}$

3. $h\mathbf{x} + k\mathbf{y} = 6\mathbf{x} + 2h\mathbf{y}$

4. $2x + h(x + y) = kx + 5y$

5. $hx + x + ky = kx + 6y$

6. $\frac{1}{3}x + h(\frac{2}{3}x + \frac{1}{3}y) = \frac{2}{3}x + (k + 1)y$

7. $h(\frac{1}{4}x + \frac{3}{4}y) = (k + \frac{1}{2})y + \frac{1}{2}x$

8. $(h - \frac{2}{5})x + y = k(x + y)$

9. $(h + 4k)x + 2ky = 2hx + y$

Given that $p = 4a + kb$ and $q = 6a + 9b$ and that p and q are parallel, find the value of k.

(As p and q are parallel, q is a multiple of p, so the coefficients of a and b are in the same ratio.)

$$\frac{4}{6} = \frac{k}{9}$$

$$\frac{9 \times 4}{6} = k$$

i.e. $k = 6$.

In each question from 10 to 16, given that p and q are parallel vectors, find k.

10. $p = 6a + 18b, \quad q = ka + 6b$

11. $p = 12a + kb, \quad q = 9a - 21b$

12. $p = (k + 1)a + (k - 1)b, \quad q = 10a + 6b$

13. $p = ka + 12b, \quad q = 3a + kb$.

14. $p = (k - 2)a + 4b; \quad q = 6a + kb$

15. $p = a + kb; \quad q = (k - 3)a + (k + 5)b$

16. $p = ka + (1 - 2k)b; \quad q = (2 - 3k)a + (k + 4)b$

$\overrightarrow{OQ} = \mathbf{q}$, $\overrightarrow{OP} = \mathbf{p}$, $\overrightarrow{OR} = \frac{1}{3}\mathbf{p} + k\mathbf{q}$, $\overrightarrow{OS} = h\mathbf{p} + \frac{1}{2}\mathbf{q}$ and R is the midpoint of QS. Find h and k.

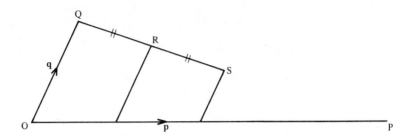

(We need to find $\overrightarrow{QR}$ and $\overrightarrow{RS}$.)

$$\overrightarrow{QR} = \overrightarrow{QO} + \overrightarrow{OR}$$

$$= -\mathbf{q} + \tfrac{1}{3}\mathbf{p} + k\mathbf{q}$$

$$= \tfrac{1}{3}\mathbf{p} + (k - 1)\mathbf{q}$$

$$\overrightarrow{RS} = \overrightarrow{RO} + \overrightarrow{OS}$$

$$= -(\tfrac{1}{3}\mathbf{p} + k\mathbf{q}) + h\mathbf{p} + \tfrac{1}{2}\mathbf{q}$$

$$= (h - \tfrac{1}{3})\mathbf{p} + (\tfrac{1}{2} - k)\mathbf{q}$$

But $\overrightarrow{QR} = \overrightarrow{RS}$ (R is the midpoint of QS.)

$\therefore$ $\tfrac{1}{3}\mathbf{p} + (k - 1)\mathbf{q} = (h - \tfrac{1}{3})\mathbf{p} + (\tfrac{1}{2} - k)\mathbf{q}$

Comparing coefficients of $\mathbf{p}$, $\tfrac{1}{3} = h - \tfrac{1}{3}$

$\therefore$ $h = \tfrac{2}{3}$

Comparing coefficients of $\mathbf{q}$, $k - 1 = \tfrac{1}{2} - k$

$$2k = 1\tfrac{1}{2}$$

$\therefore$ $k = \tfrac{3}{4}$

So $h = \tfrac{2}{3}$ and $k = \tfrac{3}{4}$.

17.

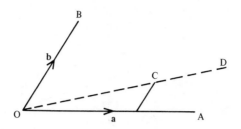

$\overrightarrow{OA} = \mathbf{a}$, $\overrightarrow{OB} = \mathbf{b}$, $\overrightarrow{OC} = \frac{2}{3}\mathbf{a} + \frac{1}{3}\mathbf{b}$. D is the point such that $\overrightarrow{OD} = k\overrightarrow{OC}$.

a) Find $\overrightarrow{OD}$ and $\overrightarrow{BD}$ in terms of **a** and **b**.

b) If BD is parallel to OA, find the value of k.

c) Find the ratio OC : CD.

18.

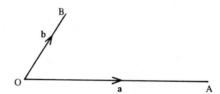

$\overrightarrow{OA} = \mathbf{a}$, $\overrightarrow{OB} = \mathbf{b}$, C and D are the points such that $\overrightarrow{OC} = \mathbf{a} - \frac{1}{2}\mathbf{b}$ and $\overrightarrow{OD} = k\mathbf{a} + \frac{3}{4}\mathbf{b}$.

a) Find $\overrightarrow{BD}$ in terms of **a** and **b**.

b) If BD is parallel to OC find the value of k.

c) Find $\dfrac{BD}{OC}$.

19.

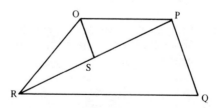

OPQR is a trapezium with OP parallel to RQ. $\overrightarrow{OP} = \mathbf{p}$, $\overrightarrow{OR} = \mathbf{r}$, $RQ = h OP$ and $PS = k PR$. Express in terms of **p** and **r**

a) $\overrightarrow{RQ}$ b) $\overrightarrow{PR}$ c) $\overrightarrow{PQ}$ d) $\overrightarrow{PS}$ e) $\overrightarrow{OS}$

f) If OS is parallel to PQ, find h in terms of k.

g) If, in addition, $\dfrac{PS}{PR} = \dfrac{1}{2}$, find k and h.

20.

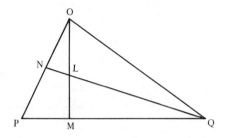

$\overrightarrow{OP} = \mathbf{p}$, $\overrightarrow{OQ} = \mathbf{q}$, M is a point on PQ such that $PM : MQ = 1 : 2$. N is the midpoint of OP. $LQ = h\,QN$.

Give in terms of **p**, **q** and h

a) $\overrightarrow{PQ}$ b) $\overrightarrow{PM}$ c) $\overrightarrow{OM}$ d) $\overrightarrow{ON}$ e) $\overrightarrow{QN}$ f) $\overrightarrow{QL}$ g) $\overrightarrow{OL}$

h) If $OL = k\,OM$, express $\overrightarrow{OL}$ in terms of **p**, **q** and k.

i) Using the two versions of $\overrightarrow{OL}$, find the values of h and k.

UNIT VECTORS

When using x and y coordinates, writing $\mathbf{a} = \begin{pmatrix} 6 \\ 4 \end{pmatrix}$ means that **a** is equivalent to a displacement of 6 units in the direction Ox (i.e. parallel to the x-axis) together with a displacement of 4 units in the direction Oy.

If we use **i** to represent a displacement of 1 unit in the direction Ox and **j** for a displacement of 1 unit in the direction Oy then we can express **a** in terms of **i** and **j**.

i.e. $\mathbf{a} = 6\mathbf{i} + 4\mathbf{j}$

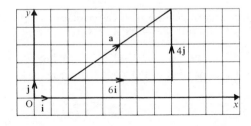

Similarly

$$\mathbf{a} = \begin{pmatrix} 4 \\ 3 \end{pmatrix} = 4\mathbf{i} + 3\mathbf{j}, \quad \mathbf{b} = \begin{pmatrix} 4 \\ -2 \end{pmatrix} = 4\mathbf{i} - 2\mathbf{j}$$

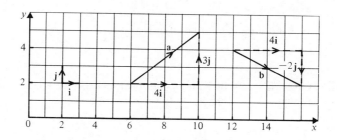

i and **j** are called *unit base vectors.*

EXERCISE 13j

$$\mathbf{p} = \mathbf{i} + 2\mathbf{j}, \quad \mathbf{q} = 3\mathbf{i} - 5\mathbf{j}, \quad \mathbf{r} = -4\mathbf{i} + 3\mathbf{j}$$

Express $\mathbf{p} + \mathbf{q}$, $\mathbf{q} - \mathbf{r}$ and $3\mathbf{r}$ in terms of **i** and **j**.

$$\mathbf{p} + \mathbf{q} = (\mathbf{i} + 2\mathbf{j}) + (3\mathbf{i} - 5\mathbf{j})$$
$$= 4\mathbf{i} - 3\mathbf{j}$$
$$\mathbf{q} - \mathbf{r} = (3\mathbf{i} - 5\mathbf{j}) - (-4\mathbf{i} + 3\mathbf{j})$$
$$= 3\mathbf{i} - 5\mathbf{j} + 4\mathbf{i} - 3\mathbf{j}$$
$$= 7\mathbf{i} - 8\mathbf{j}$$
$$3\mathbf{r} = -12\mathbf{i} + 9\mathbf{j}$$

1. $\mathbf{a} = 3\mathbf{i} + 4\mathbf{j}, \quad \mathbf{b} = -2\mathbf{i} - 5\mathbf{j} \quad \text{and} \quad \mathbf{c} = 5\mathbf{i} - \mathbf{j}$

Express in terms of **i** and **j**

a) $2\mathbf{a}$ b) $-\mathbf{b}$ c) $\mathbf{a} + \mathbf{b}$

d) $\mathbf{c} - \mathbf{b}$ e) $\mathbf{b} - \mathbf{c}$ f) $\mathbf{a} - \mathbf{b}$

g) $\mathbf{a} + \mathbf{b} + \mathbf{c}$ h) $2\mathbf{b} + \mathbf{c}$ i) $-3\mathbf{b}$

2. $a = 5i - 3j,$ $b = 4i + 3j,$ $c = 2i$

Draw diagrams to represent the following vectors.

a) **a**   b) **b**   c) **c**

d) **2c**   e) **a + b**   f) **−b**

g) **a − c**   h) **b − 2c**   i) **a + b + c**

j) Find the magnitudes of **a**, **b** and **c**.

3.   Give, in terms of **i** and **j**, the position vectors relative to the origin of the points $(2,4),$ $(-3,9)$ and $(-6,-4).$

4.

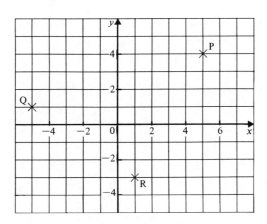

a) Give, in terms of **i** and **j**, the position vectors relative to O of the points P, Q and R.

b) Express, in terms of **i** and **j**, the vectors $\overrightarrow{PQ}$, $\overrightarrow{QR}$ and $\overrightarrow{RQ}$.

c) Find the position vectors of the midpoints of PQ and QR.

d) If $\overrightarrow{OS} = 8i + 2j,$ find the vectors $\overrightarrow{PS}$ and $\overrightarrow{QS}$.

e) Show that PS is parallel to QR and find $\dfrac{PS}{QR}.$

5.   The position vectors relative to O of A and B are $4i + 3j$ and $5j$. C is the midpoint of AB and D is the midpoint of OC.

a) Find the position vector and the coordinates of C.

b) Find the position vector of D.

c) E is the midpoint of OB. Show that EC is parallel to OA.

FURTHER PROBLEMS

EXERCISE 13k

1.

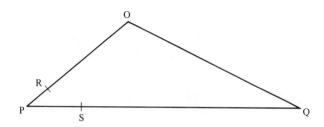

In the diagram OR $= \frac{4}{5}$OP, $\overrightarrow{OP} = $ **p**, $\overrightarrow{OQ} = $ **q** and PS : SQ $= 1 : 4$.

a) Express $\overrightarrow{OR}$, $\overrightarrow{RP}$ and $\overrightarrow{PQ}$ in terms of **p** and **q**.

b) Express $\overrightarrow{PS}$ and $\overrightarrow{RS}$ in terms of **p** and **q**.

c) What conclusion do you draw about RS and OQ ?

d) What type of quadrilateral is ORSQ ?

e) The area of $\triangle$PRS is $5\,\text{cm}^2$. What is the area of ORSQ ?

2.

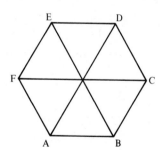

ABCDEF is a regular hexagon, $\overrightarrow{AB} = $ **a** and $\overrightarrow{AC} = $ **b**.

G is the point such that $\overrightarrow{CG} = $ **b** and H is the point such that $\overrightarrow{CH} = 2\textbf{a} - \textbf{b}$.

Find, in terms of **a** and **b**,

a) $\overrightarrow{AD}$ b) $\overrightarrow{BE}$ c) $\overrightarrow{EG}$ d) $\overrightarrow{HG}$

e) Show that HG is parallel to EF.

f) What type of quadrilateral is ADGH ?

3.

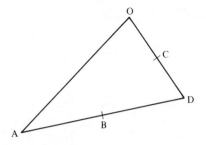

Relative to O the position vectors of A, B and C are **a**, **b** and **c**. B and C are the midpoints of AD and OD.

a) Give $\overrightarrow{OD}$ and $\overrightarrow{AD}$ in terms of **a** and **c**.

b) Find **b** in terms of **a** and **c**.

c) E is a point on OA produced such that $\overrightarrow{OE} = 4\overrightarrow{AE}$. If $\overrightarrow{CB} = k\overrightarrow{AE}$ find the value of k.

4. O, A and B are the points $(0,0)$, $(3,4)$ and $(4,-6)$ respectively.

a) C is the point such that $\overrightarrow{OA} = \overrightarrow{OC} + \overrightarrow{OB}$. Find the coordinates of C.

b) D is the point $(1,24)$ and $\overrightarrow{OD} = h\overrightarrow{OA} + k\overrightarrow{OB}$. Find the values of h and k.

5. ABCD is a parallelogram. Relative to O the position vector of A is $\begin{pmatrix} 2 \\ -1 \end{pmatrix}$, $\overrightarrow{AB} = \begin{pmatrix} 3 \\ 4 \end{pmatrix}$ and $\overrightarrow{AD} = \begin{pmatrix} -2 \\ 5 \end{pmatrix}$.

Find the coordinates of the four vertices of the parallelogram.

14 NUMBERS

INTEGERS

The numbers we use for counting are called the *natural numbers,* i.e. 1, 2, 3, 4, . . .

Whole numbers, both positive and negative, are called integers, i.e.

$$\{\text{integers}\} = \{\ldots -3, -2, -1, 0, 1, 2, 3, \ldots\}$$

SEQUENCES AND PATTERNS

The simplest sequence is the sequence of natural numbers, i.e.

$$1, 2, 3, 4, 5, 6, \ldots$$

Other simple sequences are

1, 3, 5, 7, 9, . . .	(odd numbers)
2, 4, 6, 8, 10, . . .	(even numbers)
2, 3, 5, 7, 11, 13, . . .	(prime numbers)

To continue a sequence we need to recognise a pattern. Consider, for example,

$$2, 5, 8, 11, \ldots$$

Each number in this sequence is three units greater than the preceding number. Hence the sequence continues 14, 17, 20, . . .

The pattern is not always so obvious. If you cannot immediately see a pattern, try looking for sums, products, multiples, plus or minus a number, squares, etc.

EXERCISE 14a

Find the next two terms in the following sequences.

1. 1, 4, 9, 16, . . .

2. 3, 6, 9, 12, . . .

3. 7, 14, 21, 28, . . .

4. $-6, -2, 2, 6, \ldots$

5. 3, 7, 11, 15, . . .

6. 7, 9, 11, 13, . . .

7. $6, 3, 0, -3, -6, \ldots$

8. $2, 1, \frac{1}{2}, \frac{1}{4}, \frac{1}{8}, \ldots$

9. $1 + 3, 1 + 3 + 5, 1 + 3 + 5 + 7, 1 + 3 + 5 + 7 + 9, \ldots$

10. $1, 1, 2, 3, 5, 8, \ldots$

11. $1, 2, 6, 24, 120, \ldots$

12. $0, 3, 8, 15, 24, \ldots$

13. $3, -6, 12, -24, \ldots$

14. $2, 6, 18, 54, \ldots$

15. $2, 6, 12, 20, 30, 42, \ldots$

16. $2, 5, 10, 17, \ldots$

17. $1, -\frac{1}{3}, \frac{1}{9}, -\frac{1}{27}, \ldots$

18. This is a sequence of pairs of numbers

$(1,2), (2,5), (3,10), (4,17), \ldots$

a) Find the next pair in the squence.

b) Find the 10th pair in the sequence.

19.

This triangular array of numbers is known as Pascal's triangle. Find the next three lines in the triangle.

20. Triangular numbers can be represented by dots arranged in a triangular pattern, i.e.

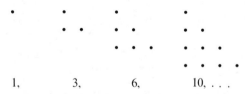

$1, \qquad 3, \qquad 6, \qquad 10, \ldots$

Write down the next four numbers in this sequence.

21.

This is a sequence of tile patterns.

Write down the first eight terms of the number sequence formed by giving the number of tiles in the first pattern, the number of tiles in the second pattern, and so on.

22.

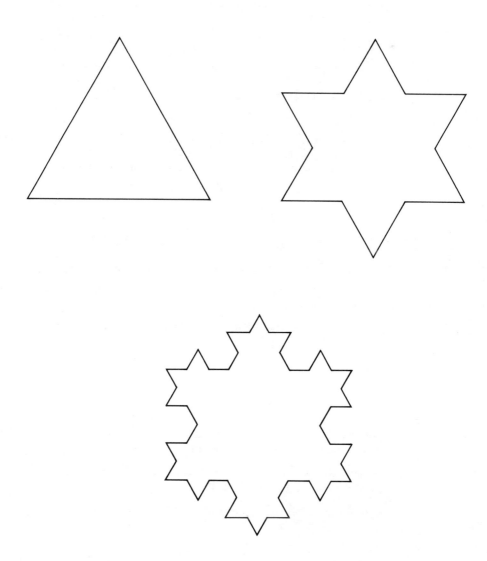

This sequence of 'snowflake' patterns is based on equilateral triangles. The second shape is made by drawing an equilateral triangle on the middle third of each side of the first shape. The third shape is obtained from the second in the same way, and so on for subsequent patterns.

Write down the first six terms of the number sequence where each term is the number of sides of the corresponding 'snowflake'.

23.

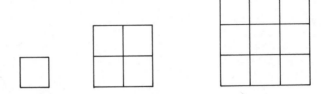

This is a sequence of square patterns.

a) Draw the next two patterns in the sequence.

b) Write down the first five terms of the number sequence formed from the number of squares in each pattern. Note that there are five squares in the second pattern, i.e. four small squares and one larger square.

24. These three triples are known as Pythagorean triples.

$$(3, 4, 5), \quad (5, 12, 13), \quad (7, 24, 25)$$

a) What is the relationship between the three numbers? (Their name should give you a clue!)

b) Find one more Pythagorean triple (not a multiple of those given above).

25. A boy is given a large bar of chocolate and decides to make it last by eating half of what is left each day. Thus he eats half the bar on the first day; he eats half of half the bar, i.e. quarter of the bar, on the second day; he eats half of quarter of the bar on the third day, and so on.

Write down the sequence giving the fraction of the bar left at the end of the first, second, third, fourth and fifth days.

In theory, how long will the bar of chocolate last?

26. Start a sequence by choosing any two integers. Continue the sequence so that each subsequent term is the difference between the previous two terms. End the sequence when zero appears as a term. Repeat this with several different pairs of integers, sometimes starting with the smaller number and sometimes starting with the larger number.

Can you find a relationship between the way a sequence ends and the pair of integers it begins with?

THE n th TERM OF A SEQUENCE

In a sequence the terms occur in a *particular order,* i.e. there is a first term, a second term and so on. The value of each term depends upon its position in that order. (This is the difference between a sequence of numbers and a set of numbers in which the members can be in any order.)

We use the notation u_1, u_1, u_3, . . . for the first, second, third, ... term in a sequence.

Thus u_{10} means the 10th term

u_{100} means the 100th term

u_n means the nth term

If we are given a formula for u_n in terms of n, e.g. $u_n = n(n+1)$, then we can find any term of the sequence.

In this case, the first term, u_1, is found by substituting 1 for n in $n(n+1)$,

i.e. $$u_1 = 1(2) = 2$$

Similarly $$u_2 = 2(2+1) = 6,$$ (substituting 2 for n)

$$u_3 = 3(3+1) = 12,$$

$$u_4 = 4(4+1) = 20$$

....

$$u_{10} = 10(10+1) = 110,$$

and so on.

Therefore the sequence defined by $u_n = n(n+1)$ is

$$2, \ 6, \ 12, \ 20, \ . \ . \ .$$

EXERCISE 14b

Give the first four terms of the sequence for which

1. $u_n = 2n + 1$

2. $u_n = 2n$

3. $u_n = 2^n$

4. $u_n = n^2$

5. $u_n = (n-1)(n+1)$

6. $u_n = n + 4$

7. A sequence is defined by $u_n = \frac{1}{2}n(n+1)$. Find

  a) the first five terms of the sequence

  b) the 20th term of the sequence

  c) an expression for the term before u_n (i.e. u_{n-1})

  d) an expression for $u_n - u_{n-1}$

  e) values of $u_2 - u_1$, $u_3 - u_2$, $u_4 - u_3$, $u_5 - u_4$

  f) the 20th term of the sequence in part (e).

8. Repeat question 7 when $u_n = \frac{1}{6}(n+1)(2n+1)$

FINDING THE *n*th TERM

When the pattern in a sequence is known, a formula for u_n can often be found.

It is helpful to start by trying to recognise the connection between each term and the number of its position order (i.e. between u_1 and 1, u_2 and 2, etc.)

This can often be done by writing the sequence in the form shown in the following example which uses the sequence of natural numbers; 1, 2, 3, . . .

$$u_1, \; u_2, \; u_3, \; u_4, \; \ldots, \; u_n, \; \ldots$$
$$\downarrow \quad \downarrow \quad \downarrow \quad \downarrow$$
$$1 \quad 2 \quad 3 \quad 4$$

In this case it is obvious that the value of a term is the same as its position number, e.g. $u_{20} = 20$ and $u_n = n$.

Now consider the sequence of even numbers:

$$u_1, \; u_2, \; u_3, \; u_4, \; \ldots$$
$$\downarrow \quad \downarrow \quad \downarrow \quad \downarrow$$
$$2 \quad 4 \quad 6 \quad 8$$

This time the value of a term is twice its position number, so the *n*th term is $2n$, i.e. $u_n = 2n$.

Finally look at the sequence of odd numbers:

$$u_1, \; u_2, \; u_3, \; u_4, \; \ldots \; u_n, \; \ldots$$
$$\downarrow \quad \downarrow \quad \downarrow \quad \downarrow$$
$$1 \quad 3 \quad 5 \quad 7$$

Comparing this sequence with the last one we see that each term here is 1 less than the corresponding term in the previous sequence.

Therefore the *n*th term of this sequence is $2n - 1$ i.e. $u_n = 2n - 1$.

Note that it is sensible to check that a formula obtained for u_n gives the correct values for $u_1, u_2, \ldots$

EXERCISE 14c

Find a formula for the nth term of the sequence

a) 2×3, 3×4, 4×5, ...   b) 5, 8, 11, 14, ...

a)
$$u_1, \quad u_2, \quad u_3, \ldots$$
$$\downarrow \qquad \downarrow \qquad \downarrow$$
$$2 \times 3 \quad 3 \times 4 \quad 4 \times 5$$

(Each term is the product of the two integers that follow the position number.)

Hence $u_n = (n+1)(n+2)$

[Check: $u_1 = (1+1) \times (1+2) = 2 \times 3$,
$$u_2 = (2+1) \times (2+2) = 3 \times 4]$$

b)
$$u_1, \; u_2, \; u_3, \; u_4, \; \ldots$$
$$\downarrow \;\; \downarrow \;\; \downarrow \;\; \downarrow$$
$$5 \quad 8 \quad 11 \quad 14$$

(Each term is 3 greater than the term before, so these terms involve multiples of 3.)

Rearranging each term gives

$$u_1, \quad u_2, \quad u_3, \quad\quad u_4,$$
$$\downarrow \qquad \downarrow \qquad \downarrow \qquad\qquad \downarrow$$
$$5 \quad 5+3 \quad 5+2(3) \quad 5+3(3)$$

$\therefore$ $u_n = 5 + (n-1)3$ from the pattern above

$$= 5 + 3n - 3$$

$$= 2 + 3n$$

[Check: $u_1 = 2 + 3(1) = 5$, $u_2 = 2 + 3(2) = 8$]

Find a formula for the nth term of each of the following sequences.

1. $2, 4, 6, 8, \ldots$ **5.** $2, 5, 10, 17, 26, \ldots$

2. $3, 5, 7, 9, \ldots$ **6.** $1 \times 2, 2 \times 3, 3 \times 4, \ldots$

3. $0, 3, 6, 9, 12, \ldots$ **7.** $1, 8, 27, 64, 125, \ldots$

4. $1, 4, 9, 16, 25, \ldots$ **8.** $2, 4, 8, 16, 32, \ldots$

9. $\frac{1}{2}, \frac{1}{3}, \frac{1}{4}, \frac{1}{5}, \ldots$ **11.** $3, 2, 1, 0, -1, -2, \ldots$

10. $0.1, 0.01, 0.001, 0.0001, \ldots$ **12.** $3, 6, 12, 24, 48, \ldots$

13.

a) Copy this pattern and write down the next three rows.

b) Add the numbers in each row to form a sequence,

i.e. $u_1 = 1, \quad u_2 = 1 + 1 = 2, \ldots$

c) Find an expression for the sum of the numbers in the nth row of Pascal's triangle.

14.

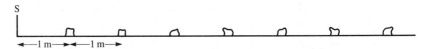

One of the races at a school sports day is set out with bean bags placed at 1 metre intervals along the track.

A competitor starts at S, runs to the first bag, picks it up and returns it to S. Then she runs to the second bag, picks it up and returns it to S, and so on.

How far has a competitor run when she has returned

a) 1 bean bag   b) 4 bean bags   c) n bean bags ?

RATIONAL NUMBERS

In the last section, the majority of the sequences involved integers only. When we include the positive and negative fractions with the integers, we have a much larger set of numbers called the *rational numbers*.

It follows that

any rational number can be expressed as $\frac{p}{q}$ where p and q are integers.

Notice that this statement includes the integers, since, for example, 2 can be expressed as $\frac{2}{1}$.

Examples of rational numbers are $\frac{3}{4}, \frac{5}{1}, \frac{12}{5}, \ldots$

This relationship becomes clearer when set notation is used.

If **N** = {integers} and **Q** = {rational numbers} then **N** ⊂ **Q**.

i.e.

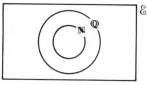

RATIONAL NUMBERS AS DECIMALS

As a rational number is a fraction, it can be expressed as a decimal if we divide the numerator by the denominator.

Any fraction can be expressed either as an exact decimal or as a recurring decimal. This is because, on division, there will at some point either be zero remainder or the remainders will form a repeating sequence. Conversely

> any decimal which is either exact or recurring is a rational number.

For some fractions, the recurring decimal is quick to obtain by division. Some of the simpler ones should be recognised, i.e.

$$\tfrac{1}{3} = 0.\dot{3}, \quad \tfrac{2}{3} = 0.\dot{6}, \quad \tfrac{1}{6} = 0.1\dot{6}, \quad \tfrac{1}{9} = 0.\dot{1}$$

Note that for practical applications it is not necessary to find the repeating pattern in a recurring decimal. The decimal should be given correct to a number of significant figures.

EXERCISE 14d

Express $\tfrac{1}{7}$ as a recurring decimal and hence write $\tfrac{1}{70}$ as a recurring decimal.

Dividing 1 by 7 gives

$$\tfrac{1}{7} = 0.\dot{1}42\,85\dot{7}$$

$$\begin{array}{r} 0.1\,4\,2\,8\,5\,7\,1\,4 \\ 7\overline{)1.0^{3}0^{2}0^{6}0^{4}0^{5}0^{1}0^{3}0 \ldots} \end{array}$$

Now $\quad \tfrac{1}{70} = \tfrac{1}{7} \div 10$

$$= 0.\dot{1}42\,85\dot{7} \div 10$$

$$= 0.0\dot{1}4\,285\,\dot{7}$$

1. Express the following fractions as recurring decimals

a) $\frac{1}{3}$ b) $\frac{1}{30}$ c) $\frac{1}{60}$ d) $\frac{1}{300}$

2. Express the following fractions as recurring decimals

a) $\frac{1}{11}$ b) $\frac{1}{22}$ c) $\frac{1}{33}$ d) $\frac{2}{11}$

3. Express the following fractions as recurring decimals

a) $\frac{1}{999}$ b) $\frac{2}{999}$ c) $\frac{1}{111}$ d) $\frac{1}{110}$

4. a) Write down $\frac{1}{9}$ as a recurring decimal.

b) Multiply your answer to (a) by 9.

c) Multiply $\frac{1}{9}$ by 9 and compare with your answer to part (b).

d) Consider the difference between 1 and $0.\dot{9}$; this may help explain the apparent paradox in part (c).

5. Express the following recurring decimals as fractions:

a) 0.1111 ...   b) 0.010 101 ...   c) 0.003 003 003 ...

With the help of your calculator find $\frac{1}{29}$ as a decimal correct to 11 decimal places.

Calculator gives $\frac{1}{29} = 0.034\,482\,8$ correct to 7 d.p.

(Clear display and enter 0.034 482, i.e. leave off the last decimal place so the number entered is *less* than $\frac{1}{29}$.)

Calculator gives $1 - (29 \times 0.034\,482) = 0.000\,022$

(This number is 29 times the difference between $\frac{1}{29}$ and 0.034 482.)

Calculator gives $0.000\,022 \div 29 = 0.000\,000\,758\,62$

(This number is the difference between $\frac{1}{29}$ and 0.034 482.)

$\therefore \frac{1}{29} = 0.034\,482\,758\,62$ correct to 11 d.p.

Questions 6 and 7 require a calculator that gives 8 s.f. in the display and has scientific notation.

6. a) Use your calculator to write down the first six decimal places when $\frac{1}{17}$ is expressed as a decimal.

b) Multiply the answer to (a) by 17.

c) Subtract the answer to (b) from 1.

d) Divide the answer to (c) by 17.

e) Use the answers to (a) and (d) to give $\frac{1}{17}$ as a decimal *correct* to 10 d.p.

7. Use a method similar to that described in question 6 to find $\frac{1}{13}$ correct to 11 d.p.

IRRATIONAL NUMBERS

We know that there are numbers which cannot be written down exactly, e.g. π, $\sqrt{2}$. It was the Greek Pythagorean philosophers who first discovered these numbers and the discovery shattered their belief that everything in the universe was exactly expressible in numbers. The Pythagoreans called these numbers 'unutterable' but we now call them *irrational*.

> An irrational number can be represented by a length but cannot be expressed as a fraction.

REAL NUMBERS

When the irrational numbers are included with the rational numbers, the combined set is called the set of *real numbers*.

The set of real numbers includes all the numbers that it is possible to represent on a number line:

i.e. for any value x, $x \in \{\text{real numbers}\}$.

If $\mathbb{R} = \{\text{real numbers}\}$ then $\mathbb{Q} \subset \mathbb{R}$.

i.e.

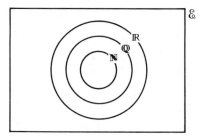

The Venn diagram suggests that there are numbers that are not real numbers, and for interest we will have a brief look at these.

Consider the quadratic equation $x^2 + 4 = 0$.

Solving this equation gives $x^2 = -4$

$$x = \pm\sqrt{-4}$$

There is no real number which, when squared, gives -4 so this equation has roots which are not real. This opens up yet another set of numbers and if you continue to study mathematics you will meet these numbers and find that they do have practical applications in the real world.

MIXED EXERCISE

EXERCISE 14e

1. A is the set $\{-4, \sqrt{5}, \pi, 0.5, -1.8, \frac{2}{7}, \sqrt{9}\}$.
List the members of A that are

a) integers   b) rational   c) irrational

Arrange the members of A in ascending order of size.

2. Insert the members of the set $\{1.6, -2, 2.5, -3, 2.501\}$ in the spaces below

_____ < _____ < _____ < _____ < _____

3. State whether the following statements are true or false.
$\sqrt{16}$ is

a) an integer   b) a real number
c) an irrational number   d) a prime number.

4. Insert the members of the set $\{0.625, 0.5, 0.125, 0.15\}$ in the spaces below

_____ < _____ < _____ < _____

5. T is the set $\{-6, 0.5, \frac{2}{5}, 3, \sqrt{5}, 1.25, -0.5\}$.
List the members of T that are

a) prime numbers   b) integers   c) rational numbers

Insert the members of T in the spaces below

_____ < _____ < _____ < _____ < _____ < _____ < _____

6. If x and y are any two integers, i.e. if $x \in \{\ldots -2, -1, 0, 1, 2, \ldots\}$
and $y \in \{\ldots -2, -1, 0, 1, 2, \ldots\}$ state whether or not the following are always integers.

a) $x + y$ b) $x - y$ c) xy d) $x \div y$

7. 'If two prime numbers are added the result is always a prime number.'
Give a pair of prime numbers which shows that this statement is false.

8. If x and y are any two rational numbers, i.e. if $x \in \mathbb{Q}$ and $y \in \mathbb{Q}$, state
whether or not the following are always rational
a) $x + y$ b) $x - y$ c) xy d) $x \div y$ e) $\sqrt{x}$

9. Goldbach's conjecture states that 'every even number is the sum of two
prime numbers'. Verify that Goldbach's conjecture is correct for
a) 8   b) 16   c) 28   d) 6

(This conjecture has been verified for all numbers up to 10000 but its
general proof still eludes mathematicians!)

Here is a method for finding a square root of a number, correct to as
many decimal places as you have patience for.

Find $\sqrt{17}$ correct to 3 d.p. using successive approximations.

(For an answer correct to 3 d.p., all working must be to 4 d.p.)

First estimate $\sqrt{17} \approx 4.1$

1.   Next find $17 \div 4.1$: $17 \div 4.1 = 4.1463$ (to 4 d.p.)
     ($\therefore$ $4.1 < \sqrt{17} < 4.1463$.)

2.   Next average 4.1 and 4.1463

$$\tfrac{1}{2}(4.1 + 4.1463) = 4.1231$$

    Use this as a better approximation for $\sqrt{17}$

Second estimate $\sqrt{17} \approx 4.1231$

    (Now repeat steps 1 and 2 using the second estimate.)

1. $17 \div 4.1231 = 4.1231$ (to 4 d.p. so step 2 is not needed)

    i.e.   to 4 d.p. $(4.1231)^2 = 17$

    $\therefore$ $\sqrt{17} = 4.123$ correct to 3 d.p.

Use the method illustrated in the worked example to find the square roots of the numbers in questions 10 to 12 giving your answers correct to 3 decimal places.

10. 20   **11.** 2.9   **12.** 0.5

13. Simplify:

a) $1 + \frac{1}{2}$ b) $1 + \dfrac{1}{2 + \frac{1}{2}}$ c) $1 + \dfrac{1}{2 + \dfrac{1}{2 + \frac{1}{2}}}$

14. Taking (a), (b) and (c) of question 13 as the first three terms in a sequence, write down the next two terms of the sequence and simplify them.

15. Using the answers to questions 13 and 14, you now have a sequence of 5 fractions. Square each of these fractions, giving your answer as a decimal to as many decimal places as your calculator permits. What do you think will happen if you continue the sequence of fractions?

(Note that these fractions are called *continued fractions.*)

15 STATISTICS AND PROBABILITY

FREQUENCY POLYGONS

The marks obtained by the candidates in an examination are given in the following table:

Mark	1–10	11–20	21–30	31–40	41–50
Frequency	1	3	6	13	14

Mark	51–60	61–70	71–80	81–90	91–100
Frequency	12	7	5	4	1

In Book 3A we saw that in a histogram the areas of the bars are proportional to the number of items in the group. The histogram representing this data is given below.

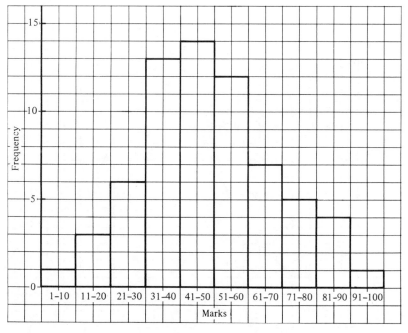

If, on a separate diagram, we plot the middle point at the top of each column of the histogram, and join these points in order, we have a frequency polygon. The frequency polygon for our histogram is given below.

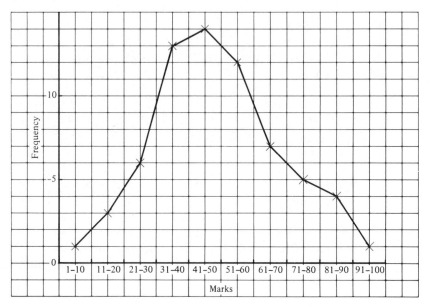

Sometimes the frequency polygon is superimposed on the histogram to give two different representations of the data in a single diagram.

The combined histogram and frequency polygon is shown below.

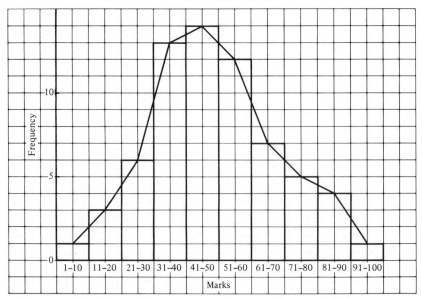

EXERCISE 15a

In questions 1 to 4, draw, on separate diagrams,   a) a histogram
b) a frequency polygon, to represent the data given in the table.

1. The table shows the results of a survey on the pocket money received by 100
ten-year-olds.

Weekly pocket money (p)	0–19	20–39	40–59	60–79	80–99	100–119
Frequency	5	15	27	33	16	4

2. In a shooting competition a competitor fired 50 shots at a target and
obtained the following results.

Score	1	2	3	4	5
Number of shots	3	4	18	16	9

3. The table shows the distribution of goals scored in the 46 football league
matches played one Saturday.

Total number of goals scored	0	1	2	3	4	5	6	7
Frequency	7	10	9	3	9	4	1	3

4. The table shows the results of counting the number of words in each of
the first 150 sentences of a novel.

Number of words	1–10	11–20	21–30	31–40	41–50	51–60	61–70
Frequency	23	53	46	21	4	2	1

5. Draw, on the same diagram, the histogram and frequency polygon for the
data given in the following table which shows the distribution of goals
scored by the twenty-six teams in a hockey league.

Number of goals scored	50–54	55–59	60–64	65–69	70–74	75–79	80–84
Frequency	3	9	4	6	0	2	2

6. Draw, on the same diagram, the histogram and frequency polygon for the
data given in the following table, which shows the distribution of shoe sizes
for the sixty teachers on a school staff.

Shoe size	3	4	5	6	7	8	9	10	11
Frequency	2	5	8	9	11	13	8	3	1

RUNNING TOTALS

After a set of examinations, the results are usually given for one subject at a time. Most pupils, however, are interested in the total of the marks they have received at any stage, i.e. their 'running total'.

The following table shows the separate subject results achieved by a certain pupil, with the running total given in the fourth column.

Lesson	Subject	Mark	Running total
1	Physics	54	54
2	French	72	126
3	Biology	62	188
4	Chemistry	45	233
5	History	78	311
6	Mathematics	64	375
7	English	45	420
8	Geography	82	502

If a similar table gives the running totals only, and not the individual subject marks, we can extract the subject marks. For example, if the running total after four subjects is 233 and the running total after five subjects is 311 then the number of marks scored in the fifth subject is $311 - 233$, i.e. 78.

EXERCISE 15b

1. The table shows the running totals of pupils staying to school lunch each day during a certain school week. Complete the table to find out how many stayed on each day.

Weekday	Number of lunches served each day	Running total of lunches served
Monday		126
Tuesday		280
Wednesday		424
Thursday		599
Friday		717

2. The mileposts along the M4 motorway show the distances, in miles, between various places as follows:

Cardiff to Newport 10,   Newport to Severn Bridge 16,
Severn Bridge to Leigh Delamere 28,   Leigh Delamere to Swindon 18,
Swindon to Reading 39,   Reading to Heathrow Airport 28,
Heathrow to Central London 15.

Make a running total of the distances along the motorway from Cardiff to Central London.

3. During a weeks' holiday a family spent the following amounts

Day	Amount spent	Running total of expenditure
Monday	£12	
Tuesday	£26	
Wednesday	£5	
Thursday	£8	
Friday	£32	
Saturday	£27	
Sunday	£4	

Complete the table and check your answer by direct addition.

CUMULATIVE FREQUENCY

It is often useful to know how many pupils have scored *less than* a certain mark.

For example, in question 1 of Exercise 15a there are $5 + 15$, i.e. 20, ten-year-olds whose weekly pocket money is less than 40 p and $20 + 27$, i.e. 47, whose weekly pocket money is less than 60 p. A simple and effective way of giving this information is to make a *cumulative frequency table*. A cumulative frequency table is constructed by adding each frequency to the sum of all those that have gone before it. The cumulative frequency table for the data given in Exercise 15a, question 2 can be set out as follows:

Score	Frequency	Score	Cumulative frequency
1	3	$\leqslant 1$	3
2	4	$\leqslant 2$	$3 + 4 = 7$
3	18	$\leqslant 3$	$7 + 18 = 25$
4	16	$\leqslant 4$	$25 + 16 = 41$
5	9	$\leqslant 5$	$41 + 9 = 50$

Even if the second column is omitted, the frequency for any score from 1 to 5 can be found from the cumulative frequencies, e.g. the number of 3s scored is given by the cumulative frequency up to 3 minus the cumulative frequency up to 2, i.e. $25 - 7 = 18$.

Notice that the last number in the cumulative frequency column can be used as a check on accuracy. It confirms that the total number of shots fired was 50.

EXERCISE 15c

1. Complete the following table which shows the distribution of goals scored by the home sides in a football league one Saturday.

Score	Frequency	Score	Cumulative frequency
0	3	⩽ 0	3
1	8	⩽ 1	$3 + 8 = 11$
2	4	⩽ 2	
3	3	⩽ 3	
4	5	⩽ 4	
5	2	⩽ 5	
6	1	⩽ 6	

2. Complete the following table which shows the distribution of the marks scored by the first year pupils in their English test.

Mark	Frequency (no. of pupils' scores within each range)	Mark	Cumulative frequency
1–10	7	⩽ 10	
11–20	14	⩽ 20	
21–30	18	⩽ 30	
31–40	33	⩽ 40	
41–50	36	⩽ 50	
51–60	43	⩽ 60	
61–70	21	⩽ 70	
71–80	15	⩽ 80	
81–90	12	⩽ 90	
91–100	8	⩽ 100	

a) How many first-year pupils are there ?

b) How many scored 50 or less ?

c) How many scored more than 60 ?

3. The table is based on a cricketer's scores in one season. Complete the table to show the cumulative frequencies.

Score	0–19	20–39	40–59	60–79	80–99	100–119	120–139
Frequency	8	14	33	6	5	3	1
Score	⩽ 19	⩽ 39	⩽ 59	⩽ 79	⩽ 99	⩽ 119	⩽ 139
Cumulative frequency							

a) How many innings did he play ?

b) In how many innings did he score less than 60 ?

c) In how many innings did he score 40 or more ?

4. The table is based on a golfer's scores on the professional circuit one summer.

Score	67	68	69	70	71	72
Frequency	2	4	9			
Score	$\leqslant 67$	$\leqslant 68$	$\leqslant 69$	$\leqslant 70$	$\leqslant 71$	$\leqslant 72$
Cumulative frequency	2	6	15	24	36	51

Score	73	74	75	76	77	78
Frequency						
Score	$\leqslant 73$	$\leqslant 74$	$\leqslant 75$	$\leqslant 76$	$\leqslant 77$	$\leqslant 78$
Cumulative frequency	64	72	77	85	91	95

Complete this table and hence find

a) the number of rounds in which he scored 73

b) the number of rounds in which he scored 75 or more.

5. A school organises a Grand Prize Draw to raise money to buy a minibus. Tickets are sold at 50 p per book and pupils are encouraged to sell as many books as possible by the award of inducement prizes, including a first prize of £20 to the pupil who sells the most books. The table shows the distribution of the numbers of books sold by the pupils in the school.

Number of books sold	0–5	6–10	11–15	16–20	21–25
Frequency	77	124			
Number of books sold	$\leqslant 5$	$\leqslant 10$	$\leqslant 15$	$\leqslant 20$	$\leqslant 25$
Cumulative frequency			383	611	775

Number of books sold	26–30	31–35	36–40	41–45	46–50
Frequency		73	32	22	9
Number of books sold	$\leqslant 30$	$\leqslant 35$	$\leqslant 40$	$\leqslant 45$	$\leqslant 50$
Cumulative frequency	867				

Complete the table and hence find

a) the number of pupils who sold more than 30 books

b) the number of pupils who sold fewer than 21 books

c) the number of pupils who sold more than 10 books but fewer than 31 books.

Was the £20 inducement prize won by one pupil or could it have been shared?

CUMULATIVE FREQUENCY CURVE

When we have cumulative tables like those given in the previous exercise we can draw a graph by plotting the cumulative frequency against the mark or score.

The data for question 2 of the previous exercise can be set out as given below.

Mark	≤ 10	≤ 20	≤ 30	≤ 40	≤ 50	≤ 60	≤ 70	≤ 80	≤ 90	≤ 100
Cumulative frequency	7	21	39	72	108	151	172	187	199	207

If we plot each cumulative frequency at the *upper value* of each interval the marks can either be joined by straight lines, in which case the graph is called the *cumulative frequency polygon,* or we can draw a smooth curve through them to obtain a *cumulative frequency curve.*

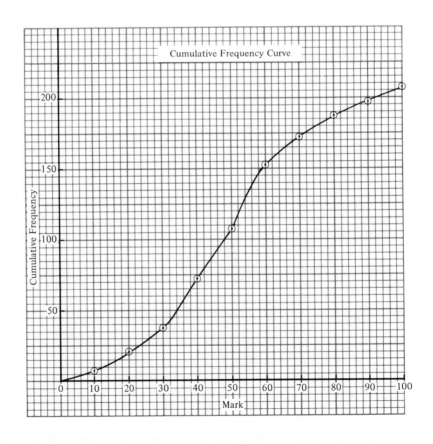

EXERCISE 15d

Use the cumulative frequency tables from Exercise 15c to draw a cumulative frequency curve for each set of data.

FINDING THE MEDIAN

Remember that when we need one number to represent a set then, depending on the circumstances, we can use either the mean, or the mode, or the median (see Book 3A, Chapter 26).

The *mean,* or arithmetic average, of a set of n numbers is the sum of the numbers divided by n.

The *mode* of a set of numbers is the number that occurs most often.

For example, given the numbers 3, 4, 5, 5, 5, 8, 9, 9

$$\text{the mean is } \frac{3+4+5+5+5+8+9+9}{8} = 6$$

the mode is 5

The *median* of a set of numbers is the number in the middle when they are arranged in order of size. If there are n numbers in the set the median is the $\left(\dfrac{n+1}{2}\right)$th number.

When n is even there is no actual $\left(\dfrac{n+1}{2}\right)$th number, so we take the average of the numbers on each side of $\dfrac{n+1}{2}$, e.g. for 10 numbers, the $\left(\dfrac{n+1}{2}\right)$th number is the $5\frac{1}{2}$th number and there is no such number, so we take the average of the 5th and 6th.

The median can easily be found from the cumulative frequency curve. At the middle value of the cumulative frequency we draw a line across to meet the curve, and read off the corresponding value of the mark.

In our example the middle value of the cumulative frequency is the $\left(\dfrac{200+1}{2}\right)$th value, i.e. the $100\frac{1}{2}$th value.

We approximate this to the 100th value.

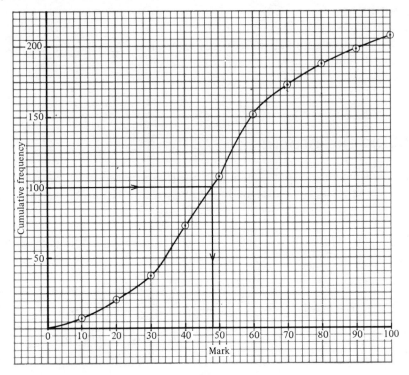

From the graph the median mark is 48.

EXERCISE 15e

1. Find the median from each cumulative frequency curve drawn for the questions in Exercise 15d.

2. Use the cumulative frequency table that follows to draw the cumulative frequency curve for the prices of all the houses advertised in a property magazine one weekend.

Price (thousands of £s)	⩽ 10	⩽ 20	⩽ 30	⩽ 40	⩽ 50
Cumulative frequency	10	22	60	128	170

Price (thousands of £s)	⩽ 60	⩽ 70	⩽ 80	⩽ 90	⩽ 100
Cumulative frequency	187	197	203	208	210

Use your graph to estimate the median 'asking price' price for a house.

3. The cumulative frequency curve for the marks in an English test is given below. Use the graph to find
a) the number of pupils sitting the test   b) the median mark.

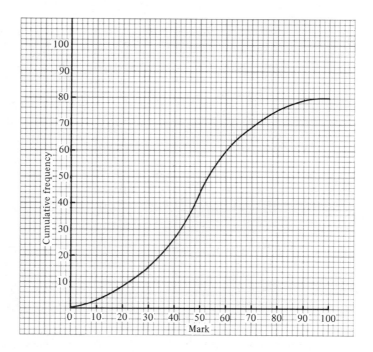

4. Use the following cumulative frequency table to draw the corresponding cumulative frequency curve for the marks obtained by candidates in an examination.

Marks	$\leqslant 9$	$\leqslant 19$	$\leqslant 29$	$\leqslant 39$	$\leqslant 49$
Cumulative frequency	7	16	28	47	80

Marks	$\leqslant 59$	$\leqslant 69$	$\leqslant 79$	$\leqslant 89$	$\leqslant 99$
Cumulative frequency	125	174	202	212	220

Use your graph to estimate the median mark.

5. The number of cars using a cross-channel ferry on each trip during a particular month was noted and the results are given in the following table.

Number of cars	40	41	42	43	44	45	46	47	48	49	50
Number of crossings	2	4	6	10	10	12	8	6	2	1	1

Give the corresponding cumulative frequency table and hence draw the cumulative frequency curve. Use your graph to estimate the median number of cars making the crossing. How many crossings did the ferry make during the month ?

6. A traffic survey counted the number of cars per hour passing Southwood Post Office each hour of the day from 8 a.m. to 6 p.m. for a week. The results are given in the table.

	8 a.m. -9 a.m.	9 a.m. -10 a.m.	10 a.m. -11 a.m.	11 a.m. -12 noon	12 noon -1 p.m.
Monday	39	37	46	36	41
Tuesday	16	31	40	39	42
Wednesday	24	39	37	45	44
Thursday	19	33	32	34	42
Friday	30	37	36	41	48
Saturday	28	38	46	39	42
Sunday	3	7	42	14	11

	1 p.m. -2 p.m.	2 p.m. -3 p.m.	3 p.m. -4 p.m.	4 p.m. -5 p.m.	5 p.m. -6 p.m.
Monday	34	33	32	22	23
Tuesday	43	39	37	24	17
Wednesday	39	38	36	29	27
Thursday	38	37	39	25	27
Friday	47	40	43	35	34
Saturday	48	42	40	31	33
Sunday	33	36	35	27	26

Use groups 0–5, 6–10, 11–15, etc. to make a frequency table and a cumulative frequency table. Draw the cumulative frequency curve and use it to estimate the median number of cars passing Southwood Post Office per hour.

RANGE

So far we have described a distribution by finding where its centre is, but we also need to know how widely the distribution is spread. One measure of spread, or *dispersion*, is the range.

The *range* is the difference between the two extreme values. It is considered to be of little use since it tells us nothing about all the other values in between, and it can be distorted by one or two extreme values.

Consider a student who scores marks of 54, 93, 86, 75, 8, 59, 73, 83, 55, 64, 73, 52, 74, 64 and 70 in fifteen examinations. The range is from 8 to 93, whereas every mark apart from the 8 is more than 50.

QUARTILES

A much better measure of spread is found by using the range of the middle 50% of the values. To find this we use the *quartiles*.

First arrange the n values in ascending order.

The *lower quartile* is the $\left(\dfrac{n+1}{4}\right)$th value, and the *upper quartile* is the $3\left(\dfrac{n+1}{4}\right)$th value.

The difference between the upper quartile and the lower quartile is called the *interquartile range*.

If Q_1 denotes the lower quartile and Q_3 the upper quartile, the interquartile range is $Q_3 - Q_1$.

When the fifteen examination marks given on page 306 are written in ascending order they are

$$8, 50, 54, 55, 59, 64, 64, 70, 73, 73, 74, 75, 83, 86, 93$$

Q_1 is the $\left(\dfrac{15+1}{4}\right)$th value, i.e. the 4th value

$\therefore \quad Q_1 = 55$

Q_3 is the $3\left(\dfrac{15+1}{4}\right)$th value, i.e. the 12th value

$\therefore \quad Q_3 = 75$

$\therefore$ the interquartile range is $Q_3 - Q_1$

$$= 75 - 55$$
$$= 20$$

Sometimes it is convenient to use the semi-interquartile range which is defined as $\dfrac{Q_3 - Q_1}{2}$.

In our example, the semi-interquartile range is $\dfrac{20}{2}$, i.e. 10.

When the number of values is large it is convenient to make certain approximations. If $n = 200$, Q_1 is the $\left(\dfrac{200+1}{4}\right)$th value, i.e. the $50\frac{1}{4}$th value, and Q_3 is the $150\frac{3}{4}$th value. We use the 50th value and the 150th value.

The cumulative frequency curve for the English test scores, given on page 302 is reproduced below.

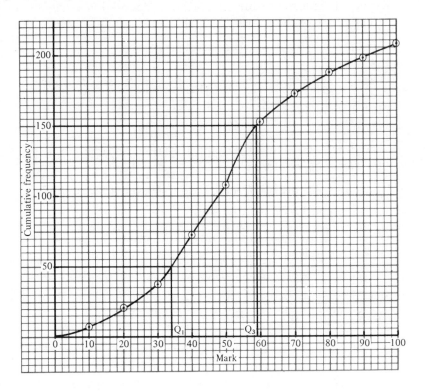

From the curve $Q_1 = 34$

and $Q_3 = 59$

The interquartile range is therefore $59 - 34 = 25$, and the semi-interquartile range is $12\frac{1}{2}$.

EXERCISE 15f

1. Use the cumulative frequency curves obtained in questions 4 to 6 of Exercise 15e to find, for each set of data,

a) the upper and lower quartiles

b) the interquartile range.

2. The cumulative frequency curve given below shows the weekly earnings, in pounds, of a group of teenagers.

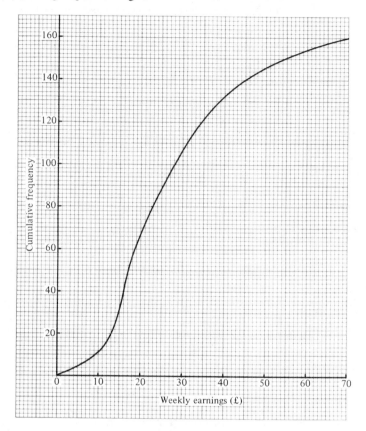

Use the graph to find   a) the median   b) the upper and lower quartiles. Hence find the interquartile range.

3. The table shows the distribution of the ages of people attending a public concert.

Age range	0–19	20–39	40–59	60–79	80–99
No. of people attending concert	8	26	110	128	56

Copy and complete the following cumulative frequency table and use it to draw a cumulative frequency curve.

Age range	< 20	< 40	< 60	< 80	< 100
No. of people attending concert					

Hence find

a) the number of people attending the concert

b) the median age

c) the upper and lower quartile ages, and the interquartile range.

4. One hundred people were chosen at random and each was allowed to fire 50 shots from a gun at a target. The number of bulls scored by each person was noted and the following frequency table was constructed.

Number of bulls	Frequency	Number of of bulls	Cumulative frequency
1–5	1	⩽ 5	
6–10	3	⩽ 10	
11–15	5	⩽ 15	
16–20	7		
21–25	15		
26–30	25		
31–35	20		
36–40	12		
41–45	8		
46–50	4		

Copy this table and complete the third and fourth columns. Draw the cumulative frequency curve using a scale of 2 cm to represent 10 bulls on one axis and 2 cm to represent a cumulative frequency of 10 on the other axis. Use your graph to estimate

a) the median   b) the upper and lower quartiles.

5. The marks obtained by the pupils sitting a test are given in the following table.

Marks	0–9	10–19	20–29	30–39	40–49	50–59	60–69	70–79
Frequency	9	13	27	43	28	20	12	8

Illustrate these figures by drawing a cumulative frequency curve. Use 2 cm to represent 10 marks on the one axis and a cumulative frequency of 20 on the other axis. Use your graph to estimate

a) the median

b) the upper and lower quartiles, and hence the interquartile range

c) the pass mark if three-quarters of the pupils pass.

6. In the first round of a golf tournament the following scores were recorded:

```
70  68  71  67  74  69  69  71  68  70
71  70  72  69  69  68  71  70  70  72
72  69  68  70  68  69  67  71  69  70
68  67  70  70  73  69  71  67  69  68
```

a) Construct a frequency table for these scores.

b) Use this to give a cumulative frequency table.

c) How many rounds of less than 70 were there ?

d) How many rounds of more than 69 were there ?

e) Use your cumulative frequency table to draw a cumulative frequency curve and use it to find the median score.

7. The table is based on a cricketer's scores in 100 innings.

Score	0–10	11–20	21–30	31–40	41–50	51–60
Frequency	7	9	11	13	16	18

Score	61–70	71–80	81–90	91–100	101–110	111–120
Frequency	11	7	4	2	1	1

a) Construct a cumulative frequency table for these scores and use it to draw a cumulative frequency curve.

b) How many scores did he have that were 50 or less than 50 ?

c) How many scores did he have that were more than 70 ?

d) Use the curve to estimate the median score, the upper and lower quartiles, and the interquartile range.

8. A botanist measured the lengths of 120 leaf specimens from a certain species of tree. The results were as follows:

Length (cm)	< 9	< 9.5	< 10	< 10.5	< 11	< 11.5	< 12
Cumulative frequency	2	6	14	23	35	50	73

Length (cm)	< 12.5	< 13	< 13.5	< 14	< 14.5	< 15
Cumulative frequency	92	103	112	116	119	120

Use this data to draw a cumulative frequency graph. From your curve find

a) the median length

b) the values of the upper and lower quartiles.

HISTOGRAMS WITH UNEQUAL INTERVALS

A histogram is constructed so that the area of each column is proportional to the number of items in that group. If the histogram has unequal intervals the height of the column must be adjusted so that its *area* is proportional to the frequency of the group. If, for example, we have a bar that is twice as wide as the others used, the height of the bar is proportional to *half* the number of items in the group.

EXERCISE 15g

Draw a histogram to illustrate the following frequency table which shows the distribution of weights, to the nearest kilogram, of seventy fourteen-year-old girls.

Weight (kg)	32–33	34–39	40–43	44–47	48–51	52–59	60–71
Frequency	1	9	10	12	17	12	9
Width of bar	2	6	4	4	4	8	12

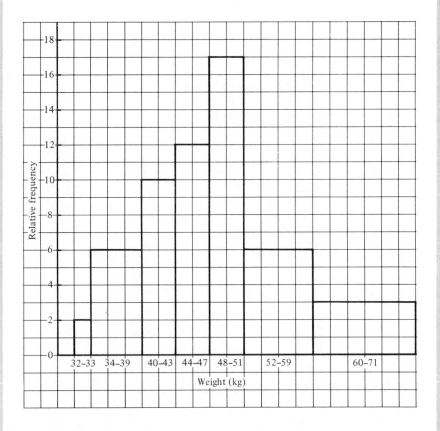

(The interval that occurs most frequently is 4. This applies to the 3rd, 4th and 5th intervals. We choose this as our 'unit' interval and extend each bar for these intervals to a height level with the number given in the frequency table.)

The width of the first interval is half that of the 'unit' interval so the height of this bar is twice its frequency, i.e. 2.

Similarly the width of the second interval is $1\frac{1}{2}$, or $\frac{3}{2}$, times the unit interval, so the height of the bar is $\frac{2}{3} \times 9$, i.e. 6.

In a similar way the height of the 6th bar is $\frac{1}{2} \times 12$, i.e. 6, and the height of the 7th bar is $\frac{1}{3} \times 9$, i.e. 3.

Draw a histogram to illustrate the frequency tables in questions 1 to 4.

1. The table gives the distribution of marks of 100 pupils in an English examination.

Mark	0–29	30–39	40–49	50–59	60–99
Frequency	9	15	24	36	16

2. The table shows the distribution of heights, to the nearest centimetre, of 65 fourteen-year-old boys.

Height (cm)	100–129	130–139	140–144	145–154	155–164	165–184
Frequency	6	11	8	14	14	12

3. The table shows the distribution of the times, to the nearest minute, taken by 85 people to get to school.

Time (minutes)	1–5	6–10	11–15	16–18	19–21	22–26	27–30	31–40
Frequency	6	11	18	10	11	16	9	4

4. The table gives the results of a survey, which was conducted in a region of France, to find the areas, to the nearest hectare, of 185 farms.

Area (hectares)	1–4	5–8	9–12	13–17	18–24	25–28	29–32	33–42
Frequency	34	22	38	30	28	13	10	10

PROBABILITY

Single event

$$P(\text{successful event}) = \frac{\text{number of successful outcomes}}{\text{total number of possible outcomes}}$$

$P(\text{certainty}) = 1$

$P(\text{impossibility}) = 0$

When an event is denoted by A, the probability that the event happens is denoted by $P(A)$ and $0 \leqslant P(A) \leqslant 1$.

If the probability that the event A does not happen is denoted by $P(\overline{A})$ then $P(\overline{A}) = 1 - P(A)$.

Two events

When we are concerned with two events (e.g. tossing two dice, together or separately) where the result of the first cannot affect the result of the second, we can use a possibility space (see Book 4A, p. 216) but when the outcome of the second event (e.g. drawing cards from a pack without replacing them) can be influenced by the outcome of the first event, we usually use a probability tree.

When the required outcome is given by more than one path in a probability tree we *add* the probabilities resulting from each path.

When we follow a path along the branches of a probability tree we *multiply* the probabilities along that path.

EXERCISE 15h

If a card is drawn from a pack of 52 find the probability that it is

a) a black card   b) a spade   c) a 5   d) the 7 of diamonds.

a) $P(\text{a black card}) = \dfrac{26}{52} = \dfrac{1}{2}$

b) $P(\text{a spade}) = \dfrac{13}{52} = \dfrac{1}{4}$

c) $P(\text{a 5}) = \dfrac{4}{52} = \dfrac{1}{13}$

d) $P(\text{the 7 of diamonds}) = \dfrac{1}{52}$

1. A bag contains two red beads, five yellow beads and eight white ones. If one bead is drawn from the bag, what is the probability that

    a) the bead is yellow

    b) the bead is red or white

    c) the bead is black

    d) the bead is not red ?

2. A letter is chosen at random from the English alphabet. Find the probability that

    a) the letter is a consonant

    b) the letter is w, x, y or z

    c) the letter is one that is used in the word MATHEMATICS.

3. A two-figure number is chosen at random. Find the probability that

    a) the number is 66

    b) the number is a multiple of 5

    c) the number is greater than 69.

4. The probabilities of three football teams A, B and C winning a competition are $\frac{1}{2}$, $\frac{1}{4}$ and $\frac{1}{5}$ respectively.

    Find the probability that

    a) either A or B will win

    b) one of these three teams will win

    c) none of these teams will win.

5. In a class of 32 pupils the probability that a pupil picked at random has completed his or her homework is $\frac{7}{8}$. How many pupils did not complete their homework ?

6. Draw a possibility space to show the outcomes when two dice are tossed. Use it to find the probability that

    a) the sum of the two numbers is 7 or more

    b) each die shows an odd number greater than 2.

    c) the product of the scores is even.

7. Fifty potatoes are selected at random from a field and the number of 'eyes' on each is recorded. The results are given in the following frequency table.

Number of 'eyes' on a potato	3	4	5	6	7	8
Frequency (number of potatoes)	5	8	10	12	10	5

a) Find the probability that a potato selected at random from the fifty potatoes will have
  i) exactly four eyes.
  ii) at least five eyes.

b) A second potato is selected from the remainder. What is the probability that
  i) both potatoes have five eyes
  ii) the first has five eyes and the second has six eyes ?

8. The table shows the distribution of the masses of 60 packages brought to the counter at Borchester Post Office and placed in a mail bag.

Mass (g)	-60	-100	-150	-200	-250	-300
Frequency	6	12	12	18	6	6

The interval -200 means $150\,\text{g} < \text{mass} \leqslant 200\,\text{g}$.

a) If one package is selected at random from the mail bag what is the probability that
  i) its mass m is such that $150\,\text{g} < m \leqslant 200\,\text{g}$
  ii) its mass is more than $150\,\text{g}$?

b) A second package is selected from the mail bag. On the assumption that the first package selected has a mass of $180\,\text{g}$ what is the probability that the mass, M, of the second package
  i) is greater than $150\,\text{g}$
  ii) satisfies the inequality $150\,\text{g} < M \leqslant 200\,\text{g}$?

9. The following cumulative frequency table is based on a cricketer's scores for a season.

Score	$\leqslant 19$	$\leqslant 39$	$\leqslant 59$	$\leqslant 79$	$\leqslant 99$	$\leqslant 119$	$\leqslant 139$	$\leqslant 150$
Cumulative frequency	7	18	32	40	46	50	53	55

a) If an innings is selected at random, what is the probability that he scored   i) a century   ii) less than 40 ?

b) Two of his scores are selected at random. What is the probability that
  i) both are centuries   ii) neither is a century ?

Two dice are tossed. What is the probability of getting
a) two prime numbers   b) a prime number on one die and
a 6 on the other ?

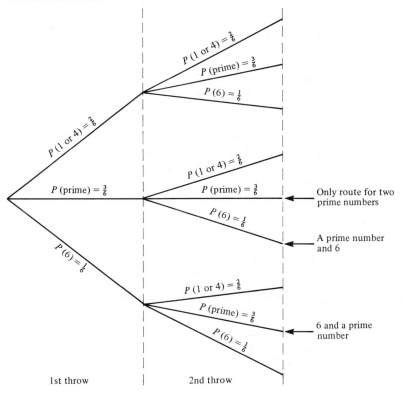

1st throw | 2nd throw

a) $P(\text{two prime numbers}) = \dfrac{3}{6} \times \dfrac{3}{6}$ (multiply the probabilities along the branches)

$$= \dfrac{1}{4}$$

b) $P(\text{a prime number and a 6 in any order})$

$$= \dfrac{3}{6} \times \dfrac{1}{6} + \dfrac{1}{6} \times \dfrac{3}{6}$$

$$= \dfrac{1}{12} + \dfrac{1}{12}$$ (add the probabilities from all the acceptable paths)

$$= \dfrac{1}{6}$$

10. The probability that my team will play as selected when I go to a football match is $\frac{9}{10}$. If I attend two games what is the probability that

a) I see two changed teams

b) I see exactly one changed team ?

11. The probability that Amer scores more than 60 in French is $\frac{4}{5}$ and that he scores more than 60 in mathematics is $\frac{3}{4}$. What is the probability that he scores

a) more than 60 in both subjects

b) 60 or less in both subjects ?

12. Tom has a very old second-hand car. The probability that he can get it started is $\frac{9}{10}$ and the probability that it will break down before he gets to work is $\frac{1}{50}$. Find the probability that he gets to work in the car without a breakdown.

13. Three coins are tossed. Find the probability of getting

a) three tails

b) two heads and one tail.

14. A die is thrown twice. Find the probability that

a) the total score is 6

b) the total score is 12

c) the first score is exactly three times the second score

d) the total is a prime number.

15. Two cards are selected at random from a pack of 52 playing cards. Find the probability that the two cards are

a) clubs

b) picture cards

c) threes

d) a heart and a spade in either order

e) the ace of spades followed by the queen of hearts.

16. The letters of the word **PREPOSSESSING** are placed in a box. A letter is then selected, replaced in the box, and a second letter is selected. Find the probability of obtaining

a) the letter S twice

b) the letter S three times

c) the letter G twice

d) a P and an E in either order.

17. The probability that Sam, who is a golfer, completes a hole in 'par' is $\frac{2}{3}$, and that he scores 'under par' is $\frac{1}{15}$. If Sam plays three holes of golf, what is the probability that

a) his score on the first hole is over par

b) he makes three par scores

c) he makes one par score and two over par scores

d) he scores par or under par on every hole ?

18. A bag contains 2 red counters, 3 white counters and 4 blue counters. Three counters are drawn from the bag without replacement. Find the probability that

a) they are all red

b) they are all blue

c) there is one of each colour

d) there are at least two of the same colour.

MIXED QUESTIONS

EXERCISE 15i

1. The following cumulative frequency table gives the percentage marks of 250 pupils in an English examination.

Mark	10	20	30	40	50	60	70	80	90	100
Number of pupils scoring up to and including this mark	5	15	29	52	89	142	197	223	240	250

a) How many pupils scored a mark of more than 70 ?

b) How many pupils scored a mark from 41 to 60 ?

c) Plot the values from the table on a graph and draw a smooth curve through your points. (Use a scale of 2 cm to represent 20 marks on the one axis and 2 cm to represent a cumulative frequency of 25 on the other axis.)

d) Use your graph to estimate
   i) the median   ii) the upper and lower quartiles.

e) State the probability that a pupil chosen at random will have a mark
   i) less than or equal to 50   ii) greater than 60.

2. The histogram illustrates the distribution of the weekly pocket money of the 240 pupils in the fourth year of a school.

a) How many pupils received from £6 to £10 inclusive ?

b) Half the pupils received more than £x per week. Estimate the value of x.

c) The line AB indicates that the value of the lower quartile of the distribution is £1.50. What does this mean ?

d) The value of the interquartile range is £2.70. What is the value of the upper quartile ? What information does the interquartile range give us about the weekly pocket money of the group ?

e) Estimate the total amount of pocket money received by the fourth year pupils.

f) Hence estimate the mean amount of pocket money received by the fourth year pupils.

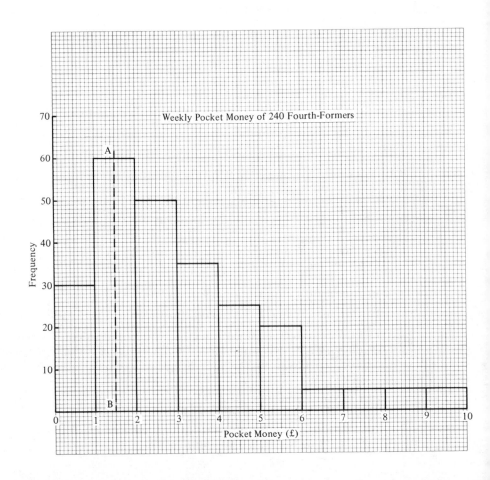

3. The following marks were obtained by the 80 candidates in an English test which was marked out of 60.

54	52	31	47	24	36	27	15	44	26
8	20	46	32	27	31	33	57	39	32
43	32	23	33	31	21	38	28	40	19
52	37	38	39	9	30	47	29	8	13
33	35	48	18	36	39	23	58	34	35
16	21	32	38	34	13	27	32	37	23
37	49	25	38	24	27	48	36	45	18
41	34	43	12	47	24	8	29	37	33

Use this data to complete the table below.

Interval	Tally	Frequency	Mark	Cumulative frequency
0–9			$\leqslant 9$	
10–19			$\leqslant 19$	
20–29			$\leqslant 29$	
30–39				
40–49				
50–60				

Use the information in your table to draw a cumulative frequency curve and from it estimate

a) the median mark

b) the upper and lower quartiles

c) the number of candidates who passed if the pass mark was 40

d) the pass mark if 70% of the candidates passed

e) the probability that a pupil selected at random scored less than 30.

4. The following cumulative frequency table gives the result of a survey into the masses (in kilograms) of 80 boys.

Mass (kg)	$\leqslant 45$	$\leqslant 50$	$\leqslant 55$	$\leqslant 60$	$\leqslant 65$	$\leqslant 70$
Cumulative frequency	7	11	14	19	25	32

Mass (kg)	$\leqslant 75$	$\leqslant 80$	$\leqslant 85$	$\leqslant 90$	$\leqslant 95$
Cumulative frequency	42	58	71	78	80

a) Calculate how many boys have a mass of more than 80 kg.

b) Calculate how many boys have a mass of 65 kg or less but more than 50 kg.

c) Using a vertical scale of 2 cm to represent 10 boys and a horizontal scale of 2 cm to represent 10 kg, plot these values on graph paper and draw a smooth curve through your points.

d) Use your graph to estimate the median mass and the upper and lower quartile masses.

e) State the probability that a boy chosen at random will have a mass in excess of 70 kg.

5. The marks of 4000 candidates in a history examination are summarised in the table.

Mark	$\leqslant 20$	$\leqslant 30$	$\leqslant 40$	$\leqslant 50$	$\leqslant 60$	$\leqslant 70$	$\leqslant 80$	$\leqslant 90$	$\leqslant 100$
Number of candidates	230	450	750	1400	2400	3300	3800	3950	4000

a) Draw a cumulative frequency diagram to represent this data. (Use 2 cm to represent 20 marks on the *x*-axis and 2 cm to represent 500 candidates on the cumulative frequency axis.)

b) Use your cumulative frequency diagram to find
   i) the number of candidates who scored 75 marks or less
   ii) the median mark
   iii) the interquartile range
   iv) the percentage of candidates who scored more than 65 %.

c) What is the probability that the score of a candidate chosen at random is
   i) 40 or less
   ii) from 41 to 60 inclusive ?

6. The ages, in completed years, of the forty applicants for a teaching post are given in the following table.

Age (years)	21–24	25–28	29–32	33–36	37–44
Frequency	3	4	8	12	13

(Note that an applicant aged 36 years and 8 months belongs to the 33–36 group.)

a) Draw up a suitable cumulative frequency table and hence draw a cumulative frequency curve. Use 2 cm to represent 2 years on the horizontal axis and 2 cm to represent 5 on the cumulative frequency axis.

b) Use your graph to find the interquartile range.

c) What is the probability that an applicant drawn at random
   i) belongs to the 29–32 group
   ii) is 33 years of age or older ?

d) If two applicants are selected at random from all the applicants, what is the probability that
   i) both come from the 29–32 age group ?
   ii) the first comes from the 29–32 age group and the second from the 37–44 age group ?

16 SINE AND COSINE FORMULAE

SINES OF OBTUSE ANGLES

In earlier work we have used sines of acute angles. Now, with the help of a calculator, we shall investigate the sines of angles from $90°$ to $180°$.

EXERCISE 16a

1. a) Copy the following table.

x	0	$15°$	$30°$	$45°$	$60°$	$75°$	$90°$
$\sin x$							

x	$105°$	$120°$	$135°$	$150°$	$165°$	$180°$
$\sin x$						

Complete the table, using a calculator to find $\sin x$ for each angle in the table, correcting the value to 2 d.p.

b) Using scales of 1 cm for $15°$ on the horizontal axis and 1 cm for 0.2 on the vertical axis, plot these points on a graph and draw a smooth curve through the points.

c) This curve has a line of symmetry. About which value of x is the curve symmetrical ?

d) From your graph, find the two angles (i.e. values of x) for which
i) $\sin x = 0.8$ ii) $\sin x = 0.6$ iii) $\sin x = 0.4$.   Find, in each case, a relationship between the two angles.

2. Use a calculator to complete the following statements.

a) $\sin 30° = \boxed{}$, $\sin 150° = \boxed{}$, $150° = \boxed{} - 30°$

b) $\sin 40° = \boxed{}$, $\sin 140° = \boxed{}$, $140° = \boxed{} - 40°$

c) $\sin 72° = \boxed{}$, $\sin 108° = \boxed{}$, $108° = \boxed{} - 72°$

These results suggest that, if two angles are supplementary (i.e. add up to $180°$), their sines are the same, i.e.

$$\sin x = \sin(180° - x)$$

323

EXERCISE 16b

Find sin x.

$$\sin x = \sin(180° - x)$$
$$= \sin A\widehat{B}C$$
$$= \frac{3}{5}$$

In questions 1 to 6, find sin x.

1.

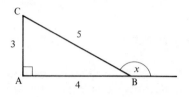

2.

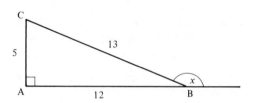

3.

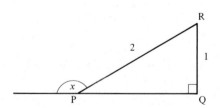

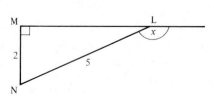

4.

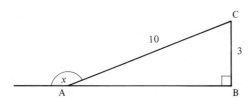

5.

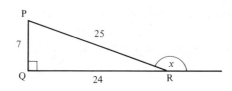

6.

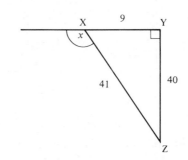

In questions 7 to 12 find the acute angle x.

7. $\sin x = \sin 165°$

8. $\sin x = \sin 140°$

9. $\sin x = \sin 152°$

10. $\sin x = \sin 100°$

11. $\sin x = \sin 175°$

12. $\sin x = \sin 91°$

COSINES OF OBTUSE ANGLES

EXERCISE 16c

1.   a) Copy and complete the following table, using a calculator to find $\cos x$ for each angle in the table, correcting the value to 2 d.p.

x	0	15°	30°	45°	60°	75°	90°
$\cos x$							

x	105°	120°	135°	150°	165°	180°
$\cos x$						

b) Using scales of 1 cm for 15° on the horizontal axis and 1 cm for 0.2 on the vertical axis, plot these points on a graph and draw a smooth curve through them.

c) This curve has a point of rotational symmetry. What is the value of x at this point ?

d) What do you notice about the sign of cosines of obtuse angles ?

e) From your graph find the angles for which
   i) $\cos x = 0.8$ ii) $\cos x = -0.8$
   What is the relationship between these two angles ?

2. Use a calculator to complete the following statements.

a) $\cos 30° = \boxed{}$, $\cos 150° = \boxed{}$, $150° = \boxed{} - 30°$

b) $\cos 50° = \boxed{}$, $\cos 130° = \boxed{}$, $130° = \boxed{} - 50°$

c) $\cos 84° = \boxed{}$, $\cos 96° = \boxed{}$, $96° = \boxed{} - 84°$

These results suggest that

1. the cosine of an obtuse angle is negative

2. the numerical value (i.e. ignoring the sign) of the cosines of supplementary angles is the same, i.e.

$$\cos x = -\cos(180° - x)$$

EXERCISE 16d

Find $\cos x$.

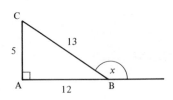

$$\cos x = -\cos(180° - x)$$

$$= -\cos A\widehat{B}C$$

$$= -\frac{12}{13}$$

In questions 1 to 4, find $\cos x$.

1.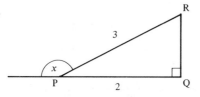

3.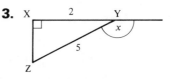

2.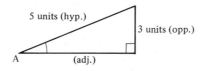

4.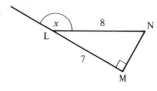

5. Find $\widehat{A}$ if   a) $\cos A = -\cos 20°$ b) $\cos A = -\cos 50°$.

TRIGONOMETRIC RATIOS AS FRACTIONS

For an angle A in a right-angled triangle, if one of $\sin A$, $\cos A$ or $\tan A$ is given as a fraction, we can draw a right-angled triangle and mark in the lengths of two sides.

For example, if $\sin A = \dfrac{3}{5}$, this triangle can be drawn.

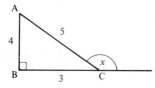

Then, using Pythagoras' Theorem, the length of the third side can be calculated. In this case it is of length 4 units.

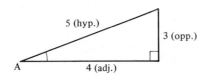

Now the cosine of angle A can be written down as a fraction,

i.e. $$\cos A = \frac{4}{5}$$

Similarly $$\tan A = \frac{3}{4}$$

EXERCISE 16e

If $\cos P = \dfrac{12}{13}$, draw a suitable right-angled triangle and hence find $\sin P$ and $\tan P$.

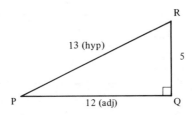

QR $= 5$ (Pythag. th., or recognising a 5, 12, 13 triangle.)

Therefore $\sin P = \dfrac{5}{13}$

and $\tan P = \dfrac{5}{12}$

---

1. If $\sin A = \dfrac{7}{25}$ find $\cos A$ and $\tan A$.

2. If $\cos A = \dfrac{5}{13}$ find $\sin A$ and $\tan A$.

3. If $\tan P = \dfrac{3}{4}$ find $\sin P$ and $\cos P$.

4. If $\cos D = \dfrac{3}{5}$ find $\tan D$ and $\sin D$.

5. If $\sin X = \dfrac{9}{41}$ find $\cos X$ and $\tan X$.

6. If $\tan A = 1$ find $\sin A$ and $\cos A$.

(Remember that $1 = \dfrac{1}{1}$ and leave the square root in your answer.)

7. If $\sin A = \dfrac{12}{13}$ find $2 \sin A \cos A$ expressing your result as a decimal.

Find $\widehat{A}$ and hence $\sin 2A$. What conclusion can you draw ?

8. If $\cos A = \dfrac{4}{5}$ find $\cos^2 A - \sin^2 A$ expressing your result as a decimal.

Find $\widehat{A}$ and hence $\cos 2A$. What conclusion can you draw?

9. If $\sin X = \dfrac{x}{y}$ find $\cos^2 X + \sin^2 X$.

10. If $\sin A = \dfrac{11}{61}$ find $\dfrac{2 \tan A}{1 - \tan^2 A}$.

MIXED QUESTIONS

EXERCISE 16f

1.

Write down, as a fraction,
a) $\cos x$ b) $\sin x$

2. Find two angles each with a sine of 0.5

3. If $\cos 59° = 0.515$ what is $\cos 121°$?

4. Find an obtuse angle whose cosine is equal to $-\cos 72°$.

5. If $\sin x = \frac{4}{5}$, find as a fraction
a) $\sin(180° - x)$ b) $\cos x$ c) $\cos(180° - x)$

6. Use your calculator to make a table of values of $\sin x$ for values of x from $0°$ to $720°$ at intervals of $45°$.

Taking scales of 1 cm to $90°$ on the x-axis and 1 cm to 1 unit on the y-axis, use the table of values, together with the knowledge of the shape drawn for $y = \sin x$ from Exercise 16a, to draw the graph of $y = \sin x$ for $0°$ to $720°$.

7. Follow the sequence of steps in question 6 to draw the graph of $y = \cos x$ for $0°$ to $720°$.

TRIANGLE NOTATION

In a triangle ABC, the sides can be referred to as AB, BC and CA. It is convenient, however, to use a single letter to denote the number of units in the length of a side and the standard notation uses

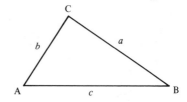

a for the side opposite to angle A
b for the side opposite to angle B
c for the side opposite to angle C.

THE SINE RULE

Consider a triangle ABC in which there is no right angle.

If a line is drawn from C, perpendicular to AB, the original triangle is divided into two right-angled triangles ADC and BDC as shown.

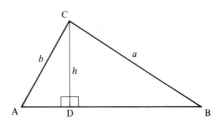

In $\triangle$ADC $\qquad$ $\sin A = \dfrac{h}{b}$ $\;\Rightarrow\;$ $h = b \sin A$

In $\triangle$BDC $\qquad$ $\sin B = \dfrac{h}{a}$ $\;\Rightarrow\;$ $h = a \sin B$

Equating the two expressions for h gives

$$a \sin B = b \sin A$$

Hence $\qquad \dfrac{a}{\sin A} = \dfrac{b}{\sin B}$ $\qquad$ (dividing both sides by $\sin A \sin B$)

Now if we were to divide $\triangle$ABC into two right-angled triangles by drawing the perpendicular from B to AC the similar result would be

$$\frac{b}{\sin B} = \frac{c}{\sin C}$$

Combining the two results gives

$$\frac{a}{\sin A} = \frac{b}{\sin B} = \frac{c}{\sin C}$$

This result is called the *sine rule* and it enables us to find angles and sides of triangles which are *not* right-angled.

It is made up of three equal fractions, but only two of them can be used at a time. When using the sine rule we choose the two fractions in which three quantities are known and only one is unknown.

In all the following exercises, unless a different instruction is given express angles correct to 1 d.p. and lengths to 3 s.f.

EXERCISE 16g

In $\triangle ABC$, $AC = 3\,\text{cm}$, $\widehat{B} = 40°$ and $\widehat{A} = 65°$.
Find the length of BC.

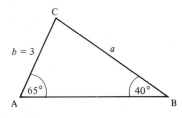

$$\frac{a}{\sin A} = \frac{b}{\sin B} = \frac{c}{\sin C}$$

(The four quantities involved are a, b, $\widehat{A}$ and $\widehat{B}$ so we use the first two fractions in the sine rule.)

$$\frac{a}{\sin 65°} = \frac{3}{\sin 40°}$$

$$a = \frac{3 \sin 65°}{\sin 40°}$$

$$= \frac{3 \times 0.9063}{0.6428}$$

$$= 4.2297$$

Therefore $BC = 4.23\,\text{cm}$ correct to 3 s.f.

1. In $\triangle ABC$, $AB = 7\,cm$, $\widehat{A} = 32°$ and $\widehat{C} = 49°$. Find BC.

2. In $\triangle PQR$, $\widehat{Q} = 56°$, $\widehat{R} = 72°$ and $PR = 11.3\,cm$. Find PQ.

3. In $\triangle XYZ$, $\widehat{X} = 28°$, $YZ = 4.9\,cm$ and $\widehat{Y} = 69°$. Find XZ.

4. In $\triangle ABC$, $BC = 8.3\,cm$, $\widehat{A} = 61°$ and $\widehat{C} = 58°$. Find AB.

5. In $\triangle LMN$, $LM = 17.7\,cm$, $\widehat{N} = 73°$ and $\widehat{L} = 52°$. Find MN.

6. In $\triangle PQR$, $\widehat{P} = 54°$, $\widehat{Q} = 83°$ and $QR = 7.9\,cm$. Find PR.

7. In $\triangle ABC$, $BC = 161\,cm$, $\widehat{A} = 41°$ and $\widehat{B} = 76°$. Find AC.

8. In $\triangle DEF$, $EF = 123\,cm$, $\widehat{D} = 66°$ and $\widehat{E} = 51°$. Find DF.

The sine rule can be used when one angle is obtuse.

In $\triangle ABC$, $AB = 6\,cm$, $\widehat{B} = 25°$ and $\widehat{C} = 110°$. Find AC.

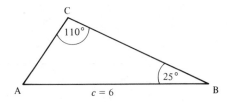

From the sine rule $\quad \dfrac{a}{\sin A} = \boxed{\dfrac{b}{\sin B} = \dfrac{c}{\sin C}}$

$$\frac{b}{\sin 25°} = \frac{6}{\sin 110°}$$

$$b = \frac{6\sin 25°}{\sin 110°}$$

$$= \frac{6 \times 0.4226}{0.9397}$$

$$= 2.6983$$

Therefore $AC = 2.70\,cm$ correct to 3 s.f.

9. In △ABC, $\widehat{A} = 124°$, $\widehat{B} = 31°$ and AC $= 41$ cm. Find BC.

10. In △PQR, $\widehat{Q} = 43°$, $\widehat{R} = 106°$ and PQ $= 7.9$ cm. Find PR.

11. In △LMN, LM $= 831$ cm, $\widehat{L} = 27°$ and $\widehat{N} = 114°$. Find MN.

12. In △DEF, EF $= 51$ cm, $\widehat{D} = 52°$ and $\widehat{F} = 98°$. Find DE.

13. In △ABC, AC $= 62.7$ cm, $\widehat{B} = 122°$ and $\widehat{C} = 35°$. Find AB.

Sometimes the third angle of a triangle must be found before two suitable fractions can be selected from the sine rule.

In △ABC, AC $= 4$ cm, $\widehat{A} = 35°$ and $\widehat{C} = 70°$. Find BC.

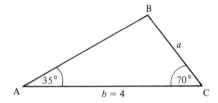

(The two sides involved are a and b so we must use $\widehat{A}$ and $\widehat{B}$ in the sine rule.)

First find $\widehat{B}$

$$\widehat{B} = 180° - 35° - 70° = 75°$$

From the sine rule $\boxed{\dfrac{a}{\sin A} = \dfrac{b}{\sin B} = \dfrac{c}{\sin C}}$

$$\frac{a}{\sin 35°} = \frac{4}{\sin 75°}$$

$$a = \frac{4\sin 35°}{\sin 75°}$$

$$= \frac{4 \times 0.5736}{0.9659}$$

$$= 2.3754$$

Therefore BC $= 2.38$ cm correct to 3 s.f.

14. In △ABC, $\widehat{B} = 81°$, $\widehat{C} = 63°$ and BC = 13 cm. Find AB.

15. In △PQR, PQ = 15.3 cm, $\widehat{P} = 106°$ and $\widehat{Q} = 21°$. Find QR.

16. In △LMN, LN = 108 cm, $\widehat{L} = 59°$ and $\widehat{N} = 44°$. Find LM.

17. In △PQR, $\widehat{P} = 61°$, $\widehat{R} = 102°$ and PR = 67 cm. Find PQ.

18. In △ABC, $\widehat{A} = 38°$, AC = 237 cm and $\widehat{C} = 94°$. Find BC.

The sine rule can be used twice if the lengths of two sides are required.

In △ABC, BC = 71 cm, $\widehat{A} = 62°$ and $\widehat{B} = 54°$. Find AB and AC.

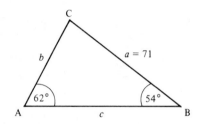

(Because $\widehat{A}$, $\widehat{B}$ and a are given we will first find b.)

From the sine rule $\dfrac{b}{\sin B} = \dfrac{a}{\sin A}$

$$\frac{b}{\sin 54°} = \frac{71}{\sin 62°}$$

$$b = \frac{71 \sin 54°}{\sin 62°}$$

$$= \frac{71 \times 0.8090}{0.8829}$$

$$= 65.05$$

Therefore AC = 65.1 cm correct to 3 s.f.

(Now to find c we need $\widehat{C}$.)

$$\widehat{C} = 180° - 62° - 54° = 64°$$

From the sine rule

$$\frac{c}{\sin C} = \frac{a}{\sin A}$$

$$\frac{c}{\sin 64°} = \frac{71}{\sin 62°}$$

$$c = \frac{71 \sin 64°}{\sin 62°}$$

$$= \frac{71 \times 0.8988}{0.8829}$$

$$= 72.27$$

Therefore AB $= 72.3$ cm correct to 3 s.f.

(Note that, to find c we equate $\dfrac{c}{\sin C}$ to $\dfrac{a}{\sin A}$ and not to $\dfrac{b}{\sin B}$. This is because a is given in the question, whereas b was calculated in the first section, and its value might not be correct.)

---

19. In $\triangle PQR$, $\widehat{P} = 61°$, $\widehat{Q} = 54°$ and PR $= 7$ cm. Find QR and PQ.

20. In $\triangle ABC$, AB $= 13$ cm, $\widehat{B} = 40°$ and $\widehat{C} = 67°$. Find BC and CA.

21. In $\triangle PQR$, $r = 191$, $\widehat{P} = 37°$ and $\widehat{Q} = 64°$. Find p and q.

22. If $\widehat{A} = 49°$, $\widehat{C} = 82°$ and $b = 27$, find a and c in $\triangle ABC$.

Complete the following table, which refers to a triangle ABC.

	AB	BC	AC	$\widehat{A}$	$\widehat{B}$	$\widehat{C}$
23.		19 cm			53°	76°
24.			146 cm	72°		69°
25.	81 cm			37°	59°	
26.			97 cm		48°	61°
27.	12 cm			54°	102°	
28.			9.9 cm	26°		121°

THE COSINE RULE

It is not always possible to use the sine rule to find the unknown facts about a triangle. For instance, if the three sides of $\triangle ABC$ are given, but no angles are known, it is impossible to select two equal fractions from the sine rule so that only one unknown quantity is involved.

$$\frac{a}{\text{unknown}} = \frac{b}{\text{unknown}} = \frac{c}{\text{unknown}}$$

For cases like this, where the sine rule fails, we need a different formula.

Consider a triangle ABC divided into two right-angled triangles by a line BD, perpendicular to AC. Let the length of AD be x so that $DC = b - x$.

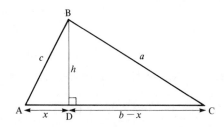

Using Pythagoras' Theorem in triangles ABD and CBD gives

$$c^2 = h^2 + x^2$$

and

$$a^2 = h^2 + (b-x)^2$$
$$= h^2 + b^2 - 2bx + x^2$$
$$= b^2 + (h^2 + x^2) - 2bx$$
$$= b^2 + c^2 - 2bx$$

In $\triangle ABD$, $x = c \cos A$. Therefore

$$a^2 = b^2 + c^2 - 2bc \cos A$$

This result is called the *cosine rule*.

If we were to draw a line from A perpendicular to BC, or from C perpendicular to AB, similar equations would be obtained, i.e.

$$b^2 = c^2 + a^2 - 2ca \cos B$$
$$c^2 = a^2 + b^2 - 2ab \cos C$$

Note that, in each version of the cosine rule, the side on the left and the angle on the right have the same letter.

When using the cosine rule it is a good idea to put brackets round the term containing the angle

i.e.

$$a^2 = b^2 + c^2 - (2bc \cos A)$$

EXERCISE 16h

In a triangle ABC, AB = 6 cm, BC = 7 cm and $\widehat{B} = 60°$. Find AC.

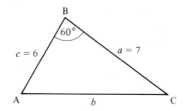

(The unknown side is b, and $\widehat{B}$ is known, so we use the version of the cosine rule that starts with b^2.)

$$b^2 = c^2 + a^2 - (2ca \cos B)$$
$$= 6^2 + 7^2 - (2 \times 6 \times 7 \cos 60°)$$
$$= 36 + 49 - (84 \times 0.5)$$
$$= 85 - 42 = 43$$
$$\Rightarrow \qquad b = \sqrt{43} = 6.557$$

Therefore AC = 6.56 cm correct to 3 s.f.

1. In $\triangle ABC$, $\widehat{A} = 54°$, $b = 7$ and $c = 5$. Find a.

2. In $\triangle ABC$, $a = 4$, $b = 11$ and $\widehat{C} = 33°$. Find c.

3. In $\triangle PQR$, PQ = 5 cm, QR = 8 cm and $\widehat{Q} = 68°$. Find PR.

4. In $\triangle LMN$, LM = 26 cm, $\widehat{M} = 45°$ and MN = 17 cm. Find LN.

5. In $\triangle PQR$, $\widehat{Q} = 51°$, $p = 9$ and $r = 5$. Find q.

6. In $\triangle ABC$, $c = 12$, $b = 13$ and $\widehat{A} = 69°$. Find a.

7. In $\triangle LMN$, $\widehat{L} = 37.6°$, $m = 21.7$ and $n = 13.8$. Find l.

8. In $\triangle DEF$, EF = 12.4 cm, DF = 9.7 cm and $\widehat{F} = 58.6°$. Find DE.

9. In $\triangle PQR$, PR = 8.9 cm, PQ = 11.2 cm and $\widehat{P} = 47.1°$. Find QR.

10. In $\triangle ABC$, $a = 40.3$, $\widehat{B} = 31.6°$ and $c = 62.6$. Find b.

If the given angle is obtuse its cosine is negative and extra care is needed in using the cosine rule; brackets are even more helpful in this case.

Triangle ABC is such that BC = 11 cm, AC = 8 cm and $\widehat{C}$ = 130°. Find AB.

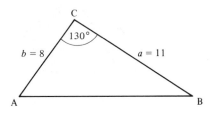

Using the cosine rule

$$c^2 = a^2 + b^2 - (2ab \cos C)$$
$$= 11^2 + 8^2 - (2 \times 11 \times 8 \cos 130°)$$
$$= 121 + 64 - (176 \times [-0.6428])$$
$$= 185 + 113.13$$
$$= 298.13$$
$$\Rightarrow \qquad c = \sqrt{298.13}$$
$$= 17.27$$

Therefore AB = 17.3 cm correct to 3 s.f.

11. In △ABC, AB = 8 cm, AC = 11 cm and $\widehat{A}$ = 113°. Find BC.

12. In △PQR, QR = 14 cm, PR = 10 cm and $\widehat{R}$ = 128°. Find PQ.

13. In △LMN, LM = 35 cm, MN = 21 cm and $\widehat{M}$ = 97°. Find LN.

14. In △ABC, $a = 19$, $b = 12$ and $\widehat{C}$ = 136°. Find c.

15. In △PQR, $q = 2.4$, $r = 3.7$ and $\widehat{P}$ = 100°. Find p.

16. In △DEF, $d = 9$, $f = 12$ and $\widehat{E}$ = 106°. Find e.

17. In △XYZ, XY = 47 cm, XZ = 81 cm and $\widehat{X}$ = 94°. Find YZ.

18. In △ABC, AC = 6.4 cm, BC = 3.9 cm and $\widehat{C}$ = 121°. Find AB.

The cosine rule can be used to find an unknown angle when three sides of a triangle are given.

In △ABC, AB = 5 cm, BC = 6 cm and CA = 8 cm. Find
a) the smallest angle   b) the largest angle,   in the triangle.

a)

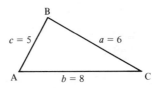

(The smallest angle is opposite to the shortest side, so we are looking for $\widehat{C}$ and will use the cosine rule starting with c^2.)

$$c^2 = a^2 + b^2 - (2ab \cos C)$$
$$5^2 = 6^2 + 8^2 - (2 \times 6 \times 8 \cos C)$$
$$25 = 36 + 64 - (96 \cos C)$$
$$96 \cos C + 25 = 100 \quad \text{(adding 96 cos C to each side)}$$
$$96 \cos C = 75 \quad \text{(subtracting 25 from each side)}$$
$$\cos C = \frac{75}{96} = 0.7813$$

Therefore $\widehat{C} = 38.6°$ to 1 d.p.

b) (The largest angle is opposite to the longest side, so we are looking for $\widehat{B}$ and will use the cosine rule starting with b^2.)

$$b^2 = c^2 + a^2 - (2ca \cos B)$$
$$8^2 = 5^2 + 6^2 - (2 \times 5 \times 6 \cos B)$$
$$64 = 25 + 36 - (60 \cos B)$$
$$60 \cos B = 61 - 64 = -3$$
$$\cos B = \frac{-3}{60} = -0.05$$

(Because cos B is negative $\widehat{B}$ must be an obtuse angle.)

Therefore $\widehat{B} = 92.9°$ to 1 d.p.

19. In $\triangle ABC$, $AB = 8\,cm$, $BC = 4\,cm$ and $AC = 5\,cm$. Find $\widehat{A}$.

20. In $\triangle PQR$, $PQ = 4\,cm$, $QR = 2\,cm$ and $PR = 5\,cm$. Find $\widehat{Q}$.

21. In $\triangle LMN$, $LM = 8\,cm$, $MN = 5\,cm$ and $LN = 6\,cm$. Find $\widehat{N}$.

22. In $\triangle ABC$, $AB = 3\,cm$, $BC = 2\,cm$ and $AC = 4\,cm$. Find the smallest angle in $\triangle ABC$.

23. In $\triangle XYZ$, $XY = 7\,cm$, $XZ = 9\,cm$ and $YZ = 5\,cm$. Find the largest angle in $\triangle XYZ$.

24. In $\triangle DEF$, $DE = 2.1\,cm$, $EF = 3.6\,cm$ and $DF = 2.7\,cm$. Find the middle-sized angle in $\triangle DEF$.

25.

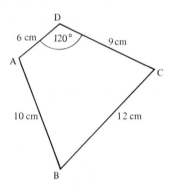

Use the information given in the diagram to find

a) the length of AC   b) $A\widehat{B}C$

MIXED QUESTIONS

If three independent facts are given about the sides and/or angles of a triangle and we are asked to find one or more of the unknown quantities, we must first decide whether to use the sine rule or the cosine rule.

The sine rule is the easier to work out so it is chosen whenever possible and that is when the given information includes a side and the angle opposite to it (remember that, if the two angles are given the third angle is also known). The cosine rule is chosen only when the sine rule cannot be used.

In some questions we are asked to find *all* the remaining information about a triangle. This is called *solving* the triangle.

EXERCISE 16i

In a triangle LMN, LM = 9 cm, MN = 11 cm and $\widehat{M} = 70°$. Find LN.

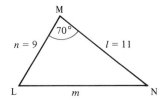

(As we are not given an angle and the side opposite to it, the sine rule cannot be used so we use the cosine rule.)

$$m^2 = l^2 + n^2 - (2ln \cos M)$$

$$= 11^2 + 9^2 - (2 \times 11 \times 9 \cos 70°)$$

$$= 121 + 81 - (198 \times 0.3420)$$

$$= 202 - 67.72 = 134.28$$

$$\Rightarrow \qquad m = \sqrt{134.28} = 11.59$$

Therefore LN = 11.6 cm correct to 3 s.f.

Fill in the blank spaces in the table.

	a	b	c	$\widehat{A}$	$\widehat{B}$	$\widehat{C}$
1.	11.7		///	39°	66°	///
2.		128	86	63°		///
3.	///		65	///	79°	55°
4.	16.3	12.7		///	///	106°
5.		263		///	47°	74°
6.	14			53°	82°	
7.		///	17.8	107°		35°
8.		16	23	81°	///	///
9.	13.2		19.6		120°	///
10.		22.6			50°	83°

Solve $\triangle PQR$ given that $PQ = 11\,\text{cm}$, $\hat{R} = 82°$ and $P = 47°$.

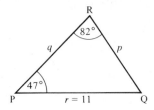

(We are given r and $\hat{R}$ so we can use the sine rule to find p.)

From the sine rule $\dfrac{p}{\sin P} = \dfrac{r}{\sin R}$

$$\frac{p}{\sin 47°} = \frac{11}{\sin 82°}$$

$$p = \frac{11 \sin 47°}{\sin 82°}$$

$$= \frac{11 \times 0.7314}{0.9903} = 8.124$$

Therefore $QR = 8.12\,\text{cm}$ correct to 3 s.f.

(Now to find q we need $\hat{Q}$ in order to use the sine rule.)

$$\hat{Q} = 180° - 47° - 82° = 51°$$

From the sine rule $\dfrac{q}{\sin Q} = \dfrac{r}{\sin R}$

$$\frac{q}{\sin 51°} = \frac{11}{\sin 82°}$$

$$q = \frac{11 \sin 51°}{\sin 82°}$$

$$= \frac{11 \times 0.7771}{0.9903} = 8.632$$

Therefore $PR = 8.63\,\text{cm}$ correct to 3 s.f.

(All sides and angles have now been found, so the triangle has been solved.)

11. Solve $\triangle ABC$ given that $a = 8.4$, $\widehat{A} = 52°$, $\widehat{B} = 74°$.

12. Solve $\triangle PQR$ given that $p = 12.6$, $q = 7.4$, $r = 16.3$.

13. Solve $\triangle LMN$ given that $m = 15$, $n = 18$, $\widehat{L} = 64°$.

14. Solve $\triangle DEF$ given that $d = 27$, $e = 19$, $f = 34$.

THE AREA OF A TRIANGLE

We already know that the area, A, of a triangle can be found by multiplying half the base, b, by the perpendicular height, h, i.e.

$$A = \tfrac{1}{2}bh$$

In some cases, however, the perpendicular height is not given, so an alternative formula is needed.

Consider a triangle ABC in which the lengths of BC and CA, and the angle C, are known.

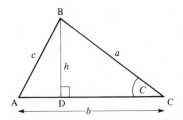

The line BD, drawn from B perpendicular to AC, is a perpendicular height of the triangle. Therefore the area of the triangle is $\tfrac{1}{2}bh$.

But, in triangle BDC, $\sin C = \dfrac{h}{a}$ i.e. $h = a \sin C$

Therefore the area of the triangle is $\tfrac{1}{2}ba \sin C$

i.e. $A = \tfrac{1}{2}ab \sin C$

If we draw perpendicular heights from A and from C, similar expressions for the area are obtained

i.e. $A = \tfrac{1}{2}bc \sin A$ and $A = \tfrac{1}{2}ac \sin B$

Note that in each of the these expressions for the area, two sides and the included angle are involved, i.e. in general

$$\text{Area} = \tfrac{1}{2} \times \text{product of two sides} \times \text{sine of included angle}$$

EXERCISE 16j

Find the area of $\triangle PQR$ if $\widehat{P} = 120°$, $PQ = 132$ cm and $PR = 95$ cm.

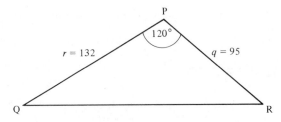

(r and q are given, and $\widehat{P}$ is the included angle.)

$$A = \tfrac{1}{2}qr\sin P$$

$$= \tfrac{1}{2} \times 95 \times 132 \times \sin 120°$$

$$= \tfrac{1}{2} \times 95 \times 132 \times 0.8660$$

$$= 5429$$

The area of $\triangle PQR = 5430$ cm^2 correct to 3 s.f.

Find the area of each of the following triangles.

1. $\triangle ABC$; $AC = 11.6$ cm, $BC = 14.2$ cm, $\widehat{C} = 80°$

2. $\triangle PQR$; $p = 217$, $q = 196$, $\widehat{R} = 117°$

3. $\triangle XYZ$; $XY = 81$ cm, $XZ = 69$ cm, $\widehat{X} = 69°$

4. $\triangle LMN$; $m = 29.3$, $n = 40.6$, $\widehat{L} = 74°$

5. $\triangle ABC$; $a = 18.1$, $c = 14.2$, $\widehat{B} = 101°$

6. $\triangle PQR$; $PQ = 234$ cm, $PR = 196$ cm, $\widehat{P} = 84°$

7. $\triangle XYZ$; $x = 9$, $z = 10$, $\widehat{Y} = 52°$

8. $\triangle ABC$; $AC = 3.7$ m, $AB = 4.1$ m, $\widehat{A} = 116°$

9. $\triangle DEF$; $EF = 72$ cm, $DF = 58$ cm, $\widehat{F} = 76°$

10. $\triangle PQR$; $q = 20.3$, $p = 16.7$, $\widehat{R} = 61°$

In triangle ABC, AB = 15 cm, $\widehat{A} = 60°$ and $\widehat{B} = 81°$. Find
a) BC   b) the area of △ABC.

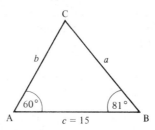

a) (In order to use the sine rule to find a we need $\widehat{C}$.)

$$\widehat{C} = 180° - 60° - 81° = 39°$$

From the sine rule, $\boxed{\dfrac{a}{\sin A}} = \dfrac{b}{\sin B} \boxed{= \dfrac{c}{\sin C}}$

$$\frac{a}{\sin 60°} = \frac{15}{\sin 39°}$$

$$a = \frac{15 \sin 60°}{\sin 39°}$$

$$= \frac{15 \times 0.8660}{0.6293} = 20.64$$

Therefore BC = 20.6 cm correct to 3 s.f.

b) (We know two sides, a and c, and the included angle is $\widehat{B}$.)

$$A = \tfrac{1}{2}ac \sin B$$

$$= \tfrac{1}{2} \times 20.64 \times 15 \sin 81°$$

$$= \tfrac{1}{2} \times 20.64 \times 15 \times 0.9877 = 152.8$$

Therefore the area of △ABC is 153 cm² correct to 3 s.f.

11. In △ABC, BC = 7 cm, AC = 8 cm, AB = 10 cm. Find $\widehat{C}$ and the area of the triangle.

12. △PQR is such that PQ = 11.7 cm, $\widehat{Q} = 49°$ and $\widehat{R} = 63°$. Find PR and the area of △PQR.

13. In △LMN, LM = 16 cm, MN = 19 cm and the area is 114.5 cm². Find $\widehat{M}$ and LN.

14. The area of △ABC is 27.3 cm². If BC = 12.8 cm and $\widehat{C} = 107°$ find AC.

PROBLEMS

In many problems a description of a situation is given which can be illustrated by a diagram. Our aim is to find, in this diagram, a triangle in which three facts about sides and/or angles are known. A second diagram, showing only this triangle, can then be drawn and the appropriate rules of trigonometry applied to it.

EXERCISE 16k

From a port P a ship Q is 20 km away on a bearing of 125 ° and a ship R is 35 km away on a bearing of 050 °. Find the distance between the two ships.

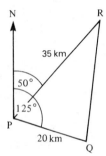 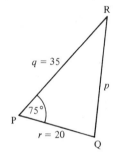

(PQR is a suitable triangle since PR and PQ are known and $Q\hat{P}R$ can be found.)

$$Q\hat{P}R = 125° - 50° = 75°$$

(The distance between the ships is QR, so we must calculate p in $\triangle PQR$.)

In $\triangle PQR$, using the cosine rule gives

$$p^2 = q^2 + r^2 - (2qr\cos P)$$
$$= 35^2 + 20^2 - (2 \times 35 \times 20 \times \cos 75°)$$
$$= 1225 + 400 - (1400 \times 0.2588)$$
$$= 1625 - 362.3 = 1262.7$$

$\Rightarrow \qquad p = 35.53$

Therefore the distance between the ships is 35.5 km correct to 3 s.f.

In a quadrilateral ABCD, AB = 5 cm, BC = 6 cm, CD = 7 cm, $\widehat{ABC} = 120°$ and $\widehat{ACD} = 90°$. Find
a) the length of the diagonal AC   b) the area of $\triangle ABC$
c) the area of $\triangle ADC$ d) the area of the quadrilateral.

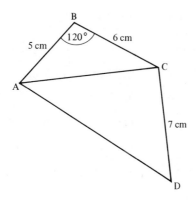
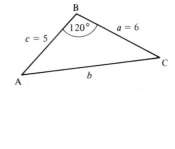

a) In $\triangle ABC$, using the cosine rule gives

$$b^2 = c^2 + a^2 - (2ca \cos B)$$
$$= 5^2 + 6^2 - (2 \times 5 \times 6 \cos 120°)$$
$$= 25 + 36 - (60 \times [-0.5])$$
$$= 61 + 30$$
$$= 91$$

$\Rightarrow \qquad\qquad b = \sqrt{91}$
$$= 9.539$$

Therefore AC = 9.54 cm correct to 3 s.f.

b) For $\triangle ABC$, $\qquad A = \frac{1}{2}ac \sin B$
$$= \frac{1}{2} \times 5 \times 6 \times \sin 120°$$
$$= 15 \times 0.8660$$
$$= 12.99$$

Area $\triangle ABC = 13.0 \text{ cm}^2$ correct to 3 s.f.

c) ($\triangle ADC$ is a right-angled triangle, so we use $\frac{1}{2}$ base × perpendicular height for its area.)

$$A = \tfrac{1}{2}bh$$

$$= \tfrac{1}{2} \times AC \times CD$$

$$= \tfrac{1}{2} \times 9.539 \times 7$$

$$= 33.38$$

Area $\triangle ADC = 33.4\,cm^2$ correct to 3 s.f.

d) Area quadrilateral $ABCD$ = area $\triangle ABC$ + area $\triangle ADC$

$$= 12.99\,cm^2 + 33.38\,cm^2$$

$$= 46.37\,cm^2$$

$$= 46.4\,cm^2$$ correct to 3 s.f.

1. Starting from a point A, an aeroplane flies for 40 km on a bearing of 169° to B, and then for 65 km on a bearing of 057° to C. Find the distance between A and C.

2.

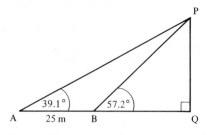

From two points A and B, on level ground, the angles of elevation of the top of a radio aerial PQ are found to be 39.1° and 57.2°. If the distance between A and B is 25 m, find

a) $A\widehat{P}B$

b) the height of the radio aerial.

3. A children's slide has a flight of steps of length 2.7 m and the length of the straight slide is 4.2 m. If the distance from the bottom of the steps to the bottom of the slide is 4.9 m find, to the nearest degree, the angle between the steps and the slide.

4.

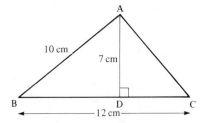

Using the information given in the diagram

a) find the area of $\triangle ABC$

b) use the formula   area $\triangle ABC = \frac{1}{2}ac \sin B$ to find $\widehat{B}$

c) find AC.

5.   A helicopter leaves a heliport A and flies 2.4 km on a bearing of 154° to a checkpoint B. It then flies due east to its base C. If the bearing of C from A is 112° find the distances AC and BC. The helicopter flies at a constant speed throughout and takes 5 minutes to fly from A to C. Find its speed.

6.   P, Q and R are three points on level ground. From Q, P is 60 m away on a bearing of 325° and R is 94 m away on a bearing of 040°.

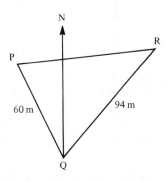

a) Find $\cos Q$.

b) Using the cosine rule, or otherwise, find the distance between P and R.

7.   ABCD is a quadrilateral in which AB = 4.1 cm, BC = 3.7 cm, CD = 5.3 cm, $\widehat{ABC} = 66°$ and $\widehat{ADC} = 51°$. Find

a) the length of the diagonal AC

b) $\widehat{CAD}$

c) the area of quadrilateral ABCD, considering it as split into two triangles by the diagonal AC.

8.

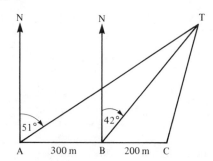

The diagram shows three survey points, A, B and C, which are on an east–west line on level ground. From point A the bearing of the foot of a tower T is 051°, while from B the bearing of the tower is 042°. Find

a) $\widehat{TAB}$ and $\widehat{ATB}$ b) AT   c) CT.

9.

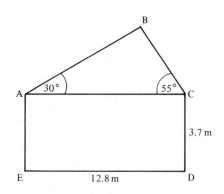

The diagram shows the end wall of a bungalow. The roof has one face inclined at 30° to the horizontal and the other inclined at 55° to the horizontal. Find

a) the length of AB   b) the length of BC
c) the area of the end wall   d) the height of B above ED.

10.   In a quadrilateral ABCD, DC is of length 3 cm, the length of the diagonal BD is 10 cm, $\widehat{BAD} = 30°$, $\widehat{BDA} = 45°$ and $\widehat{BDC} = 60°$. Calculate
a) the length of (i) AB (ii) BC   b) the area of the quadrilateral.

11.   The points A, B and C are on the circumference of a circle with centre O and radius 10 cm. The lengths of the chords AB and BC are 8 cm and 3 cm respectively. Calculate

a) $\widehat{AOB}$ b) $\widehat{BOC}$ c) the length of the chord AC
d) the area of quadrilateral ABCO.

  (Remember that OA = OB = OC = 10 cm.)

MIXED EXERCISES

EXERCISE 16I

1. If $\cos A = \frac{5}{13}$ and A is acute, find $\sin A$ and $\tan A$.

2. Find two angles between $0°$ and $180°$ for which the sine is 0.3667 (a calculator may be used).

3. If $\cos P = 0.43$ find, without using a calculator, $\cos(180° - P)$.

4. In $\triangle ABC$, $BC = 161\,cm$, $\widehat{B} = 109°$ and $\widehat{A} = 51°$. Find AC.

5. A triangle PQR is such that $QR = 7.6\,cm$, $PQ = 5.9\,cm$ and $\widehat{Q} = 107°$. Find PR.

6. Find the area of $\triangle ABC$ if $AB = 8.2\,cm$, $BC = 11.3\,cm$ and $\widehat{B} = 125°$.

7. A boat sails $11\,km$ from a harbour on a bearing of $220°$. It then sails $15\,km$ on a bearing of $340°$. How far is the boat from the harbour?

EXERCISE 16m

1. If $\sin x = \sin 37°$ and x is obtuse, find x.

2. $\cos x = 0.123$ and $\cos y = -0.123$. If x is acute and y is obtuse find y in terms of x.

3. If A is an acute angle, complete the following statements.
a) $\cos(180° - A) = \ldots$ b) $\sin(180° - A) = \ldots$ c) $-\cos A = \ldots$

4. In $\triangle PQR$, $\widehat{Q} = 61°$, $PR = 14.7\,cm$ and $\widehat{R} = 84°$. Find PQ.

5. Find the lengths of AB and BC in $\triangle ABC$ if $AC = 7.3\,cm$, $\widehat{A} = 49°$ and $\widehat{C} = 78°$.

6. In $\triangle LMN$, $MN = 14.2\,cm$, $LN = 17.3\,cm$ and $LM = 11.8\,cm$. Find
a) the smallest angle   b) the largest angle,   in the triangle.

7. A flower bed in a park is in the form of a quadrilateral ABCD in which $AB = 4\,m$, $\widehat{B} = 90°$, $BC = 6.2\,m$, $CD = 7.3\,m$ and $DA = 5\,m$. Find
a) the angle $A\widehat{D}C$ b) the area of the flower bed
c) the amount of peat needed to cover the bed evenly to a depth of $8\,cm$.

17 FUNCTIONS

RELATIONSHIPS

Consider the set $\{-2, -1, 0, 1, 2\}$.

If each member of this set is doubled we get the set $\{-4, -2, 0, 2, 4\}$.

The set that we start with, $\{-2, -1, 0, 1, 2\}$ is called the *domain* of the relation.

The set that results from the rule 'double each member', i.e. $\{-4, -2, 0, 2, 4\}$, is called the *image set* or the *range* of the relationship.

The relationship between the domain and range can be expressed more briefly using symbols as follows

$$x \mapsto 2x \quad \text{for } x \in \{-2, -1, 0, 1, 2\}$$

(the symbol $\mapsto$ means 'maps to')

and can be illustrated on an arrow diagram:

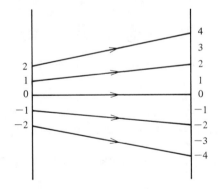

EXERCISE 17a

Give the range of the relation $x \mapsto \pm\sqrt{x}$ for $x \in \{4, 1, 0\}$.

(The domain is $\{4, 1, 0\}$. The positive and negative square roots of each member give the range.)

The range is the set $\{\pm 2, \pm 1, 0\}$, i.e. $\{-2, -1, 0, 1, 2\}$.

Give the range of each of the following relationships.

1. $x \mapsto 4x$ for $x \in \{-1, 0, 1\}$

2. $x \mapsto -x$ for $x \in \{-1, 0, 1, 2\}$

3. $x \mapsto 2x - 1$ for $x \in \{-2, -1, 0, 1, 2\}$

4. $x \mapsto \dfrac{1}{x}$ for $x \in \{\frac{1}{2}, 1, 2, 3\}$

5. $x \mapsto$ 'the largest prime factor of x' for $x \in \{2, 6, 12, 14\}$

6. $x \mapsto \sin x$ for $x \in \{0\,°, 30\,°, 60\,°, 90\,°\}$

Draw an arrow diagram to illustrate the relationship

$$x \mapsto \pm\sqrt{x} \quad \text{for } x \in \{0, 1, 4\}$$

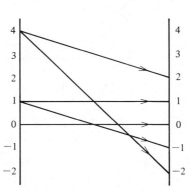

Draw arrow diagrams to illustrate the following relationships.

7. $x \mapsto x^2$ for $x \in \{-1, 0, 1, 2\}$

8. $x \mapsto 2x + 3$ for $x \in \{-2, 0, 2\}$

9. $x \mapsto \cos x$ for $x \in \{0, 30\,°, 60\,°, 90\,°\}$

10. $x \mapsto$ 'the integer from 1 to 6 which has x as its smallest factor' for $x \in \{2, 3\}$

11. $x \mapsto$ 'the largest odd factor of x apart from x itself' for $x \in \{2, 4, 6, 9\}$

Find, in the form $x \mapsto \ldots$, a relationship between the domain and image set shown in each of the following diagrams.

12.

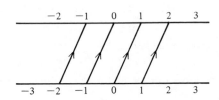

13.

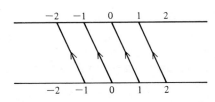

14.

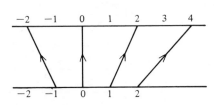

15.

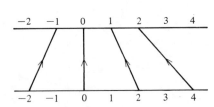

16.

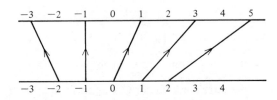

17.

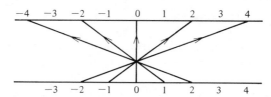

18.

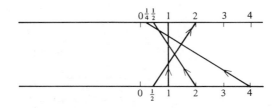

FUNCTIONS

A relationship is called a function when each member of the domain maps to one, and not more than one, member of the range.

Consider for example $x \mapsto x^2$.

Without specifying a domain it is clear that $x \mapsto x^2$ is a function, because for any value of x there is only one corresponding value of x^2.

The arrow diagram illustrates $x \mapsto x^2$ for $x \in \{-1, 0, 1, 2\}$.

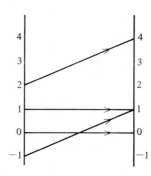

Note that the converse of the condition does not have to be satisfied, i.e. it does not matter if a member of the image set arises from more than one member of the domain.

Now consider $x \mapsto \pm\sqrt{x}$ for $x \geqslant 0$.

Any positive value of x corresponds to two members of the image set, so this relationship is not a function.

The arrow diagram illustrates $x \mapsto \pm\sqrt{x}$ for $x \in \{0, 1, 4\}$.

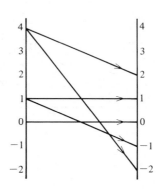

EXERCISE 17b

State whether or not each of the following relationships is a function.

1. $x \mapsto (x+2)^2$ for $x \in \{-1, 0, 1\}$

2. $x \mapsto \dfrac{4}{x}$ for $x \in \{1, 2, 4\}$

3. $x \mapsto \pm x$ for $x \in \{1, 2, 3\}$

4. $x \mapsto 2x - 3$ for $x \in \{-2, -1, 0, 1, 2\}$

5. $x \mapsto$ multiples of x, $x \in \{1, 2, 3\}$

6. $x \mapsto$ lowest prime factor of x, $x \in \{2, 3, 4, 5, 6\}$

7. $x \mapsto \sin x$ for $x \in \{0, 30°, 60°, 90°\}$

8. $x \mapsto \sqrt{x}$ for $x \in \{0, 1, 4, 9\}$

The domain of a relationship need not be a set of distinct and separate numbers. The domain may be specified as a range of values, e.g. $\{x : x \geqslant 0\}$. If the domain is not specified, assume that the domain contains all possible values of x.

Determine whether $x \mapsto \dfrac{2}{x}$, $x \neq 0$, is a function.

($x = 0$ is excluded, so we assume that all values of x are possible except $x = 0$.)

For any value of x, there is only one value of $\dfrac{2}{x}$

so $x \mapsto \dfrac{2}{x}$, $x \neq 0$, is a function.

State whether or not each of the following relationships is a function.

9. $x \mapsto \frac{1}{2}(x-2)^2$

10. $x \mapsto 3x + 4$

11. $x \mapsto 1 \pm \sqrt{x}$ for $x \geqslant 0$

12. $x \mapsto \dfrac{1}{x+1}$, $x \neq -1$

13. $x \mapsto$ the largest integer less than or equal to x

14. $x \mapsto$ the perimeter of a rectangle whose area is x^2, $x > 0$

15. $x \mapsto$ the roots of the equation $y^2 = x$ for $x \geqslant 0$

16. $x \mapsto \sqrt{x}$ for $x > 0$

FUNCTION NOTATION

The relationship that maps x to $5x$ is a function and we write this as
$$f(x) = 5x \quad \text{or, sometimes,} \quad f : x \mapsto 5x$$

The notation $f(x) = 5x$ is read as 'the function f of x is $5x$'.

The notation $f : x \mapsto 5x$ is read as 'the function f such that x maps to $5x$'.

If 4 is a value of x we can write $f(4) = 20$ (or $f : 4 \mapsto 20$)

EXERCISE 17c

If $f(x) = \frac{1}{2}x^2 - 4$ find $f(-2)$.

$$f(x) = \frac{1}{2}x^2 - 4$$
$$\therefore \qquad f(-2) = \frac{1}{2}(-2)^2 - 4$$
$$= \frac{1}{2}(4) - 4$$
$$= -2$$

1. Given that $f(x) = 2x - 1$ find
a) $f(0)$ b) $f(4)$ c) $f(-1)$ d) $f(-5)$

2. The function f is defined by $f(x) = \dfrac{4}{x}$, $x \neq 0$. Find

a) $f(\frac{1}{2})$ b) $f(2)$ c) $f(8)$ d) $f(-2)$

3. If $f: x \mapsto 3 - x$ find
a) $f(1)$ b) $f(-1)$ c) $f(5)$ d) $f(-3)$

4. If $f(x) = 5 - 3x$ find
a) $f(2)$ b) $f(-1)$ c) $f(0)$ d) $f(4)$

5. If $f(x) = (1-x)^2$ find
a) $f(1)$ b) $f(0)$ c) $f(-1)$ d) $f(-2)$

6. The function f is defined by $f(x) = \dfrac{3}{1+x}$, $x \neq -1$. Find

a) $f(2)$ b) $f(-2)$ c) $f(0)$ d) $f(-4)$

7. The function f is given by $f(x) = x^2 + 4$. Find
a) $f(1)$ b) $f(2)$ c) $f(0)$ d) $f(-3)$

8. If $f(x) = x^2 - 2x + 3$ find
a) $f(4)$ b) $f(0)$ c) $f(3)$ d) $f(-1)$

9. If $f(x) = 1 - \dfrac{1}{x}$, $x \neq 0$, find

a) $f(2)$ b) $f(3)$ c) $f(-1)$ d) $f(-2)$

10. The function f is defined by $f(x) = 2x - \dfrac{4}{x}$, $x \neq 0$. Find

a) $f(2)$ b) $f(4)$ c) $f(-1)$ d) $f(-8)$

11. If $f(x) = \sin x$ find

    a) $f(30°)$ b) $f(90°)$ c) $f(0)$ d) $f(75°)$

If $f(x) = x^2 - 4$ find the values of x for which f maps x to 5

$$f(x) = x^2 - 4$$

If f maps x to 5 then $f(x) = 5$.

Therefore

$$x^2 - 4 = 5$$
$$x^2 = 9$$

$x = 3$ or $x = -3$

Find the value(s) of x for which the given functions have the given value.

12. $f(x) = 5x - 4$, $f(x) = 2$

13. $f(x) = \dfrac{1}{x}$, $f(x) = 5$

14. $f(x) = 3 - x$, $f(x) = -4$

15. $f(x) = 2x + 1$, $f(x) = -9$

16. $f(x) = x^2$, $f(x) = 9$

17. $f(x) = \dfrac{1}{x^2}$, $f(x) = 1$

18. $f(x) = x^2 - 2x$, $f(x) = 3$

19. $f(x) = x + \dfrac{1}{x}$, $f(x) = 2$

20. $f(x) = (x + 1)(x - 2)$, $f(x) = 0$

21. $f(x) = \sin x$ for $0 \leqslant x \leqslant 90°$, $f(x) = \frac{1}{2}$

22. If f is the function that maps x to $f(x)$ where $f(x) = 2 - 4x$ find

    a) $f(3)$ b) $f(-4)$ c) the value of x for which f maps x to 6.

23. The function f maps x to $f(x)$ where $f(x) = x^2 + x - 6$. Find

    a) $f(1)$ b) $f(-1)$ c) the values of x for which f maps x to 0.

24. The function f is given by $f(x) = x^3 - 8$.
Find $f(0)$, $f(1)$ and the value of x for which f maps x to 0.

25. Find the value of k if f maps 3 to 3 where
a) $f : x \mapsto kx - 1$ b) $f : x \mapsto x^2 - k$

26. The function f is defined by $f(x) = x^2 + x - 4$.
a) Find $f(0)$, $f(1)$ and $f(2)$.
b) Is $f(x)$ increasing or decreasing in value as x increases in value from zero ?
c) *Estimate* the value of x, greater than zero, for which $f(x) = 0$.

27. The function f is defined by $f(x) = 6 - x$.
a) Find $f(-2)$, $f(0)$, $f(4)$.
b) Is $f(x)$ increasing or decreasing in value as x increases in value ?
c) Give the value of x for which $f(x) = 0$.
d) Give the range of values of x for which $f(x)$ is negative.

28. The function f is given by $f : x \mapsto x^2 + 2$.
a) Find $f(0)$, $f(2)$, $f(4)$, $f(10)$.
b) Is $f(x)$ increasing or decreasing in value as x increases in value from zero ?
c) Find $f(0)$, $f(-1)$, $f(-3)$, $f(-10)$.
d) What is the least value in the image set ?

GRAPHICAL REPRESENTATION

While arrow diagrams can be used to represent functions when there is a finite number of members in the domain, they are not practical when the domain is an infinite set of values of x. A different form of representation is needed for these cases.

Consider the function $f(x) = 4x - 2$ for all values of x.

If we write $y = f(x)$ then the equation $y = 4x - 2$ can be used to give a graphical representation of $f(x)$.

Now $y = 4x - 2$ is the equation of a straight line with gradient 4 and y intercept -2.

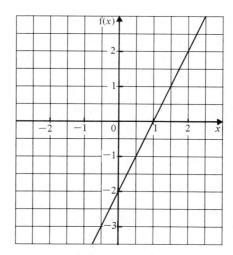

The vertical axis can be labelled either $f(x)$ or y, since $y = f(x)$. A graphical representation like this gives a picture of how $f(x)$ changes in value as the value of x varies.

In this case the graph shows that, as x increases in value the corresponding values of $f(x)$ also increase.

EXERCISE 17d

Draw a sketch graph to represent each of the following functions.

1. $f(x) = 2x$

2. $f(x) = x + 1$

3. $f(x) = x^2$

4. $f(x) = \dfrac{1}{x}, \quad x \neq 0$

5. $f(x) = x^3$

6. $f(x) = 1 - x$

7. $f(x) = 1 - x^2$

8. $f(x) = \dfrac{2}{x^2}, \quad x \neq 0$

9. $f(x) = (x - 1)(x - 2)$

10. $f(x) = x^3 + 2$

11. On the same set of axes draw a sketch graph of
   a) $y = x$ b) $y = x + 1$ c) $y = x - 2$
   Give the transformation that maps the graph of (a) to the graph of (b).

12. On the same set of axes draw a sketch graph of
   a) $y = x^2$ b) $y = x^2 + 1$ c) $y = x^2 - 2$
   Give the transformation that maps the graph of (a) to the graph of (c).

13. Repeat question 12 for each of the following, for $x > 0$.

a) $y = \dfrac{1}{x}$ b) $y = \dfrac{1}{x} + 1$ c) $y = \dfrac{1}{x} - 1$

14. On the same set of axes draw the graph of
a) $y = x^2$ b) $y = -x^2$
Name the transformation that maps the curve for (a) to the curve for (b).

15. Repeat question 14 for the graphs of
a) $y = x - 1$ b) $y = 1 - x$

16. Repeat question 14 for the graphs of
a) $y = x^2 + 2$ b) $y = -x^2 - 2$

THE GRAPHS OF $y = f(x)$ AND $y = f(x) + k$

Consider the curves whose equations are

$$y = f(x) \quad \text{and} \quad y = f(x) + k$$

where k is a number.

Questions 11 to 13 of the last exercise show that the curve $y = f(x) + k$ is a translation, parallel to the y-axis, of the curve $y = f(x)$.

If k is positive the translation is k units upwards and if k is negative the translation is k units downwards.

This fact can be very useful when sketching curves. For example, to sketch the curve $y = x^3 - 4$, we can start with the known shape and position of $y = x^3$ and then move it 4 units downwards.

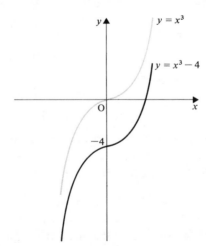

THE GRAPHS OF $y = f(x)$ AND $y = -f(x)$

Questions 14 to 16 of the last exercise show that the curve $y = -f(x)$ is the reflection in the x-axis of the curve $y = f(x)$.

Hence to sketch the curve $y = -\dfrac{1}{x}$, we can start with the known curve

$y = \dfrac{1}{x}$ and reflect it in the x-axis.

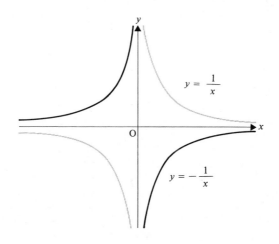

EXERCISE 17e

1. This is a sketch of the graph of $f(x) = 2^x$.

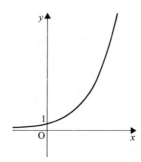

Copy this sketch and, on the same diagram, sketch the graph of
a) $f(x) = -2^x$ b) $f(x) = 1 - 2^x$.

The function f is given by $f(x) = x^3 + x - 5$.

Find $f(0), f(1)$ and $f(2)$ and hence estimate the value of x for which $f(x) = 0$.

$f(x) = x^3 + x - 5$

$f(0) = -5, \quad f(1) = -3, \quad f(2) = 5$

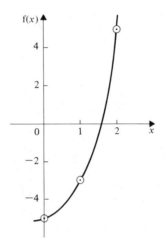

(A rough graph of $y = f(x)$ shows that $f(x) = 0$ for a value of x between $x = 1$ and $x = 2$.)

From the sketch, $f(x) = 0$ when $x \approx 1.6$.

2. If $f(x) = x^3 + x - 3$ find $f(0), f(1), f(2)$ and hence estimate the positive value of x for which $f(x) = 0$.

3. If $f(x) = x^3 - 12$ find $f(1), f(2), f(3)$ and hence estimate the value of x for which $f(x) = 0$.

4. Draw a rough graph of $f(x) = x^2 - 7$. Hence estimate the value of $\sqrt{7}$.

5. Are there any values of x for which f maps x to zero when $f(x) = x^2 + 1$?

6. Draw a rough graph of $f(x) = x^2 + 2$ for $-4 \leqslant x \leqslant 4$.

a) Give the equation of the line of symmetry of the curve.

b) What are the values of x for which $f(x) = 6$?

c) In general how many values of x map to each value of $f(x)$?

d) Fill in the blanks in each of the following.

$f(2) = f(\ )$, $f(4) = f(\ )$, $f(-1) = f(\ )$, $f(x) = f(\ )$

7. Draw a rough graph of $f(x) = (x - 2)^3$ for $0 \leqslant x \leqslant 4$.

a) Give the coordinates of the point about which the curve has rotational symmetry.

b) Give the relationship between

i) $f(4)$ and $f(0)$

ii) $f(3)$ and $f(1)$

iii) $f(x)$ and $f(4 - x)$

INVERSE FUNCTIONS

The function f defined by $f(x) = 4x$ maps 1 to 4, 2 to 8, and so on.

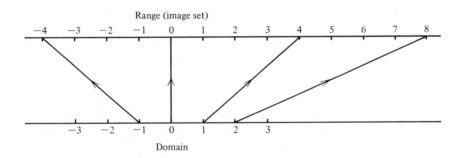

Range (image set)

Domain

The relationship which maps a member of the image set back to its corresponding member in the domain is such that $4 \mapsto 1$, $8 \mapsto 2$, and so on. Clearly this relationship is $x \mapsto \frac{1}{4}x$ which is a function; it is called the *inverse function* of f and is denoted by f^{-1}

i.e.   if f is defined by $f(x) = 4x$ (or $f : x \mapsto 4x$)

then f^{-1} is given by $f^{-1}(x) = \frac{1}{4}x$ (or $f^{-1} : x \mapsto \frac{1}{4}x$)

We may think of the inverse function of f as the function which interchanges the domain and image set of f.

Not all functions have inverses. Consider the function defined by $f(x) = x^2$.

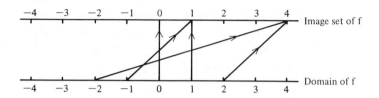

The relationship that maps the members of the image set of f to their corresponding members in the domain of f is such that, for example

$$1 \mapsto 1 \text{ and } -1$$

This relationship is not a function (remember that, for a relationship to be a function, one member of the domain must map to one and only one member of the range). Therefore we say that $f(x) = x^2$ does not have an inverse.

EXERCISE 17f

In this exercise it is important to realise that f represents an *operation*. For example, in $f(x) = x^2$, f means 'square' and this instruction can be applied to any variable. Thus $f(x) = x^2$, $f(z) = z^2$, $f(t) = t^2$, etc, all mean the same thing.

> Find the inverse of the function f defined by
>
> $$f(x) = 2x - 1$$
>
> Let $y = f(x)$, i.e. $y = 2x - 1$.
>
> (To find the inverse function we need to find the relationship that maps y to x, i.e. we need to find $g(y)$ such that $g(y) = x$.)
>
> From $\qquad\qquad y = 2x - 1$
>
> we have $\qquad\qquad y + 1 = 2x$
>
> giving $\qquad\qquad \dfrac{y + 1}{2} = x$
>
> $\therefore \qquad\qquad g(y) = \dfrac{y + 1}{2} \text{ maps } y \text{ to } x$
>
> i.e. g is f^{-1}, hence $f^{-1}(x) = \dfrac{x + 1}{2}$.

Each of the following functions has an inverse. Find the inverse function in the form $f^{-1}(x)$.

1. $f(x) = 2x$

2. $f(x) = 3x$

3. $f(x) = \dfrac{1}{x}, \quad x \neq 0$

4. $f(x) = \frac{1}{5}x$

5. $f(x) = x + 1$

6. $f(x) = x - 2$

7. $f(x) = 4 - x$

8. $f(x) = 2x + 1$

9. $f(x) = \dfrac{12}{x}, \quad x \neq 0$

10. $f(x) = 4 - 2x$

11. $f(x) = x^3$

12. $f(x) = 3x - 5$

Determine whether the function f has an inverse when
a) $f(x) = x^3 + 1$ b) $f(x) = (x - 3)^2$.

(A sketch graph of $y = f(x)$ can be used to find out whether or not an inverse exists.)

a) $f(x) = x^3 + 1$

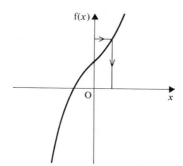

One value of $f(x)$ corresponds to one and only one value of x. Therefore $f(x)$ does have an inverse.

b) $f(x) = (x - 3)^2$

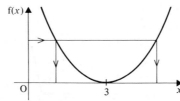

One value of $f(x)$ corresponds to *two* values of x, so $f(x)$ does not have an inverse.

Draw a sketch graph of each of the following functions and find, when it exists, the inverse function.

13. $f(x) = x^2 + 1$

14. $f(x) = x^3$

15. $f(x) = 3x - 1$

16. $f(x) = (3x - 1)^2$

17. $f(x) = 2(x - 1)$

18. $f(x) = (2 + x)(x - 1)$

19. $f(x) = \dfrac{1}{x^2}, \quad x \neq 0$

20. $f(x) = \dfrac{1}{x + 1}, \quad x \neq -1$

21. $f(x) = 7 - 3x$

22. $f(x) = \dfrac{12}{x}, \quad x \neq 0$

23. $f(x) = \dfrac{1}{x}, \quad x \neq 0$

24. $f(x) = x(x - 1)(x - 2)$

25. The function f is defined by $f : x \mapsto 3x$.
Define f^{-1} and find $f^{-1}(2)$, $f^{-1}(0)$ and $f^{-1}(-1)$.

26. The function f is given by $f(x) = x + 4$. Find
a) $f^{-1}(x)$
b) $f^{-1}(-2)$
c) the value of x for which $f^{-1}(x) = 3$

27. Given that $f(x) = 4 - 2x$ find
a) $f(2)$
b) $f^{-1}(2)$

28. If $f(x) = 2x + 4$ find the value of x for which $f(x) = f^{-1}(x)$.

INVERSE FUNCTIONS AND REFLECTIONS

Consider the function f, where $f(x) = y$ where $y = 5x$.

The graph of $y = 5x$ is

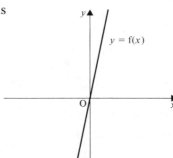

Now the inverse of f maps y to x, i.e. $f^{-1}(y) = x$, and so we can sketch the graph of f^{-1} by interchanging the x and y axes:

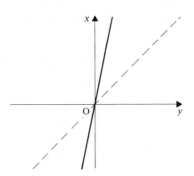

The axes can then be returned to their conventional positions by reflecting the diagram in the line $y = x$:

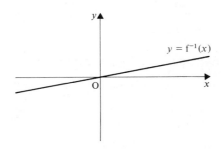

Therefore the graphs of $f(x)$ and $f^{-1}(x)$ are reflections of each other in the line $y = x$.

EXERCISE 17g

Sketch the graph of $f(x)$ and, without finding $f^{-1}(x)$, sketch the graph of $f^{-1}(x)$ on the same diagram if

1. $f(x) = 2x$ **6.** $f(x) = 3 - x$

2. $f(x) = \frac{1}{4}x$ **7.** $f(x) = -\frac{1}{2}x$

3. $f(x) = x + 1$ **8.** $f(x) = 1 - 2x$

4. $f(x) = x - 2$ **9.** $f(x) = 3x + 1$

5. $f(x) = x^3$ **10.** $f(x) = \frac{1}{2}x - 1$

11. Write down two functions for which $f \equiv f^{-1}$.

COMBINING FUNCTIONS

If more than one function is being dealt with at the same time, we use f for one function and different letters for other functions. For example, we can have two functions f and g where $f(x) = 2x$ and $g(x) = x + 1$

Two (or more) functions can be combined in a variety of ways, e.g. they

can be multiplied $f(x)g(x) = (2x)(x + 1)$

$$= 2x^2 + 2x$$

they can be added $f(x) + g(x) = (2x) + (x + 1)$

$$= 3x + 1$$

EXERCISE 17h

The functions f and g are defined by

$$f(x) = x + 1   \text{and}   g(x) = 2x - 1$$

The function h is given by $h(x) = f(x)g(x)$ and the function j is given by $j(x) = f(x) - g(x)$.

a) Find $h(-2)$.   b) Define j in the form $j(x) = \ldots$

a) $h(x) = f(x)g(x)$

$\therefore$ $h(-2) = f(-2)g(-2)$

$$= (-2 + 1)[2(-2) - 1]$$

$$= (-1)(-5)$$

$$= 5$$

b) $j(x) = f(x) - g(x)$

$$= (x + 1) - (2x - 1)$$

$$= x + 1 - 2x + 1$$

$$= 2 - x$$

$\therefore$ $j(x) = 2 - x$.

In the following questions the functions f, g and h are given by

$$f(x) = 4x, \quad g(x) = x + 1, \quad h(x) = \frac{1}{x}$$

1. The function j is defined by $j(x) = f(x) + h(x)$. Find $j(4)$ and $j(-1)$.

2. The function k is defined by $k(x) = f(x)g(x)$. Find $k(2)$ and $k(-3)$.

3. The function m is defined by $m(x) = \dfrac{f(x)}{g(x)}$. Find $m(4)$ and $m(-2)$. Does $m(-1)$ have any meaning ?

4. Express the function j (as given in question 1) in the form $j: x \mapsto \ldots$ Hence find the values of x for which j maps x to 5.

5. Express the function k (as given in question 2) in the form $k: x \mapsto \ldots$ Hence find the values of x for which k maps x to 3.

6. The function q is defined as $q(x) = g(x)h(x)$.
a) Find $q(2)$ and $q(-1)$.
b) Express $q(x)$ in terms of x.
c) Find the value of x for which q maps x to $\frac{1}{2}$
d) Does $q(0)$ have any meaning ?

7. The function m is defined by $m(x) = f(x) + g(x)$. Find $m(x)$ in terms of x and hence find $m^{-1}(x)$.

8. The functions j and q are defined in questions 1 and 6. Find, if they exist, j^{-1} and q^{-1}.

9. What value of x must be excluded from the domain of the function h ? Give the most general domain possible for each of the functions j, k and q^{-1}.

10. If the function $g(x) + k[f(x)]$, where k is a number, maps 3 to 8, find the value of k.

11. The function j is defined by $j(x) = \dfrac{4(x + 1)}{x}$

a) Find (i) $j(4)$ (ii) $j(-1)$
b) Find the value of x for which j maps x to 2.
c) Is there any value of x for which $j(x)$ is meaningless ?
d) If $aj(x)$ maps 4 to 10, find the value of the number a.

COMPOSITE FUNCTIONS

Consider the functions f and g defined by

$$f(x) = 2x \quad \text{and} \quad g(x) = x + 1$$

If the domain of f is $\{-1, 0, 1\}$, the image set is $\{-2, 0, 2\}$. Applying the function g *to the image set* of f gives the set $\{-1, 1, 3\}$.

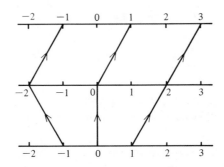

The set $\{-1, 1, 3\}$ is the result of taking the function g of the function f.

In more general terms, if x is any member of the domain of f,

then f maps x to $2x$

and if $2x$ is any member of the domain of g,

then g maps $2x$ to $2x + 1$

The operation that maps x directly to $2x + 1$ is called the *composite function* gf (or $g \circ f$) and we write

$$gf(x) = 2x + 1 \qquad \text{where } gf(x) \text{ means } g\{f(x)\}$$

Note that we read gf from left to right but we operate the opposite way, i.e. gf means apply f first (because it is closest to x), then apply g to the result.

It is interesting to compare functions and transformations: like a transformation, a function is an operation and not the object or the image, e.g. 'double' is a function. The notation used for compound transformations, e.g. TR, and the notation used for composite functions, e.g. fg, are similar in that they both require the operations to be performed in the order right to left.

EXERCISE 17i

The functions f and g are defined by

$$f(x) = \frac{1}{x}, \quad x \neq 0, \quad \text{and} \quad g(x) = x + 2$$

Find, in terms of x,   a) $gf(x)$ b) $fg(x)$

a) $f(x) = \frac{1}{x}$

$g(x) = x + 2 \quad \Rightarrow \quad g\{f(x)\} = \frac{1}{x} + 2$

$$= \frac{1 + 2x}{x}$$

i.e. $gf(x) = \frac{1 + 2x}{x}$.

b) $g(x) = x + 2$

$f(x) = \frac{1}{x} \qquad \Rightarrow \quad f\{g(x)\} = \frac{1}{x + 2}$

i.e. $fg(x) = \frac{1}{x + 2}$.

(Note that $fg \neq gf$.)

1. If $f(x) = x + 1$ and $g(x) = 2x$ find $fg(x)$ and $gf(x)$ in terms of x.

2. Given that $f(x) = 4x$ and $g(x) = 3 - x$ find $fg(x)$ and $gf(x)$ in terms of x.

3. If $f(x) = 3x + 1$ and $g(x) = 5 - x$ find $fg(x)$ and $gf(x)$ in terms of x.

4. The functions f and g are defined by $f(x) = 3 - 2x$ and $g(x) = 4x - 1$. Find $fg(x)$ and $gf(x)$ in terms of x.

5. The functions f and g are defined by $f(x) = \frac{1}{2}(x - 5)$ and $g(x) = 2x + 5$. Find $fg(x)$ and $gf(x)$ in terms of x. Comment on the results.

The functions f, g, h and i are defined by

$$f(x) = 3x, \quad g(x) = x - 1, \quad h(x) = \frac{2}{x}, \quad i(x) = x^2$$

Express the following composite functions in terms of x.

6. fg	**8.** fh	**10.** gh	**12.** fi	**14.** hi
7. gf	**9.** hf	**11.** hg	**13.** ig	**15.** gi

The functions f and g are defined by $f(x) = 2x$ and $g(x) = x^2$.
Find $fg(2)$.

$$g(2) = (2)^2$$
$$= 4$$

and

$$f(4) = 2(4)$$
$$= 8$$

∴

$$fg(2) = 8$$

---

16. The functions f and g are defined by $f(x) = x^2$ and $g(x) = x - 1$. Find $fg(2)$ and $gf(2)$.

17. If $f(x) = 3x + 1$ and $g(x) = \dfrac{1}{2x^2}$, find $fg(3)$ and $gf(-1)$.

18. If $f(x) = \dfrac{1}{x+2}$ and $g(x) = x^2$ find $fg(-1)$ and $gf(0)$.

19. The functions f and g are defined by $f: x \mapsto x + 1$ and $g: x \mapsto 3x$. Find $gf(-2)$ and $fg(-2)$.

20. The functions f and g are defined by $f(x) = \dfrac{1}{2x}$ and $g(x) = x^2$. Find $gf(3)$ and $fg(3)$.

21. The functions f, g and h are defined by

$$f: x \mapsto x + 1, \quad g: x \mapsto 3x, \quad h: x \mapsto x^2$$

Find   a) $fgh(4)$ b) $gfh(1)$ c) hgf,   as a function of x.

The functions f and g are defined by

$$f(x) = \frac{9}{1+x} \quad \text{and} \quad g(x) = 2x^2 - 1$$

Find the values of x for which $fg(x) = 2$.

$$g(x) = 2x^2 - 1$$

$$f\{g(x)\} = f(2x^2 - 1) = \frac{9}{1 + (2x^2 - 1)}$$

$$= \frac{9}{2x^2}$$

$$\therefore \qquad fg(x) = \frac{9}{2x^2}$$

When $fg(x) = 2$,

$$\frac{9}{2x^2} = 2$$

$$9 = 4x^2$$

$$4x^2 - 9 = 0$$

$$(2x - 3)(2x + 3) = 0$$

$$x = \tfrac{3}{2} \quad \text{and} \quad x = -\tfrac{3}{2}$$

22. If $f(x) = x + 1$ and $g(x) = x^2$ find the values of x for which $gf(x) = 4$.

23. If $f(x) = 2x^2$ and $g(x) = x - 3$ find the values of x for which $gf(x) = 5$.

24. If f and g are the functions defined by

$$f: x \mapsto 2x \quad \text{and} \quad g: x \mapsto 3 - x$$

find the value of x such that
a) fg maps x to 4   b) gf maps x to 2

25. The functions f and g are given by

$$f(x) = x^2 + 1 \quad \text{and} \quad g(x) = \frac{2}{x}, \quad x \neq 1.$$

Find the values of x for which gf maps x to 1.

THE SINE AND COSINE FUNCTIONS

We first meet sines and cosines when dealing with angles in right-angled triangles and so tend to think of sines and cosines in relation to acute angles only. However, $\sin x$ has a value for any value of x and the same is true of $\cos x$.

We will now investigate the relationships $x \mapsto \sin x$ and $x \mapsto \cos x$ for $0 \leqslant x \leqslant 360°$.

EXERCISE 17j

1. Copy and complete the following table, using a calculator to find $\sin x$ correct to 2 d.p. for each value of x.

x	0	15°	30°	45°	60°	75°	90°	105°	120°
$\sin x$	0	0.26							

x	135°	150°	165°	180°	195°	210°	225°	240°
$\sin x$								

x	255°	270°	285°	300°	315°	330°	345°	360°
$\sin x$								

2. Use the values in the table in question 1 to draw the graph of $y = \sin x$. Draw the y-axis from -1 to 1 using a scale of 2 cm to 0.5. Draw the x-axis from 0 to 360° using 1 cm to 30°.

3. Make another table, using the values of x given in question 1, for values of $\cos x$ correct to 2 d.p., i.e.

x	0	15°	30°	...	345°	360°
$\cos x$	1	0.97				

4. Use the values in the table in question 3 to draw the graph of $y = \cos x$. Use the same ranges and scales for the x and y axes as in question 2.

f(x) = sin x

The graph drawn for question 2 in the last exercise should look like this.

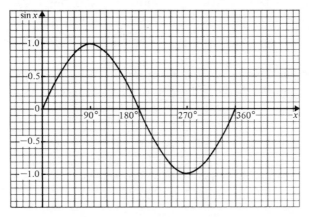

From this graph we can see that the relationship $x \mapsto \sin x$ is a function.

The curve has a distinctive shape: it is called a *sine wave*.

f(x) = cos x

The graph drawn for question 4 in the last exercise should look like this.

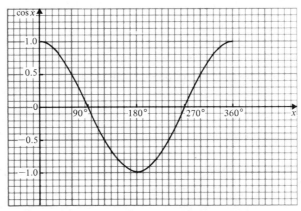

This graph also shows that $x \mapsto \cos x$ is a function.

Notice that the cosine curve looks quite different from the sine curve for the domain $0 \leqslant x \leqslant 360°$.

However, if both curves are drawn for a larger domain, the relationship between the curves becomes obvious. Questions 6 and 7 in the next exercise explore this relationship further.

EXERCISE 17k

1. a) Draw a sketch of the curve $y = \sin x$ for $0 \leqslant x \leqslant 360°$.
   b) On the same diagram sketch the curve $y = -\sin x$.
   c) On the same diagram sketch the curve $y = 1 + \sin x$.

2. a) Draw a sketch of the curve $y = \cos x$ for $0 \leqslant x \leqslant 360°$.
   b) On the same diagram sketch the curves
      i) $y = 1 + \cos x$ ii) $y = -\cos x$

3. a) Draw a sketch of the curve $f(x) = \sin x$ for $0 \leqslant x \leqslant 360°$.
   b) From your sketch, find the values of x for which $f(x) = 0$.
   c) On the same axes draw a line to show how the value(s) of x can be found for which $f(x) = 0.4$.

4. a) Draw a sketch of the curve $f(x) = \cos x$ for $0 \leqslant x \leqslant 360°$.
   b) For what values of x is $f(x) = 0$?
   c) On the same axes draw lines to show how the values of x can be found for which   i) $f(x) = 0.5$ ii) $f(x) = -0.8$.

5. For the domain $0 \leqslant x \leqslant 360°$, give the range of
   a) $f(x) = \sin x$ b) $f(x) = \cos x$.

6. Draw the graph of $y = \sin x$ for $0 \leqslant x \leqslant 720°$ using the following steps.
   a) Make a table of values of $\sin x$ for values of x from $0°$ to $720°$ at $30°$ intervals. Use a calculator and give values of $\sin x$ correct to 2 d.p.
   b) Draw the y-axis on the left-hand side of a sheet of graph paper and scale it from -1 to 1 using 2 cm to 0.5 units. Draw in the x-axis and scale it from $0°$ to $720°$ using 1 cm to $60°$.
   c) Plot the points given in the table made for (a) and draw a smooth curve through them.

7. Draw the graph of $y = \cos x$ for $0 \leqslant x \leqslant 720°$ using the same sequence of steps as in question 6.
   Compare the graphs of $y = \sin x$ and $y = \cos x$. What do you notice ?

8. The function f is given by $f(x) = \sin x$ and the function g is given by $g(x) = 2x$.
   a) Express the function $fg(x)$ in terms of x.
   b) Use a calculator to find $fg(80°)$ correct to 3 s.f.
   c) Express the function $gf(x)$ in terms of x.
   d) Use a calculator to find $gf(80°)$ correct to 3 s.f.

In questions 9 to 11 use the functions f, g and h, where

$$f(x) = \sin x, \quad g(x) = x - 1, \quad h(x) = x^2.$$

9. Express $hf(x)$ in terms of x and find $hf(15°)$ correct to 3 s.f.

10. Express $gf(x)$ in terms of x and sketch the graph of $gf(x)$ for $0 \leqslant x \leqslant 360°$.

11. a) Express $g^{-1}(x)$ in terms of x.
b) Express $g^{-1}f(x)$ in terms of x and sketch the graph of $y = g^{-1}f(x)$ for $0 \leqslant x \leqslant 360°$.

In questions 12 to 15 use the functions f, g and h where

$$f(x) = \cos x, \quad g(x) = 1 + x, \quad h(x) = 3x.$$

12. Express $fh(x)$ in terms of x and find $fh(50°)$ correct to 3 s.f.

13. Express $gf(x)$ in terms of x and sketch the graph of $y = gf(x)$ for $0 \leqslant x \leqslant 360°$.

14. Find $hf(x)$ in terms of x and hence write down the value of
a) $hf(30°)$ b) $hf(60°)$.

15. Find $g^{-1}(x)$ in terms of x and hence find $g^{-1}f(x)$ in terms of x. Sketch the graph of $y = g^{-1}f(x)$ for $0 \leqslant x \leqslant 360°$.

In questions 16 to 20 use the functions f, g, h and i, where

$$f(x) = \sin x, \quad g(x) = \cos x, \quad h(x) = 4x, \quad i(x) = 1 - x$$

16. Express $ig(x)$ in terms of x. Sketch the graph of $y = ig(x)$. Draw a line on your sketch from which the solutions of the equation $ig(x) = 0.5$ can be found.

17. Sketch the graph of $y = f(x)$ for $0 \leqslant x \leqslant 360°$. On the same axes sketch the graph of $y = g(x)$. Hence estimate the values of x for which $\sin x = \cos x$.

18. Express $gh(x)$ in terms of x. Evaluate $gh(0)$, $gh(15°)$, $gh(22.5°)$, $gh(30°)$, $gh(45°)$, $gh(60°)$, $gh(75°)$, $gh(90°)$. Use these values to *sketch* the graph of $y = gh(x)$ for $0 \leqslant x \leqslant 90°$.

19. Express $hf(x)$ in terms of x. On the same axes sketch the graphs of $y = f(x)$ and $y = hf(x)$, for $0 \leqslant x \leqslant 360°$.

20. On the same axes, sketch the graphs of $y = f(x)$ and $y = g(x)$ for $0 \leqslant x \leqslant 90°$.

a) What, roughly, is the value of x at the point of intersection of the two graphs ?

b) For which equation is the value of x in (a) a solution ?

21.

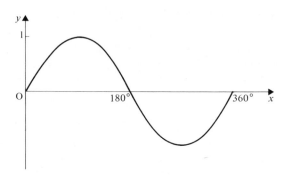

This is a sketch of the graph of $y = \sin x$ for $0 \leqslant x \leqslant 360°$. Copy this sketch, and on the same axes sketch the graph of

a) $y = 2 \sin x$ b) $y = \sin 2x$.

22.

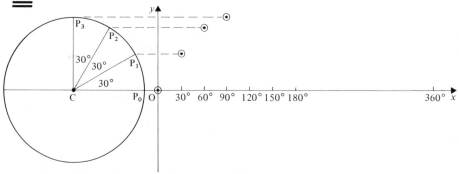

Copy this diagram on graph paper. Make the circle of radius 3 cm. Scale the x-axis from 0 to 360° using 1 cm to 30°.

a) Starting at P_0, mark positions of P at 30° intervals right round the circle. (The diagram shows P_0 to P_3.)

b) Draw a line from each position of P parallel to the x-axis, stopping the line above (or below) the point on the x-axis corresponding to the angle that CP makes with CP_0. Mark the end of the line with a circle. The first four lines are drawn in the diagram.

c) Draw a smooth curve through the marked points. What is this curve ?

MIXED EXERCISES

These exercises do not involve the trigonometric functions.

EXERCISE 17I

In this exercise each question is followed by several alternative answers. Write down the letter that corresponds to the correct answer.

1. The function f is given by $f: x \mapsto x^2 - 1$. The value of $f(-1)$ is

  A 0   **B** 2   **C** -2 **D** -3

2. If $f(x) = 3x$ and $g(x) = 2x$ then $f(x)g(x)$ is

  A $6x$ **B** $6x^2$ **C** $5x$ **D** $5x^2$

3. If $f(x) = x + 1$ then $f^{-1}(x)$ is

  A $x - 1$ **B** $x + 1$ **C** $\dfrac{1}{x+1}$ **D** $1 - x$

4. The function g is such that g maps x to $2x - 1$. The value of x which g maps to 7 is

  A 7   **B** 3   **C** 13   **D** 4

5. If $f(x) = 2x$ and $g(x) = x + 1$, the value of x for which $f(x) = g(x)$ is

  A 3   **B** $\frac{1}{3}$ **C** 1   **D** -1

6. When $f(x) = x^2$, the inverse of this function

  A is $\sqrt{x}$ **B** is x **C** does not exist   **D** is $\dfrac{1}{x^2}$

7. Given $f(x) = x + 1$ and $g(x) = 5x$ then fg maps x to

  A $5x + 1$ **B** $5(x + 1)$ **C** $6x + 1$ **D** $5x(x + 1)$

8. If $f(x) = x^2$ and $g(x) = x + 1$ then gf maps x to

  A $x^2(x + 1)$ **B** $x^2 + 1$ **C** $(x + 1)^2$ **D** $x^2 + x + 1$

9. The function f is defined by $f(x) = 3x^2$ and the function g is defined by $g(x) = x - 2$. $gf(x)$ is defined by

  A $3(x - 2)^2$ **B** $3(x^2 - 2)$ **C** $3x^2(x - 2)$ **D** $3x^2 - 2$

EXERCISE 17m

1. The function f is defined by $f(x) = x + x^2$ for $x \in \{-2, -1, 0, 1, 2\}$. Write down the image set of f.

2. State whether or not each of the following relationships is a function.

a) $x \mapsto \dfrac{1}{x^2}$, $x \neq 0$ b) $x \mapsto \dfrac{1}{x}$, $x \neq 0$ c) $x \mapsto 4$

3. The function f is given by $f(x) = 5x - 2$. Find:

a) $f(0)$ b) $f(6)$ c) $f(-4)$

4. The function f is defined by $f : x \mapsto (x + 1)^2$. Find:

a) $f(0)$ b) $f(-1)$ c) $f(-3)$

5. The function f is given by $f(x) = \frac{1}{2}x$ and the function g is given by $g(x) = x - 2$. Find:

a) the value of x for which $f(x) = 3$
b) the value of x for which $g(x) = 3$
c) the value of x for which $f(x) = g(x)$.

6. The function f is defined by $f(x) = x + k$. Find the value of k if f maps 2 to 8.

7. Find, as a function of x, the inverse of

a) $f(x) = x + 1$ b) $f(x) = 2 - x$ c) $f(x) = 3x - 2$

8. The function f is defined by $f : x \mapsto 5x$. Find:

a) $f^{-1}(x)$ b) $f^{-1}(10)$ c) the value of x for which $f^{-1}(x) = 2$.

9. Given that $f(x) = 6x$, $g(x) = x + 2$ and $h(x) = \dfrac{1}{x}$, $x \neq 0$, find

a) $f(x)g(x)$ b) $g(x) + h(x)$ c) $fg(x)$ d) $hg(x)$

10. Given that $f : x \mapsto 4x$, $g : x \mapsto 2 - x$ and $h : x \mapsto x^2$, find

a) $f(2)g(2)$ b) $fg(2)$ c) $gh(2)$ d) $f(2) - h(2)$

11. The functions f, g and h are defined by

$$f(x) = 2x^2, \quad g(x) = 3 - x, \quad h(x) = 6x$$

a) Find the composite functions fg and gh.
b) Find the inverse functions g^{-1} and h^{-1}.
c) Find the composite functions $g^{-1}f$ and gh^{-1}.
d) Find the value of x for which $h^{-1}(x) = g(x)$.

12. The functions f and g are given by

$$f(x) = 3x - 4 \quad \text{and} \quad g(x) = \frac{2}{x}, \quad x \neq 0$$

a) Find the value of x that g maps to 4.
b) Find the value of x that fg(x) maps to 8.
c) Find the values of x for which $f(x) = 2g(x)$.
d) If h is the function gf, find, in terms of x, the function h^{-1}.

13. The function f maps x to $f(x)$ where $f(x) = 6 - 2x$. Find:
a) $f(-2)$ b) $f^{-1}(-2)$ c) $ff(-2)$.

14. The function f is defined by $f(x) = 10x$ and the function g is defined by $g(x) = 2x - 5$. Find:
a) the values of x for which $f(x)g(x) = 0$
b) h^{-1} where $h = gf$
c) $fgf(2)$.

CHAPTER 1

Exercise 1a   Page 2

No answers are provided for this exercise because there is more than one way to write out an acceptable solution. Also some of the criticisms of the given solutions are in the category of 'bad practice' and the perception of bad practice varies from one individual to another.

CHAPTER 2

Test 2A   Page 10

1. 82   **2.** 14   **3.** 32   **4.** 645   **5.** 1677   **6.** 43   **7.** 1791   **8.** 66   **9.** 48

Test 2B   Page 10

1. -5 **3.** 10   **5.** -12 **7.** -2 **9.** -10 **11.** 22   **13.** 6   **15.** -5 **17.** -8
2. -12 **4.** -3 **6.** -2 **8.** 9   **10.** 7   **12.** 6   **14.** -4 **16.** -18 **18.** 11

Test 2C   Page 11

1. 2   **3.** 7   **5.** 14   **7.** 6   **9.** 30   **11.** 100   **13.** $2 \times 3 \times 5$ **15.** $2 \times 3^2 \times 5^2 \times 7$
2. 4   **4.** 7   **6.** 8   **8.** 12   **10.** 60   **12.** 120   **14.** $2^2 \times 3^3$

Exercise 2a   Page 12

1. 111   **4.** 2820   **7.** 68   **10.** 3   **13.** 752   **16.** 51459   **19.** 27   **22.** 504
2. 217   **5.** 72   **8.** 2271   **11.** 6   **14.** 1518   **17.** 19   **20.** 126   **23.** 120
3. 1144   **6.** 12   **9.** 80   **12.** 13   **15.** 1782   **18.** 31   **21.** 30   **24.** 60

Exercise 2b   Page 12

1. 5   **4.** 15   **7.** -2 **10.** 48   **13.** 15   **16.** 6   **19.** 2   **22.** -12
2. 1   **5.** 8   **8.** -7 **11.** 16   **14.** 30   **17.** 19   **20.** 5   **23.** 0
3. -10 **6.** 3   **9.** -4 **12.** 2   **15.** -2 **18.** -8 **21.** 4   **24.** 16

Exercise 2c   Page 13

1. 6   **8.** 6   **15.** $2^2 \times 3 \times 7^2$ **22.** $2^2 \times 7^2$; 14
2. 4   **9.** 30   **16.** $2^2 \times 3 \times 5^2$ **23.** $2^6 \times 7^2$; 56
3. 5   **10.** 20   **17.** $2^2 \times 3 \times 11$ **24.** $3^2 \times 13^2$; 39
4. 12   **11.** 24   **18.** $2 \times 3 \times 13$ **25.** 3×5^2; 3
5. 18   **12.** 42   **19.** $2 \times 3^2 \times 13$ **26.** 2×3^4; 2
6. 12   **13.** $2^3 \times 3$ **20.** $2^3 \times 3^2 \times 5^2$ **27.** $2^2 \times 3 \times 5$; 15
7. 15   **14.** $2^3 \times 3^2$ **21.** $2^2 \times 3^2 \times 5^2$; 30   **28.** $2^4 \times 3^2 \times 7 \times 11$; 77

Test 2D   Page 15

1. $1\frac{1}{12}$ **3.** $\frac{7}{20}$ **5.** $1\frac{17}{35}$ **7.** $3\frac{3}{4}$ **9.** $\frac{65}{108}$ **11.** $\frac{4}{5}$ **13.** $\frac{13}{22}$ **15.** $3\frac{7}{16}$
2. $\frac{1}{4}$ **4.** $1\frac{7}{8}$ **6.** $\frac{9}{35}$ **8.** $1\frac{21}{40}$ **10.** $\frac{16}{39}$ **12.** $2\frac{1}{20}$ **14.** $16\frac{2}{3}$

Exercise 2d   Page 16

1. $1\frac{5}{24}$
2. $4\frac{26}{35}$
3. $\frac{3}{20}$
4. $1\frac{1}{6}$

5. $\frac{3}{5}$
6. 5
7. $\frac{9}{77}$
8. $\frac{25}{48}$

9. $\frac{5}{36}$
10. $\frac{9}{20}$
11. $1\frac{4}{5}$
12. $\frac{17}{30}$

13. $-\frac{1}{6}$
14. $\frac{3}{5}$
15. $\frac{13}{20}$
16. $\frac{15}{16}$

17. 62
18. $12\frac{3}{8}$
19. $3\frac{1}{3}$
20. $1\frac{1}{2}$

21. $75\frac{3}{5}$
22. $\frac{3}{5}$
23. $2\frac{14}{65}$
24. $\frac{1}{128}$

Test 2E   Page 17

1. 0.3
2. 1.23

3. 4.21
4. −1.39

5. 0.72
6. 0.017

7. 0.45
8. 0.016

9. 50
10. 0.3

11. 4
12. 40

Test 2F   Page 17

1. 0.47
2. 9400

3. 1 120 000
4. 17.8

5. 0.0705
6. 2.8

7. 0.2572
8. 5.033

9. 0.232

Exercise 2e   Page 17

1. 60
2. 0.0054

3. 0.064
4. 0.79

5. 1.423
6. 800

7. 1.44
8. −3.4

9. 12.58
10. 0.125

11. −0.73
12. 0.458

13. 1
14. 4

15. 1
16. 50

17. 0.0675
18. 0.9

Exercise 2f   Page 19

1. 200
2. 2000
3. 0.005
4. 0.1
5. 3

6. 1000
7. 93.0
8. 3.54
9. 0.627
10. 5.25

11. 0.000 006 33
12. 0.182
13. 164.06, 1669.09, 0.01, 0.12,
    2.96, 914.48

Test 2G   Page 19

1. a) 24%   b) 29 %
   c) 175 %   d) 1.5 %
2. a) $\frac{21}{50}$ b) $\frac{7}{25}$
   c) $\frac{1}{40}$ d) $1\frac{1}{5}$

3. a) 0.12   b) 1.15
   c) 6.4   d) 0.45
4. £ 57
5. 34 m³

6. 400
7. 288 m²
8. $\frac{3}{5}$
9. 65 %

Test 2H   Page 20

1. 300 g

2. 4.4 lb

3. 20 % decrease

4. 5 p

5. 800

Exercise 2g   Page 21

1. a) $\frac{3}{10}$ b) $\frac{41}{50}$
   c) $1\frac{3}{10}$ d) $1\frac{1}{4}$

2. a) 93 %   b) 9.5 %
   c) 37.5 %   d) 46.7 %

3. a) 0.54   b) 2.6
   c) 1.38   d) 0.364

4. 28 %, $\frac{2}{7}$, 0.3, $\frac{1}{3}$
5. $\frac{4}{7}$, 55 %, 0.48, $\frac{5}{11}$
6. a) £ 2.20   b) 700 g
7. a) £ 9   b) 35 miles
8. a) $8\frac{1}{8}$ (exactly) or 8.13 m
   to 3 s.f.
   b) £ 54.60

9. a) 48 litres   b) 0.6 m
10. a) $\frac{4}{13}$ b) 30.8 %
11. 40 %
12. 38
13. 57.7 %
14. 2849
15. 4 %

Exercise 2h   Page 23

Answers are given exact, or corrected to the nearest penny, cm, etc. depending on the context of the question.

1. a) £88   b) £92
c) £86

2. a) £139.50   b) £105
c) £131.25

3. a) 25%   b) 20%
c) 10%   d) $12\frac{1}{2}$%

4. £5600

5. £41.40

6. £4.52

7. 2.70 m (2.63 m is the calculated answer but dress material is usually sold in 10 cm units)

8. Loss of 11.1%

9. £100 000

10. 37.5%

11. 162 cm

12. 25%

13. 70

14. £1400

15. 460 g

16. 660 tons

17. £150

18. £850.67

19. a) 5%   b) 5%

Test 2I   Page 24

1. 2:3   **2.** 1:3.5   **3.** 2.5 km   **4.** 2:5   **5.** £28, £8

6. 6 m, 18 m, 24 m   **7.** 30 cm

Exercise 2i   Page 26

1. 16:3

2. 15:2

3. 2:4:7

4. 1:1.75

5. 1:3.125

6. 1:4000

7. 4:7

8. 1:50

9. 1:10 000

10. 1:200 000

11. 1:2500

12. 1:24

13. 6 cm

14. 1:40 000

15. 2.6 km

16. £20, £25

17. £14, £21, £21

18. 5 cm

19. 1:4

20. 1:26 000

Exercise 2j   Page 27

1. a) 14.3 cm³   b) 70

2. £297.50

3. 215

4. 2614.4

5. a) 5, 7   b) 9, 12, 15
c) 1, 4, 5, 12, 15
d) 1, 4, 9

6. £900

Exercise 2k   Page 27

1. $3^2 \times 7^2$, 21

2. £510

3. 19 full glasses with some left over.

4. $\frac{11}{100}$

5. £114 000

6. 1:192 000

Exercise 2I   Page 28

1. a) 36   b) 24
c) 3

2. £14 675

3. £12.50, £15, £22.50

4. £575.13

5. $3\frac{5}{6}$

6. £186.67

Exercise 2m   Page 28

1. C   **3.** D   **5.** A   **7.** A   **9.** A
2. B   **4.** B   **6.** B   **8.** B   **10.** C

CHAPTER 3

Exercise 3a   Page 30

1. $5x^2 - 2x - 3$
2. $6x^2$
3. x^3
4. $3x$
5. $x^2 - 3x$
6. $-a + 3b + c$
7. $42xy$
8. $x^2 - x$
9. x^3
10. x
11. $6x - 8$

12. $6a + 4b + 8$
13. $10x - 1$
14. $x^2 - 4x - 8$
15. $ac - ab$
16. $12 + 9x - 4x^2$
17. $x^2 + 7x + 12$
18. $x^2 + x - 12$
19. $x^2 - 25$
20. $x^2 - 10x + 25$
21. $2x^2 - 5x - 12$
22. $ac - bc - ad + bd$

23. $4x^2 + 20x + 25$
24. $1 + x - 6x^2$
25. $x^2 - xy - 2y^2$
26. $6 - x - x^2$
27. $4x + 9$
28. $a^2 + b^2$
29. $4x^2 - 4x - 8$
30. $2x^2 + 4x + 10$
31. $4x - x^2$
32. $-12x^2 + 42x - 18$

Exercise 3b   Page 31

1. $2(x - 2)$
2. $3(2a - 3c)$
3. $2x(x - 3)$
4. $6(2a + 3b - 4c)$
5. $xy(x + y)$
6. $x(x - 1)$

7. $4(x + 5y)$
8. $9xy(x + 2y)$
9. $(a + b)(x + 2y)$
10. $(p - r)(q + s)$
11. $(2m - p)(n - q)$
12. $(x^2 + 1)(x + 1)$

13. $(v - 3)(4u + 3)$
14. $(a - 2x)(b - 2v)$
15. $(ax - y)(x + a)$
16. $(a - 2c)(b - 3d)$

Exercise 3c   Page 32

1. $(x + 4)(x + 8)$
2. $(x - 3)(x - 4)$
3. $(x - 3)(x + 3)$
4. $(x - 3)^2$
5. $(a - 7)^2$
6. $(y + 9)(y - 2)$
7. $(x + 5)(x - 3)$
8. $(x - 11)(x + 7)$
9. $(x + 6)(x + 5)$
10. $x(x - 9)$
11. $3(x + 1)^2$
12. $4(x + 3)(x - 1)$
13. $9(y - 2)(y + 2)$
14. $5(y - 1)(y - 7)$
15. $6x(2x - 3)$
16. $2(b - 5)(b - 2)$

17. $8(x + 3)(x - 2)$
18. $3(x + 4)(x - 4)$
19. $(2x + 1)(x + 3)$
20. $(3x - 1)(x + 4)$
21. $(3x - 2)(3x + 2)$
22. $(3x + 1)(2x + 3)$
23. $(3x - 5)(4x - 3)$
24. $(5x - 2)(x + 1)$
25. $(2x + 3)(2x - 1)$
26. $(6x - 5)(2x - 3)$
27. $(12x - 5)(x - 3)$
28. $(5x - 3)(5x + 3)$
29. $x(x - 25)$
30. $(5 - x)(5 + x)$
31. $(x - 5)^2$
32. $(x - 25)(x - 1)$

33. $(x + 5)(x + 5a)$
34. $(1 - 4x)(1 + 3x)$
35. $5(2x - 3)(x - 1)$
36. $5(x + 5)^2$
37. $(101 + 99)(101 - 99)$
    $= 200 \times 2 = 400$
38. $(2x - 3)(2x + 3)$.
    $391 = 20^2 - 3^2 = 23 \times 17$
39. $3(x - 4)(x + 4)$.
    $252 = 3(10^2 - 4^2)$
    $= 3 \times 3 \times 2 \times 2 \times 7$
40. $k = 6.24$
41. 57.2
42. $30^2 - 1^2 = 29 \times 31$

Exercise 3d   Page 34

1. 11
2. 45
3. 72
4. 3

5. -6

6. 12
7. $3\frac{1}{3}$
8. -9
9. $-\frac{1}{6}$

10. 2

11. 7
12. 0
13. a) ± 13 b) ± 3
14. a) 18 b) 2
    c) -9
15. $\pm \frac{6}{5}$

Exercise 3e   Page 35

1. 32	**10.** 5^3	**19.** a^7	**28.** x	**37.** $a^{1/2} \times b^{1/3}$
2. 1	**11.** 4^0	**20.** 1	**29.** 1	**38.** $x^{1/4}$
3. 10	**12.** 2^{-2}	**21.** a^{-10}	**30.** $x^{1/4}$	**39.** x^2
4. 27	**13.** a^7	**22.** 6	**31.** $x^{1/2} + x^{1/3}$	**40.** 4
5. $\frac{1}{25}$	**14.** a^3	**23.** 2	**32.** $18x^6$	**41.** 4
6. $\frac{1}{9}$	**15.** a^{10}	**24.** 5	**33.** x^2	**42.** 6
7. 18	**16.** a^3	**25.** $\frac{3}{2}$	**34.** $x^{1/2}$	**43.** 2
8. $1\frac{2}{3}$	**17.** $a^5 + a^2$	**26.** $\frac{3}{2}$	**35.** x^2	**44.** 3
9. 2^6	**18.** $7a$	**27.** 3	**36.** x	**45.** 3

Exercise 3f   Page 37

1. 8	**9.** $\frac{1}{10\,000}$	**17.** $x^{1/2}$
2. 27	**10.** x^2	**18.** $x^{3/2}$
3. 1	**11.** $x^{3/2} + x^{1/2}$	**19.** a) 6×10^{10} b) 1.5×10^2
4. $\frac{1}{2}$	**12.** x	c) 3.02×10^6 d) 2.98×10^6
5. 25	**13.** x	**20.** a) 3.84×10^{-7} b) 1.5×10
6. 81	**14.** x^2	c) 2.56×10^{-3} d) 2.24×10^{-3}
7. $\frac{3}{2}$	**15.** x^{-2}	
8. 100	**16.** $x^{1/2}$	

Exercise 3g   Page 38

1. 2	**6.** 1	**11.** -3
2. $1\frac{2}{3}$	**7.** 3	**12.** $5\frac{1}{2}$
3. 5	**8.** -8	**13.** 4
4. 1	**9.** $\frac{3}{2}$	**14.** 12
5. $\frac{3}{4}$	**10.** 2	**15.** $34\frac{1}{2}°$, $69°$, $76\frac{1}{2}°$

Exercise 3h   Page 39

1. $x = 2, y = 1$	**9.** $x = 9, y = -6$	**17.** $x = -2, y = 4$
2. $x = -1, y = 4$	**10.** $x = 1, y = -2$	**18.** $x = 3, y = -2$
3. $x = 2, y = -3$	**11.** $x = 4, y = 1$	**19.** 6, 19
4. $x = 0, y = -1$	**12.** $x = -7, y = 3$	**20.** 27
5. $x = 12, y = 5$	**13.** $x = 4\frac{1}{2}, y = 1\frac{1}{2}$	**21.** length 17 cm, width 9 cm
6. $x = -4, y = -24$	**14.** $x = -1, y = 1$	**22.** 10 p
7. $x = 8, y = 3$	**15.** $x = 6, y = 1$	
8. $x = 3, y = 1$	**16.** $x = 2, y = 1$	

Exercise 3i   Page 42

1. $x = 2$ or $x = -3$	**8.** 2 or -3	**15.** 1 or 2
2. $x = 0$ or $x = 4$	**9.** 7	**16.** -1 or $\frac{1}{2}$
3. $x = 1$ or $x = 5$	**10.** 0 or -3	**17.** 3.83 or -1.83
4. $x = -\frac{1}{4}$ or $x = \frac{6}{7}$	**11.** -2 or 4	**18.** $\frac{1}{2}$ or -1
5. 4 or 3	**12.** 5 or 2	**19.** -0.38 or -2.62
6. ± 5	**13.** ± 1	**20.** 1.18 or -0.85
7. 0 or 6	**14.** $1\frac{1}{2}$ or $-2\frac{1}{2}$	**21.** 2.56 or -1.56

22. 3.64 or -0.14
23. 1.45 or -3.45
24. -0.29 or -1.71
25. 3.30 or -0.30
26. 1.43 or 0.23
27. 1.16 or -5.16
28. 1.62 or -0.62
29. ± 4.47

30. 3.19 or 0.31
31. -5.16 or 1.16
32. 1.59 or -1.26
33. ± 2
34. $p = -10, q = 25$
35. $x^2 - 5x + 6 = 0$
36. 2.70 m
37. 4

38. 4 and 5
39. 9 or -8
40. a) $\dfrac{90}{x}$ hours   b) $\dfrac{90}{x+5}$ hours
c) $\dfrac{90}{x} - \dfrac{90}{x+5} = \dfrac{1}{4}$; 40
41. 30 mph
42. 8

Exercise 3j   Page 46

1. $\dfrac{xy}{10}$

2. $\frac{5}{3}$

3. $\dfrac{1}{ab}$

4. $\dfrac{b}{a}$

5. $\dfrac{8p}{qr}$

6. $\dfrac{12}{c}$

7. $\frac{3}{4}$

8. $\frac{9}{10}$

9. $3(x-1)$

10. $\dfrac{x}{2}$

11. $\dfrac{2}{x}$

12. -1

13. $\dfrac{x-2}{x+2}$

14. $\dfrac{x-2}{x-1}$

15. $\dfrac{x+3}{3}$

16. $\dfrac{a+b}{a-b}$

17. $-\dfrac{x+1}{2}$

18. $\frac{4}{9}$

19. 2

20. $\dfrac{1}{(x+a)^2}$

21. $\dfrac{(x+3)(x-4)}{(x-3)(x+4)}$

22. $\dfrac{(a+b)^2}{a^2+b^2}$

Exercise 3k   Page 47

1. $\dfrac{11x}{15}$

2. $\dfrac{3x+1}{21}$

3. $\dfrac{7x-1}{12}$

4. $\dfrac{19x-19}{10} = \dfrac{19(x-1)}{10}$

5. $\dfrac{x+3}{12}$

6. $\dfrac{2+x^2}{2x}$

7. $\dfrac{11}{2x}$

8. $\dfrac{-5}{(x+2)(x-3)}$

9. $\dfrac{26-3x}{20}$

10. $1\frac{4}{11}$

11. -5

12. $2\frac{4}{7}$

13. 2

14. 81

15. 1 or 6

16. 2

17. 4, -2

18. a) $\dfrac{5x+y}{x^2-y^2}$ b) $\dfrac{a^2-b^2+4}{2(a-b)}$

19. a) 11   b) $\frac{1}{15}$

20. a) 9 or 1   b) $\frac{1}{2}$

21. a) $\dfrac{1}{1-x}$ b) $\dfrac{1}{y-x}$

Exercise 3l   Page 49

1. $x = 1, y = 2$
or $x = 2, y = 1$
2. $x = 3, y = -1$
or $x = 1, y = -3$

3. $x = 2, y = 2$
or $x = 1\frac{1}{2}, y = 2\frac{1}{2}$
4. $x = 4, y = -2$
or $x = -\frac{8}{3}, y = \frac{14}{3}$

5. $x = 0, y = 3$
or $x = 3, y = 0$
6. $x = 3, y = 4$
or $x = 4, y = 3$

7. $x = -2, y = 6$
 or $x = 2\frac{1}{2}, y = -3$
8. $x = 0, y = -3$
 or $x = 2, y = 1$
9. $x = 4, y = 4$
 or $x = 0, y = 0$
10. $x = 2, y = 4$
 or $x = 12, y = \frac{2}{3}$
11. $x = 2, y = 3$
 or $x = 3, y = 2$

12. $x = 5, y = 1$
 or $x = -7, y = -5$
13. $x = 1, y = -1$
 or $x = \frac{17}{13}, y = \frac{-7}{13}$
14. $x = 1, y = 1$
 or $x = 5, y = -5$
15. $x = 3, y = 2$
 or $x = -\frac{3}{2}, y = -4$
16. $x = -1, y = 2$
 or $x = 7\frac{1}{2}, y = -1\frac{2}{5}$

17. $x = 1, y = -2$
 or $x = -\frac{1}{3}, y = -2\frac{8}{9}$
18. $x = 3, y = 2$
 or $x = \frac{5}{4}, y = 3\frac{3}{4}$
19. $x = 3, y = 1$
 or $x = -1, y = 5$
20. $x = 3, y = -2$
 or $x = 1, y = 2$

Exercise 3m   Page 52

1. a) $\dfrac{18a^2}{bc}$ b) $\dfrac{2c}{b}$

 c) $\dfrac{b + 2c}{6a}$

2. $6 - 3x$

3. a) 5 or 0   b) $\frac{1}{4}$

4. $\dfrac{5}{(x-2)(x+3)}$

5. a) 1   b) 2
 c) $\frac{1}{64}$

6. 1.81 or -3.31

Exercise 3n   Page 52

1. ± 9

2. a) x^5 b) x^{-1} or $\dfrac{1}{x}$

 c) x^6

3. $x = 5, y = 3\frac{1}{2}$

4. a) $(x+2)(x-3)$

 b) $(x^2+1)(x-1)$

5. $\dfrac{14x - 26}{15}$ or $\dfrac{2(7x-13)}{15}$

6. -6 or 7

Exercise 3p   Page 53

1. $\frac{1}{3}$ or 3
2. a) $4(x-2)(x+2)$
 b) $4x(x-4)$
 c) $(4x-1)(x-1)$

3. $-3\frac{4}{5}$
4. a) 4
 b) $\frac{1}{8}$
 c) 64

5. $x = -5, y = -4$
6. 5.83 or 0.17

CHAPTER 4

Exercise 4a   Page 56

1. 483 cm^2
2. 69.7 cm^2
3. 56 cm^2
4. 96 cm^2
5. 216 cm^2
6. 24 cm^2
7. 38.5 cm^2
8. 32.5 cm^2
9. 226 cm^2
10. 2120 cm^2
11. 9.02 cm^2
12. 86.6 cm^2
13. a) 101 cm^2 i.e. 32π cm^2

 b) 3.27 cm^2 i.e. $\dfrac{25\pi}{24}$ cm^2

14. 603 cm^2 (192π),
 1010 cm^2(320π)

15. a) i) 29.4 cm by 21.6 cm
  ii) 0.294 m by 0.216 m
 b) i) 635.0 cm^2 ii) 635 cm^2
16. 188 cm^2 (60π), 8 cm
17. 2.82 cm
18. 2.44 cm
19. a) i) 25.1 cm   ii) 13.7 cm^2
  iii) 78.5%
 b) e.g.

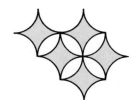

20. a) 95 hectares
 b) 0.95 km^2
21. a) 33.7 °   b) 3.15 m
 c) 1.05 m   d) 1.65 m^2
22. a) 85.8 cm^2 b) 114 cm^2
 75.2%
23. a) 75 m^2 b) 175 m^2
 c) 80 m
24. 67.7 mm
25. a) 4630 cm^2 b) 253 cm
 c) £ 3.70   d) 61 p
26. 400 cm^2

Exercise 4b   Page 63

1. $1470\,\text{cm}^3$, 681
2. $720\,\text{cm}^3$
3. $121\,\text{cm}^3$
4. $14\,100\,\text{cm}^3$
5. a) $3.4\,\text{cm}^3$ b) $3400\,\text{mm}^3$
6. $251\,\text{ml}$
7. $6\frac{2}{3}\,\text{cm}$
8. $4.30\,\text{cm}$
9. a) $25\,\text{cm}$ b) $1250\,\text{cm}^2$
   c) $900\,\text{cm}^2$ d) $7800\,\text{cm}^2$
   e) $45\,000\,\text{cm}^3$
10. $20.5\,\text{cm}$
11. $27\,\text{cm}$
12. $20.4\,\text{mm}$

13. $3.04\,\text{cm}$
14. $579\,\text{cm}^3$
15. a) $12\,\text{cm}^3$ b) $55.8\,\text{cm}^3$
    c) $48\,\text{cm}^3$
16. a) $402\,\text{ft}^2$ (128π)
    b) $37.4\,\text{m}^2$ c) $34.2\,\text{m}^3$
17. $6.93\,\text{cm}$
18. a) $9\pi\,\text{cm}^3$ b) $3\pi\,\text{cm}^3$
    c) $20\pi\,\text{cm}^3$
19. $1.5\,\text{cm}$
20. a) $\frac{80}{9}\,\text{cm}$ b) $\frac{8}{3}$
21. a) i) $3850\,\text{cm}^3$ ii) 3.85 litres
    b) $0.223\,\text{m}^2$

22. a) i) $216\,\text{cm}^3$ ii) $72\,\text{cm}^3$
       iii) $144\,\text{cm}^3$
    b) i) $216\,\text{cm}^2$ ii) $116\,\text{cm}^2$
23. a) $25.0\,\text{cm}$ b) $50\,\text{cm}$
    c) $26.6°$ d) $1490\,\text{cm}^2$
24. a) $11\,\text{m}^2$ b) $2.5\,\text{m}$
    c) $15\,\text{m}^2$ d) $66\,\text{m}^3$
    e) $24\,\text{m}^2$ f) $0.48\,\text{m}^3$
25. $740\,\text{g/sec}$, $800\,\text{cm}^3/\text{sec}$, 36π sec
26. a) $9\,\text{cm}$
    b) i) $18\,\text{cm} \times 36\,\text{cm}$ ii) 8
       iii) 18   iv) $20\,\text{cm}$
       v) 21.5% wasted in each
          layer
27. $8.25\,\text{cm}$

Exercise 4c   Page 70

1. to 10. Any acceptable answer
11. 18
12. 5
13. 1.2
14. 4
15. 2
16. 5.6
17. 180

18. 2
19. $\frac{1}{16}$
20. 10
21. $\frac{10}{3}$
22. $\frac{1}{2}$
23. 0.04
24. 4

25. 3
26. $\frac{1}{60}$
27. 405
28. 1000
29. 12
30. 10
31. 1

Exercise 4d   Page 73

1. a) $46\,499$ b) $45\,500$
2. 1450
3. 550 and 649

4. a) 2049 millions
   b) 1950 millions
5. a) £47.49   b) £42.50

6. £754 999 and £745 000
7. 2049 and 1950
8. 999

9.

Number	Correct to nearest	Smallest possible value	Largest possible value
4560	10	4555	4564
1800	100	1750	1849
5000	1000	4500	5499
80 000	10 000	75 000	84 999
30 000	1000	29 500	30 499
66 700	100	66 650	66 749
4500	100	4450	4549
4000	1000	3500	4499

Exercise 4e   Page 75

1. Nearly $9.65\,\text{cm}$, $9.55\,\text{cm}$
2. Nearly $5.65\,\text{cm}$, $5.55\,\text{cm}$
3. Nearly $31.5\,\text{cm}$, $30.5\,\text{cm}$
4. Nearly $8250\,\text{mm}$, $8150\,\text{mm}$

5. Nearly $126.5\,\text{m}$, $125.5\,\text{m}$
6. Nearly $50.5\,\text{kg}$, $49.5\,\text{kg}$
7. Nearly 350 miles, 250 miles
8. Nearly $8.355\,\text{m/s}$, $8345\,\text{m/s}$

9. $3.5 \leqslant l < 4.5$
10. $2.5 \leqslant D < 3.5$
11. $165.5 \leqslant h < 166.5$
12. $0.488 < D < 0.512$

Exercise 4f   Page 76

1. a) just less than 10.5 cm
   b) 9.5 cm
   c) just less than 110.25 cm^2
   d) 90.25 cm^2
2. a) Upper limits: just less
      than 20.5 cm and 15.5 cm
      Lower limits: 19.5 cm and
      14.5 cm
   b) just less than 317.75 cm^2,
      282.75 cm^2
3. just less than 58.0325 cm^2,
   56.5125 cm^2
4. just less than 18.0625π cm^2,
   14.0625π cm^2
5. just less than 20.5^3 cm^3
   i.e. 8615.125 cm^3,
   19.5^3 cm i.e. 7414.875 cm^3
6. just less than 580.125 cm^3,
   391.875 cm^3
7. just less than 68.25, 52.25
8. just less than 169, 121

9. just less than $\frac{4}{3}\pi \times 12.5^3$,
   $\frac{4}{3}\pi \times 11.5^3$
10. just less than 22.75, 21.25
11. just less than 86.625, 39.375
12. just less than 14, 13
13. just less than 45π, 43π
14. just less than 306.25π,
   272.25π
15. just less than $\dfrac{12.5^2}{10.5}$, $\dfrac{11.5^2}{9.5}$
16. just less than 16, just above 4
17. i) just less than 11.45, 10.95
   ii) just less than 7.2, just
      above 6.8
   iii) just less than 8.265, 7.425
18. 8.14, 7.99
19. a) 11.5, 16.5   b) 1.5, 6.5
   c) 30, 63   d) 1.25, 2.625
20. a) 5.5, 10.2   b) 5.16, 23.12
   c) $5\frac{2}{3}$, $\dfrac{4.3}{3.4}$ d) 19.93, 57.8
   i.e. 1.264

21. a) 4.5 cm, 5.5 cm
   b) 121.5 cm^2, 181.5 cm^2
   c) 91.125 cm^3, 166.375 cm^3
22. a) 1%   b) 4.04 m
   c) 3663 cm to the nearest cm
23. 873 mm × 485 mm
24. the exact amount
25. £ 5.10
26. a) 360   b) 40 p
   c) £ 14.40
27. 150 g butter, 60 g sugar,
   210 g flour
28. 21%
29. a) 130 °F   b) 38 °C
   c) 60 °C   d) 130 °F
30. 200 miles
31. a) i) 150 cm^2 ii) 1700 cm^2
   b) i) 4 cm   ii) 28 cm
32. a) i) 30 m.p.g.   ii) 36 m.p.g.
   iii) 56 m.p.g.
   b) i) 9 l/100 km
   ii) 7 l/100 km
   iii) 5.6 l/100 km

Exercise 4g   Page 81

1. C   **2.** A   **3.** C   **4.** B   **5.** D   **6.** B   **7.** A   **8.** A

CHAPTER 5

Note that where questions require reasons, the main steps in the arguments are given: they are *not* full answers but may help if you are stuck.

Exercise 5a   Page 85

1. a) 25
   b)

p	q	x
3	4	5
5	12	13
7	24	25

   c) 250 mm
2. $p = 74°$, $q = 68°$, $r = 38°$
3. a) $x = 108°$
   b) $y = 45°$, $z = 67.5°$
   c) $p = 120°$, $q = 60°$,
      $r = 30°$
4. a) $\triangle$AEB ||| $\triangle$CDB
      equiangular (||| means 'is
      similar to')
   b) $\triangle$ABD $\equiv$ $\triangle$ACD r.h.s.

   c) $\triangle$ABF $\equiv$ $\triangle$BAC s.a.s. and
      $\triangle$AGF $\equiv$ $\triangle$BGC a.a.s.
   d) ABC ||| $\triangle$ADB ||| $\triangle$BCD
      equiangular
   e) $\triangle$ABF $\equiv$ $\triangle$DCE a.a.s. or
      s.a.s. or s.s.s. and
      $\triangle$BGC ||| $\triangle$FGD ||| $\triangle$DCE
      a.a.a.
   f) $\triangle$LMR ||| $\triangle$PQR a.a.a.
5. a) 1:2.5 = 2:5
   b) 0.72 cm   c) 25:4
6. $\triangle$AMB $\equiv$ $\triangle$ALC (s.a.s.)
   $\Rightarrow$ $\triangle$BLN $\equiv$ $\triangle$CNM (a.a.s.)
   where N is the intersection
   of LC and
   BM $\Rightarrow$ B$\hat{L}$C = B$\hat{M}$C
7. 36°

8. a) 15 cm   b) 16 cm^2
9. a) In $\triangle$BCX, $\hat{B}$ = $\hat{C}$ = 60°
   $\therefore$ $\hat{X}$ = 60°.
   Similarly $\hat{Y}$ = $\hat{Z}$ = 60°
   b) 24 cm
10. a) A$\hat{F}$E = A$\hat{B}$E
   (opp $\angle$'s ||gram)
   = B$\hat{D}$C
   (alt $\angle$'s);
   F$\hat{A}$G = A$\hat{D}$B
   = D$\hat{B}$C;
   $\triangle$AFC ||| $\triangle$BDC
   b) a.a.s.
   c) 1:4
   d) 1:8

Exercise 5b   Page 90

1. $x = 70°$
2. $p = 90°, q = 56°, r = 34°$
3. $x = 41°, y = 27°, z = 112°$
4. $x = 39°, y = 51°, w = 78°$
5. $e = 48.5°, f = 71°, g = 60.5°$
6. $p = 36°, q = 24°, r = 78°$
7. AC is a diameter

as $A\widehat{B}C = 90°, \therefore A\widehat{D}C = 90°$

8. $A\widehat{O}B = 120° (2\widehat{C})$,

$\therefore A\widehat{O}D = 60°, OA = OD$

$\Rightarrow O\widehat{A}D = O\widehat{D}A = 60°$
9. $25°$
10. 5 cm
11. a) $\triangle AEC \parallel\mid \triangle DBE$ (a.a.a.)

b) $\dfrac{AE}{ED} = \dfrac{CE}{EB}$ c) 12 cm

12. $\triangle RMP \parallel\mid \triangle QMS$
13. $p = 30°, q = 120°, r = 30°$
14. $x = 20°, y = 50°, z = 110°$
15. $p = 35°, q = 35°, r = 75°$
16. $x = 60°, y = 90°$

17. $A\widehat{C}B = A\widehat{B}C$ (isos $\triangle$)

$= T\widehat{A}C$ (alt. seg thm)

$\Rightarrow$ alt $\angle$'s equal

Exercise 5c   Page 93

1. 3
2. 4.5
3. 12
4. 4

5. 4
6. 3
7. 3

8. 9 or 1
9. 5
10. 0.536 correct to 3 s.f.

Exercise 5d   Page 94

These answers have been calculated. Your answers should be within $+0.1$ or -0.1 of the given answer.

1. 5 cm
2. 6.2 cm

3. 7 cm
4. 10 cm

5. 5.8 cm
6. 10.4 cm

Exercise 5e   Page 98

1.

a) 6 cm   b) 64 cm^2

2.

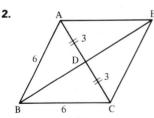

a) a rhombus (diagonals bisect at rt $\angle$'s)

b) 10.4 cm   c) 31.2 cm^2

3.

a) In $\triangle$ PRS, $\widehat{P} = 30°$,

$\widehat{R} = 120°, \widehat{S} = 30°$

b) 12 m   c) 10.4 m

4.

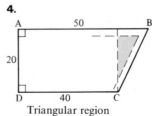

Triangular region

5.

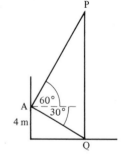

a) $\widehat{A} = 90°, \widehat{P} = 30°$,

$\widehat{Q} = 60°$

b) 8 m   c) 16 m

Exercise 5f   Page 99

1. a) i) PQ as 9.2 cm and
   QR as 7.6 cm
   ii) PQ as 11.5 cm and
   QR as 9.5 cm
   b) 68 m   c) 57.6 m
   d) 1 cm to 10 m is easier than
   1 cm to 8 m as multiplication and division by 10 is
   easier than by 8.
   1 cm to 5 m is easier than
   1 cm to 8 m as it is easier
   to multiply and divide by
   5 than by 8.

2. a) 0.5 km   b) 72 cm
   c) 21 km

3. a) about 25 km

b) about 55 km
c) about 50 km
(By the nature of the
problem, those answers can
only be approximate: any
attempt to be more accurate
is ridiculous!)

4. a) a bit more than 5 miles
   b) about $12\frac{1}{2}$ miles
   c) about 8 miles.

5. a) 10 km   b) 1.25 %
   c) 20 km   d) 5 km
   Choose a scale that involves
   straightforward calculations,
   i.e. $1:10^n$ or $1:5 \times 10^n$ or
   $1:2 \times 10^n$ and then select

from these the scale that
gives the largest practical
drawing.

6. a) 4.06 cm calculated correct
   to 3 s.f.
   b) 4.21 cm calculated correct
   to 3 s.f.; 3.7 % (2 s.f.)
   c) A difference less than
   0.1 cm means that your
   drawing is as accurate as
   you can expect without
   using expensive drawing
   equipment. A difference
   greater than 0.1 cm
   probably indicates an
   error.

Exercise 5g   Page 101

These answers are calculated. You can consider your drawings accurate if your answers are within the
given tolerances.

1. b) 22.8 cm, 25 m, 13.2 m:
   all ± 0.1 m
   c) area AEFD $= 126.9$ m^2,
   area EBCF $= 133.1$ m^2:
   ± 0.5 m^2; 126.9 : 133,1
   $\approx 25:26$

2. a) Q is nearest to P on
   calculated distance, (64 m).
   However a small error in

the drawing will give
another answer – all three
gates are **about** 65 m
walking distance from P.
   c) 5170 m^2 correct to 3 s.f.:
   ± 200 m^2

3. b) 9.5 m (± 0.2 m)

4. a) 479 m:, ± 1 m, 12 600 m^2;
   ± 100 m^2

b)

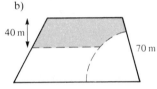

5. a) 20.4 m; ± 0.1 m
   b) 500 m^3; exact

Exercise 5h   Page 104

These answers are calculated. You can consider your drawings accurate if your answers are within the
given tolerances.

1. a) 118°
   b) 7.73 km: ± 0.1 km

2. 36.4 km; ± 0.7 km

3. 38.1 nautical miles; ± 0.7;
   320° ± 1°

4. a) $A\hat{B}C = 100°$,
   $B\hat{C}D = 142°$,
   $A\hat{D}C = 70°$
   b) 290°
   d) 76.3 m ± 1.5 m
   e) 100° ± 1°; 330.5° ± 1°

5. a)

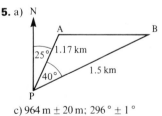

c) 964 m ± 20 m; 296° ± 1°

CHAPTER 6

Exercise 6a   Page 107

1 . 48 in
2. 8 oz
3. 3 ft

4. 5 yd
5. 91 lb
6. 20 pints

7. $1\frac{1}{4}$ lb
8. $1\frac{1}{2}$ gallons
9. 6 in

10. 288 sq in
11. 12 oz
12. 3 miles
13. 60 gallons
14. 20 sq yd
15. $2\frac{1}{2}$ fl oz
16. 1760 yd $= 1$ mile

17. a) OED: A furlong (from a furrow long) is the length of a furrow that can be ploughed across a 'common' field.
A 'common' field has an area of 10 acres.
b) 48 400 sq yd
c) 1 square furlong $= 10$ acres

18. In the days when a tythe of 10% was taken, a weight of 112 lb would yield about 100 lb after the deduction of the tythe. (In the U.S. 1 cwt is 100 lb.)

Exercise 6b   Page 108

Exact answers are not possible when using approximate equivalents. It is therefore pointless and misleading to give several significant figures: The answers given here are considered accurate enough in the context of the questions. If you give more significant figures than are given here, reread the question and consider again what would be a reasonable answer.

1. 54 litres
2. 47 p
3. 16 mm (some patterns say use $\frac{5}{8}''$ or 15 mm)

4. 10
5. 1.8 m
6. 350 acres
7. Apples sold loose. (1 kg for 98 p is about 45 p per lb)

8. 3 metres 90 cm. (i.e. 13 units)
9. 12
10. £ 9.60 per m² is cheaper:
£ 8.75/sq yd $\simeq$ £ 10.25 m²

Exercise 6c   Page 110

1. 22.2 m/s
2. 360 km/h
3. 58.7 ft/s
4. 25 m.p.h.
5. 110 km/h

6. 1 800 000 litres/hour
7. 18 km/litre
8. 2000 sq ft/gallon
9. 330 gal/minute
10. 55 lb/sq in

11. about 200 miles
12. 1600 m²
13. 48 lb/sq in
14. 3 min
15. About 1 h 4 min

Exercise 6d   Page 111

1. a) 08.30   b) 20.30
   c) 05.42   d) 14.36
2. a) 3 a.m.   b) 7.42 p.m.
   c) 8.51 a.m.   d) 10.43 p.m.
3. a) 11 h 48 min   b) 6 h 12 min
   c) 13 h 5 min   d) 21 h 12 min
4. 08.35, 2 h 3 min

5. 11.48, 1 h 23 min
6. 08.35 or 10.35 and both are through trains.
7. 1 hour 26 min
8. 1 hour 12 min
9. a) Reading; 6 min
   b) Bristol Parkway

10. 09.24, 51 min
11. 09.22, 45 min
12. 12.15
13. 15.29
14. No   a) Partick or Hyndland
   b) 5 min   c) 10.36
15. 21 min, 3.44 p.m.

Exercise 6e   Page 117

1. a) 69 p   b) 45 p
   c) 24 p   d) £ 2.96

2. a) 12 p   b) 28 p
   c) 18 p

3. 750 g

4.

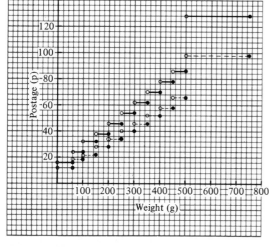

——— 1st class
– – – 2nd class

5. a) 16 p   b) 52 p
   c) 73 p
6. a) 36 p   b) 99 p
   c) 130 p
7. and **8.**

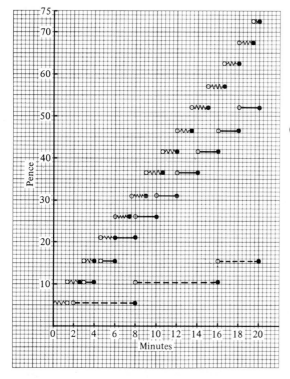

(Note that the horizontal
scale here has been halved.)

– – – Cheap
——— Standard
⋀⋀⋀ Peak

9. a) 40 p   b) 80 p
   c) £ 2.80
   To discourage long-term
   parking
10. a) £ 3.20
   b) Probably because that is
      the maximum weight that
      one machine will hold
   c) Two loads of 9 lb each:
      £ 7.60

Exercise 6f   Page 120

1. £ 420
2. £ 240
3. a) £ 78
   b) the piano and the ring
4. a) £ 251.40   b) £ 243.78
5. a) £ 525   b) £ 252
6. £ 2078, about £ 40 (£ 39.96)
7. About £ 160 (£ 158.23)

CHAPTER 7

Exercise 7a   Page 122

1. 62.3°
2. 43.4°
3. 4.57 cm
4. 7.71 cm

5. 25.8°
6. 4.69 cm
7. 2.83 cm
8. 19.1 cm

9. $\hat{Q} = 55.8°$, $\hat{R} = 34.2°$
10. YZ = 4.06 cm, $\hat{Y} = 52.0°$
11. $\hat{B} = 36°$, AC = 2.53 cm
12. $\hat{N} = 48.6°$, $\hat{L} = 41.4°$

Exercise 7b   Page 125

1. AC = 6, $A\hat{C}B = 45°$
2. 94.2 m
3. a) 54°   b) 5.51 cm
   c) 110 cm²
4. 6.71 cm

5. 82.8°
6. a) 6.5 m
   b) 45.2°   (or 134.8° for the obtuse angle)
7. 5.77 m

8. 6.40 miles on a bearing 231.3°
9. a) 36.9°   b) 26.6°
   c) 63.4°
10. 25.7 m

Exercise 7c   Page 126

1. a) 10 m   b) 21.5 m
   c) 36.9°   d) 15.6°
   e) 528 m²

2. a) 10 m   b) 13.4 m
   c) 14.4 m   d) 22.6°
3. a) 13.0 cm   b) 45°
   c) 49.1°

4. a) 21.2 m   b) 27.9°
   c) 22.6 m   d) 70.6°
5. a) 3.20 cm   b) 51.3°
   c) 38.7°   d) 32.7°

Exercise 7d   Page 131

1. a)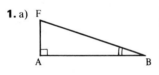

b)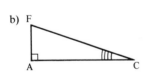

c)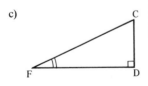

d)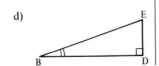

2. a)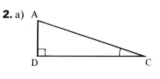

b)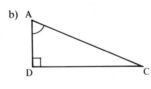

c)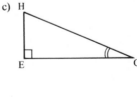

d)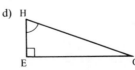

3. a)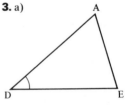
   E (midpoint of CB)

b)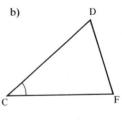
   F (midpoint of AB)

c)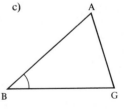
   G (midpoint of DC)

Exercise 7e   Page 132

1. a) AF = 5 cm, FC = 23.9 cm
b) F

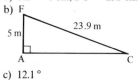

5 m   23.9 m

A   C

c) 12.1°
2. a) 14.1 m   b) 12.2 m
c) 54.7°   d) 11.2 m
e)   V   f) 63.4°

11.2 m   11.2 m

X   Y
14.1 m

3. a) i) 3 m   ii) 2.60 m
b)   A

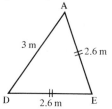

3 m   2.6 m

D   2.6 m   E

d) 54.7°, 70.5°, 2.45 m
(allow ±2% for drawing)

4. a) 5 m, 9.43 m
b) 36.9°, 18.5°
5. a) 107.5 m, 152 m
b) i) 249 m   ii) 272 m
c) i) from the midpoint of a
side of the base, e.g. P;
ii) from a corner of the
base, e.g. A.

CHAPTER 8

Exercise 8a   Page 135

1. a) 8.2   b) 7.5
c) 0.006
2. a) 462   b) 3.5

3. a) $7\frac{1}{2}$ b) −10

4. a) 25 °C   b) 59 °F

5. a) 14   b) 20
c) 12

Exercise 8b   Page 136

1. a) $T = a + b + c$
b) $a = T - b - c$
2. $C = 6p + 8q$
3. $y = 20(N + 1)$

4. a) $x - 6$
b) $P = 6(x - 6)$

5. $T = N(S - C)$

6. $A = \dfrac{xy}{10\,000}$
7. a) 25
b) $x = 4 + 3(n - 1)$
or $x = 1 + 3n$

Exercise 8c   Page 137

1. $\dfrac{c}{a + b}$

2. $\dfrac{c - b}{a}$

3. $\dfrac{c}{a + b}$

4. $\dfrac{c}{a + 1}$

5. $\dfrac{c}{b - 1}$

6. $\dfrac{b - a}{c}$

7. $\dfrac{c - 2a}{a}$

8. $\dfrac{ab}{a + b}$

9. $\dfrac{bc - a}{2}$

10. $\dfrac{a + b}{b - a}$

11. $\dfrac{c}{a + b}$

12. $\dfrac{ab + a - b}{a + b}$

13. 0   (If $a \neq b$)

14. 1

15. $u = v - at$

16. $h = \dfrac{A - 2\pi r^2}{2\pi r}$

17. $m = \dfrac{y - k}{x - h}$

18. $a = \dfrac{v - u}{t}$

19. $a = \dfrac{2A - bh}{h}$

20. $l = \dfrac{S - 4a^2}{4a}$

21. $d = \dfrac{l - a}{n - 1}$

22. $n = \dfrac{l - a + d}{d}$

23. $t = \dfrac{2S}{u + v}$

24. $f = \dfrac{uv}{u + v}$

25. $F = \dfrac{12}{6 - n}$

Exercise 8d   Page 138

1. a) $AB = 30 - 2x$,
  $BC = 40 - 2x$
b) $V = x(30 - 2x)(40 - 2x)$
  or $V = 4x(15 - x)(20 - x)$

2. a) $A = x(300 - x)$
b) $120\,m$ by $180\,m$

3. a) $(x^2 - 5)\,cm^2$
b) $(x^3 - 5x)\,cm^3$ c) 5

4. a) $\dfrac{144}{x^2}$ b) $5.66\,cm$

5. 10

6. 3

7. a) $12x°$
b) Hour hand $(60 + x)°$,
  minute hand: $12x°$
c) $\frac{60}{11}$

8. $5x + y = (x + y) + 4x$.
Both $(x + y)$ and $4x$ are
divisible by 4.

Exercise 8e   Page 140

1. $x \geqslant 2$
2. $x < 2$
3. $x \geqslant \frac{1}{5}$
4. $x > 1\frac{2}{3}$
5. $1 < x < 5$
6. $x < -2\frac{1}{4}$

7. $8\frac{1}{2} < x < 11\frac{1}{2}$
8. $-7 < x < 2$
9. $-5 \leqslant x \leqslant 1$
10. $x < 3\frac{1}{2}, 3$
11. $-2 < x < \frac{1}{2}, -1, 0$
12. $x > 9, 10$

13. $2x + y \geqslant 5, x > 0$ and $y > 0$
14. For example $(2, 3)$, $(1, 1)$,
  $(-10, -12)$
15. For example $(10, 1)$, $(3, -2)$,
  $(3\frac{3}{4}, \frac{1}{4})$
16. a) $x \geqslant 2, x \leqslant 2$ so $x = 2$ only
b) $x > 2, x < 2$ so there are
  no values of x

Exercise 8f   Page 141

1. a) $x + 3 > 2x, x < 3$
b) 2, 1
2. a) $a > 2m, m > s + 4$
b) 15, 7, 2 or 1; 14, 6, 1;
  15, 6, 1; 13, 6, 1
3. a) $(15x + 20y)$ pence
b) $3x + 4y \leqslant 40, x \geqslant 0$,
  $y \geqslant 0, x \leqslant 13, y \leqslant 10$

c) 0, 1, 2, 3 or 4
d) $x = 5, y = 2$
  or $x = 1, y = 5$

4. a) $C = 6x + 2\frac{1}{2}y$
b) $12x + 5y \leqslant 200, x \geqslant 0$,
  $y \geqslant 0, x \leqslant 16, y \leqslant 40$
c) 9 hardbacks, 14 paperbacks

5. a) 5 bottles   b) $C = x + 6y$
c) $x + 6y = 30; x \geqslant 0, y \geqslant 0$,
  $x \leqslant 30, y \leqslant 5$
d) $x = 30, y = 5$
6. a) $P = x + 2y$
b) $x > 3y$
c) $x + 2y \geqslant 22$
d) 5 desks and 11 tables

Exercise 8g   Page 143

1. $\dfrac{x + y}{1000}\,kg$

2. a) 4   b) 25
  c) 144   d) 4

3. $x > -\frac{2}{3}$

4. $p = \dfrac{A - 3\pi q}{\pi}$

5. a) $b = \dfrac{2a}{3}$

b) $\dfrac{b}{a} = \dfrac{2}{3}$

Exercise 8h   Page 143

1. -4
2. ± 2.85

3. a) $1\frac{1}{3}$ b) 4
  c) 4

4. $h = \dfrac{A - \pi r^2}{\pi r}$

5. $\frac{1}{2}x + 2y \leqslant 70$

Exercise 8i   Page 144

1. $x = \dfrac{5p + 3q}{p + q}$

2. $10x + y$

3. -2

4. a) $-1, 1\frac{1}{2}$ b) Roughly $1\frac{1}{2}$

5. $y = \dfrac{ab^2}{z^2}$

CHAPTER 9

Exercise 9a   Page 146 ─────────────────────────────

1. a) 5, 2　　　b) 5, −3.5

c) $\frac{2}{3}, \frac{1}{5}$　　d) 1, −2

e) $\frac{1}{2}, 2$　　f) −2, $\frac{3}{2}$

g) $-\frac{1}{2}, 2$　　h) $\frac{3}{5}, -\frac{2}{5}$

2. B

3. B

4. C

5. C

6. A

7. B

8. B

9. C

10. A

11. a)
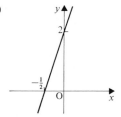

b)
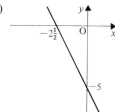

c)
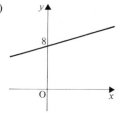

d)
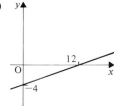

e)
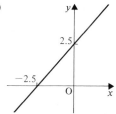

f)
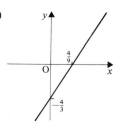

g)
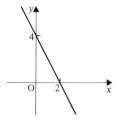

h)
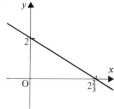

i)
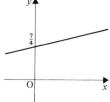

j)
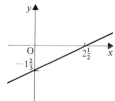

12. a) Yes　　b) No

c) Yes　　d) Yes

e) Yes　　f) Yes

13. a) $y = 2x + 3$

b) $y = -3x + 4$

c) $y = \frac{1}{2}x - 2$

or $x - 2y - 4 = 0$

d) $y = -\frac{2}{3}x + \frac{1}{3}$

or $2x + 3y - 1 = 0$

e) $y = \frac{2}{5}x - \frac{4}{5}$

or $2x - 5y - 4 = 0$

f) $y = -\frac{1}{2}x + \frac{2}{3}$

$3x + 6y - 4 = 0$

14. a) 3　　b) −3

c) 18

15. a) 0　　b) 1

c) −2

16. $a = 0, b = 8, c = 0$

17. a) $\frac{1}{2}$ b) 7
   c) $-\frac{1}{3}$ d) $\frac{3}{4}$
   e) $\frac{2}{9}$ f) $\frac{4}{7}$

18. a) A(1,0), B(4,6)
   b) 2   c) −2
   d) $y = 2x - 2$

19. a) A(−6,0), B(3,3)
   b) $\frac{1}{3}$ c) 2
   d) $y = \frac{1}{3}x + 2$

20. a) A(−2, 6), B(2, 2)
   b) −1   c) 4
   d) $y = -x + 4$

21. a) A(−4, −1), B(4, −3)
   b) $-\frac{1}{4}$ c) −2
   d) $y = -\frac{1}{4}x - 2$

22.

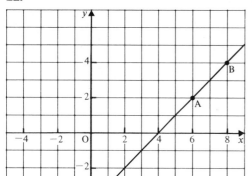

gradient is 1
y intercept is −4
$y = x - 4$

23.

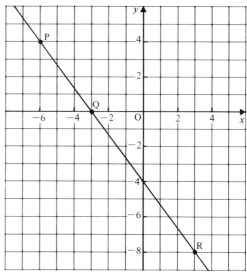

gradient is $-\frac{4}{3}$,
y intercept is −4
$4x + 3y + 12 = 0$

24.

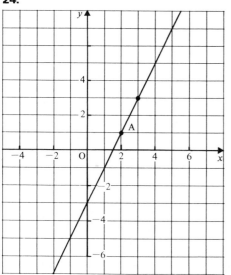

25.

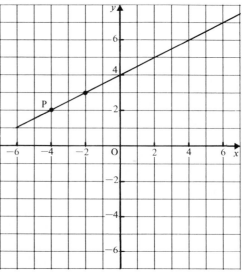

26.

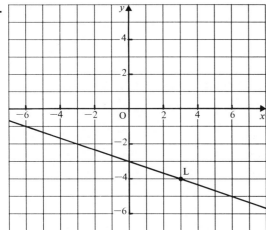

Exercise 9b   Page 152

1. a) $y = 3x + 2$
 b) $y = \frac{1}{2}x + \frac{5}{2}$
 c) $y = -2x + 10$
 d) $y = \frac{2}{3}x + \frac{7}{3}$
 e) $y = -\frac{3}{4}x - \frac{23}{4}$

2. a) $y = \frac{1}{2}x + 2$
 b) $y = -x + 4$
 c) $y = -\frac{1}{3}x - \frac{4}{3}$
 d) $y = \frac{5}{2}x - \frac{11}{2}$

 e) $y = \frac{5}{2}x + 3$
 f) $y = -\frac{2}{7}x + \frac{26}{7}$
 g) $y = \frac{2}{5}x - \frac{19}{5}$
 h) $y = 3$

3. a) Yes   b) Yes
 c) No   d) No
 e) Yes   f) No
 g) Yes

4. a) i) A(3,0)   (ii) B(0,4)
 b) $-\frac{4}{3}$ c) 5 units
 d) 6 sq units

5. a) P(−12,0)   b) Q(0,5)
 c) $\frac{5}{12}$ d) 13 units
 e) 30 sq units

6. a) 3
 b)

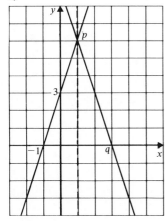

 c) $y = -3x + 9$

7. a) $-\frac{1}{2}$
 b)

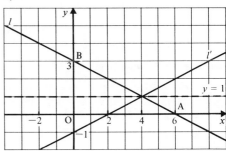

 c) $\frac{1}{2}$, C(0,−1)
 d) $y = \frac{1}{2}x - 1$
 e) 12 sq units

8. a) A(0, 4)

b)

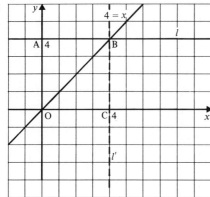

c) Equation of l is $y = 4$
   Equation of l' is $x = 4$
d) C(4, 0)   e) 8 sq units

Exercise 9c   Page 156

Note that values read from a graph can only be approximate.

1. $x = 2.6, y = 2.3$
2. $x = 2.2, y = 4.5$
3. $x = 3.1, y = -2.2$

4. $x = 2, y = -2.8$
5. $x = 2, y = 3.3$
6. $x = -1.9, y = -2.2$

7. $x = 3.51, y = 0.85$

Exercise 9d   Page 158

Remember that most of these answers are approximations.

1. a) £76   b) 65
   c) £12
   d) gradient $= \frac{88}{240} = \frac{11}{30}$
      $= 0.37$ pounds per therm.
      The gradient gives the
      cost of one therm.
2. a) £37.50   b) 760
   c) 1000

d) Tariff A   Gradient $= \frac{55}{1000}$
   $= 0.055$, i.e. electricity costs
   £0.055 per unit i.e. 5.5 p/unit.
   Vertical intercept is 10
   i.e. standing charge is £10
   Tariff B   Gradient $= \frac{40}{1000}$
   $= 0.04$, i.e. electricity costs
   £0.04 per unit i.e. 4 p/unit.
   Vertical intercept is 25
   i.e. standing charge is £25

3. a) £20   b) £64.80
   c) 1143
   d) 0.07, 7 p ∴ cost is 7 p
      per unit.
4. a) £420   b) £314
   c) 1.4   Rate in the pound
      is 140 p.

Exercise 9e   Page 161

These are calculated answers. You cannot expect yours to be as accurate.

1. a) $11.05   b) £13.85
   c) £3.80   d) $19

2. a) 78 hectares   b) 116 acres
   c) 18.6 hectares d) 217 acres

3. a) 3740 lbs   b) 886 kg

Exercise 9f   Page 162

These answers are approximate.

1. b) i) 73 kg   ii) 169 cm
2. a) i) 80   ii) 100   iii) 42
   b) No – appears inversely
      related.   iii) 80

3. b) i) £2.10   ii) 305
4. b) i) £35   ii) 7 years
   c) The bicycle was in very
      poor condition

5. a) 40 gallons   b) 17.3 °C

Exercise 9g   Page 166

In questions 1 to 8 the shaded area is the area that does **not** satisfy the inequality.

1.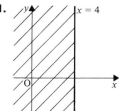

4.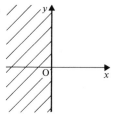

7.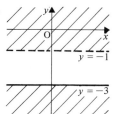

2.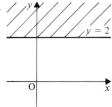

5.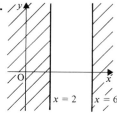

8.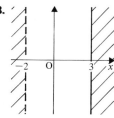

3.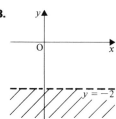

6.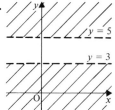

9. $-1\frac{1}{2} \leqslant y \leqslant 5$

10. $2 < x \leqslant 5$

11. $-3 \leqslant y < 3\frac{1}{2}$

12. $-3 < x \leqslant -1$

13.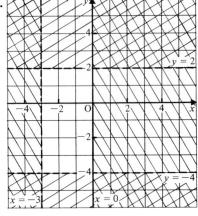

14.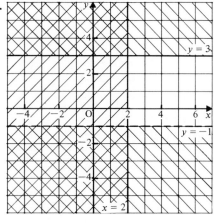

15. $-2 \leqslant x \leqslant 4, y \leqslant 2$

16. $-3 \leqslant x \leqslant 2, -4 < y < -1$

17.

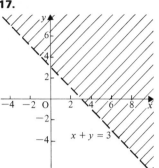

$x + y = 3$

19.

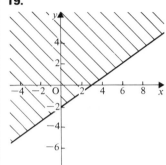

21.

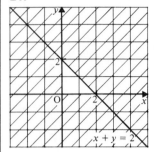

$x + y = 2$

18.

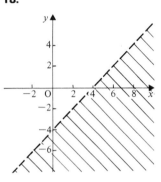

20.

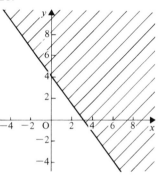

22.

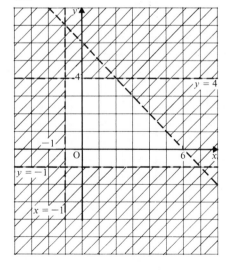

$y = 4$

$y = -1$

$x = -1$

23.

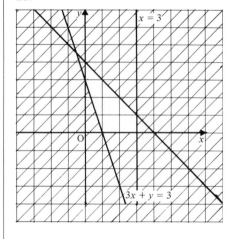

$x = 3$

$3x + y = 3$

24.

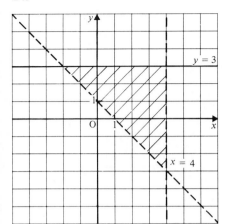

25.

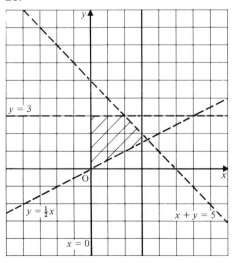

26.

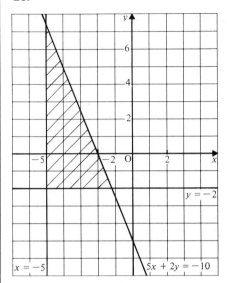

27. $y \geqslant -1.5$, $x \leqslant 3\frac{1}{2}$, $x - y \leqslant -3$

28. $y > -4$, $2x - y < -2$ or $y > 2x - 2$,
  $x + y \leqslant 4$

29. a) $x + y \leqslant 3$, $y \geqslant 0$,
    $x - 2y \leqslant -6$
  b) $y \leqslant 5$, $y \geqslant 0$, $x + y \geqslant 3$,
    $x - 2y \leqslant -6$
  c) $y \leqslant 5$, $x + y \geqslant 3$,
    $x - 2y \geqslant -6$
  d) $y \geqslant 0$, $y \leqslant 5$, $x - 2y \leqslant -6$

CHAPTER 10

Exercise 10a   Page 174 ─────────────────────────────

1. a)
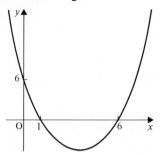

b)
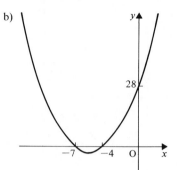

c)
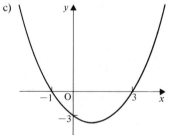

d)
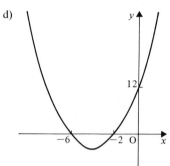

e)
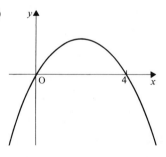

f)
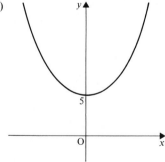

g)
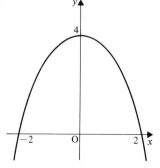

h)
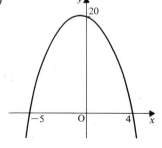

2. a) $A(0,6)$, $B(2,0)$, $C(3,0)$
b) $2x + y - 6 = 0$
c)

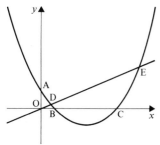

$x^2 - 6x + 6 = 0$

3. a) $p = -1$, $q = -20$
b) $(0, -20)$

4. a) $A(-3,0)$, $B(0,9)$
b)

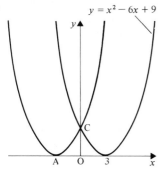

$y = x^2 - 6x + 9$

5.

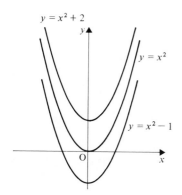

$y = x^2 + 2$

$y = x^2$

$y = x^2 - 1$

6.

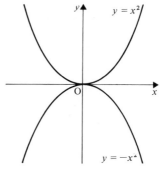

$y = x^2$

$y = -x^2$

A reflection in the x-axis.

7.

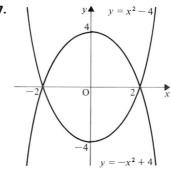

$y = x^2 - 4$

$y = -x^2 + 4$

A reflection in the x-axis.

8. a) $A(-3,0)$, $B(0,9)$, $C(3,0)$
b)

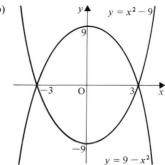

$y = x^2 - 9$

$y = 9 - x^2$

9. a) $A(-4,0)$, $B(0,-8)$, $C(2,0)$
b)

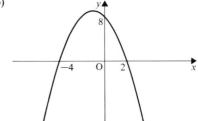

10.

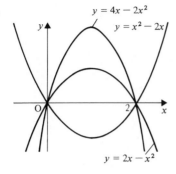

$y = 4x - 2x^2$

$y = x^2 - 2x$

$y = 2x - x^2$

11. a)

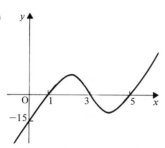

b)

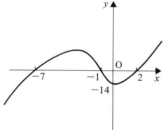

c)

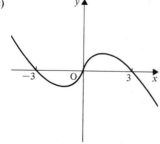

d)

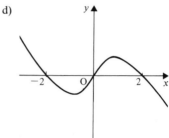

e)

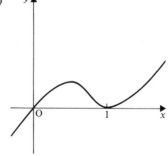

12. A($-3,0$), B($4,0$)

13. A($-2,0$), B($2,0$),
      C($3,0$), D($0,12$)

14. $a = 0, b = -16$

15.

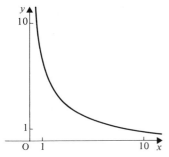

16.

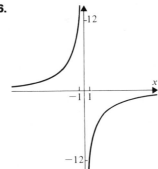

17. C   **18.** C   **19.** B   **20.** C   **21.** D

Exercise 10b   Page 182

1. a) -2.76
   b) 1.38 and 3.62
   c) -3.25 when x is 2.5
   d) 0.68 and 4.30

2. a) -3.24 and 1.24
   b) -3.24 and 1.24
   c) -2.73 and 0.73
   d) -4.74 and 2.74
   e) -4.43 and 2.32

3.

x	-1	0	2
$3x^2 - 3x - 2$	4	-2	4

   a) -0.46 and 1.46
   b) -0.88 and 1.88
   c) -1.30 and 2.30

4. a) -6.31
   b) -0.37 and 4.37
   c) 0.44 and 4.56,
      $x^2 - 5x + 2 = 0$
   d) $0.44 < x < 4.56$

5. a) -3.45 and 1.45,
      $x^2 + 2x - 5 = 0$
   b) -2.41 and 0.41

6. a)

x	-1	0	4
$4(x^3 - x^2 - 6x)$	16	0	96

   b) Graph
   c) 3.1 (only value)
   d) For all values of x
      above 3.1
   e) $4x^3 - 4x^2 - 19x - 20 = 0$

7. a) -1.73, 0, 1.73
   b) approx. values -1.5,
      -0.45, 1.85
   c) approx. values -0.88,
      0.35, 1.53
   Yes, when the graph crosses
   the x-axis.
   $\sqrt{3} \approx \pm 1.73$

8. a) -3.42 b) 3.27
   c) -2.71

9. a) -3.3, 0.5, 2.8
   b) $x^3 - 9x + 4 = 0$
   c) One root

10. a) 2.2 and 10.8
   b) $x^2 - 13x + 20 = 0$
   c) $2.2 \leqslant x \leqslant 10.8$
   d) Because we cannot find a
      value of y when $x = 0$.

11. a) 1.51
   b) 0.23 and 8.77
   c) $x^2 - 9x + 2 = 0$

12.

x	0.5	1
$\dfrac{1}{x^2}$	4	1

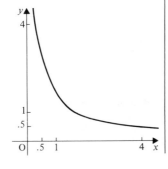

13.

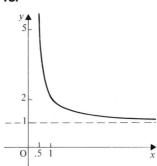

14.

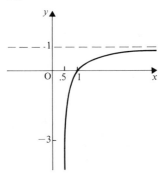

15.

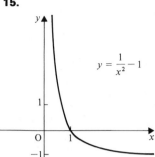

$$y = \frac{1}{x^2} - 1$$

16.

x	2	3.5
$1 + \dfrac{2}{x^2}$	1.50	1.16

a) Graph

b)

x	1	2	3
$y = \frac{1}{2}(x+3)$	2	2.5	3

c) $x^3 + x^2 - 4 = 0,\ x \approx 1.32$

17. a) and b) Graphs
c) 1.71
d) $2x^2 - 4x + 1 = 0$

Exercise 10c   Page 190

1. a) £ 10 250 when 3500
   switchboards are produced
 b) 300 (approx)
 c) More than 1440 but less
   than 5560
2. a) 57.2 m
 b) 1.44 s and 5.56 s
 c) 3.5 s
 d) 0 m

3.

x	20	40	70
A	2000	2400	0

 a) 13.2 m or 56.8 m
 b) Max. area 2450 m² when
   $x = 35$ m
 c) 1568 m²
 d) 2352 m²

4. a) $A = 4x^2$

 b)

x	2	2.3
C	16	21.2

 c) Graph
 d) 3.16 ft
 e) $2.75 \leqslant x < 3$

5. a) i) $(16 - 8x)$ cm
   ii) $(4 - 2x)$ cm
   iii) $3x^2$ cm²
   iv) $16 - 16x + 4x^2$ cm²
 c) Graph
 d) $\frac{8}{7}$ or 1.14
6. a) $(10 - \frac{1}{2}x)$ cm

 c)

x	0	2	4	6
C	0	36	128	252

x	8	10	12	14
C	384	500	576	588

x	16	18	20
C	512	324	0

 d) Graph   e) 114 cm³
 f) 8.3 and 17.4
 g) 592 cm² when $x = 13.3$

7. a) Beginning of week 1 to
   the end of week 3 and
   from beginning of week 7
   to the early part of
   week 11
 b) Beginning of week 4 to
   the end of week 6 and
   from the middle of week
   11 on.
 c) i) End of week 6
   ii) Early in week 11
 d) Probably a fall in price
 e) False
8. a) 4 a.m.   b) 3 hours
 c) 40 °C   d) 1.1 °C
 e) About 2 p.m.

Exercise 10d   Page 196

1. a) after 4 s   b) 5 m/s
 c) 2.5 m/s   d) 10 m/s
 e) −10 m/s   f) −4.8 m/s
2. a)

t	0	2	3
h	10	20	10

 b) Graph
 c) i) 15 m/s   ii) −10 m/s
 d) $21\frac{1}{4}$ m   e) 3.56 s

3. a) 3.025 m   b) 0.5 m/s²
 c) When $t = 5.5$ s
 d) 22 m approx

4. a) At the beginning of the
   race.
 b) 200 m.p.h.   c) 1 minute
 d) 75 m.p.h.   e) 2.8 miles.

5. a) 24 m/s²   b) 17 m/s
 c) 3.3 min   d) $15\frac{3}{4}$ km
 e) 13 km

Exercise 10e   Page 199

1. He is
a) accelerating from rest
b) running at a constant speed
c) accelerating as he runs towards the winning post.

2. a) He accelerates from rest until he attains his steady running speed
b) He runs at a steady speed
c) He increases his speed until he is running 'flat out' at the end.

3. B

4. C

5. C

6.

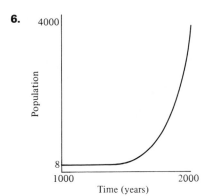

7.

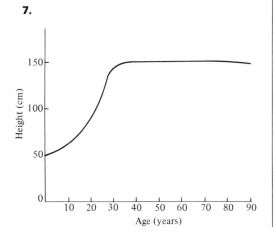

8.

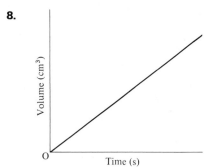

9. a) true   b) false   c) true

10.

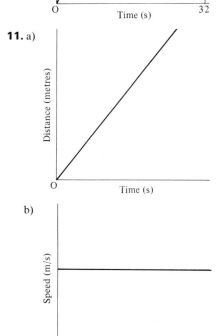

11. a)

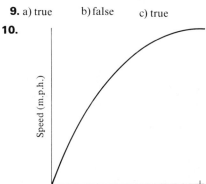

b)

12.

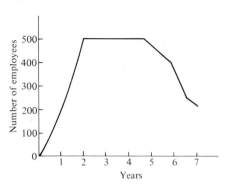

13.

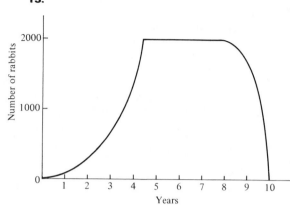

CHAPTER 11

Exercise 11a   Page 204

1. and **2.**

a) 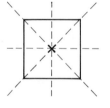   4

b) 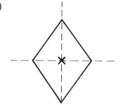   2

c)    2

d) 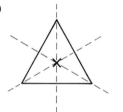   3

e) 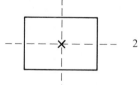   2

f)    1

3.

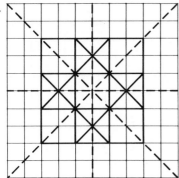

Rotational symmetry of order 4.

b)

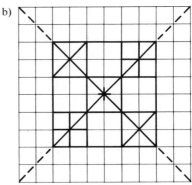

Rotational symmetry of order 2.

4. a)

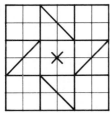

Rotational symmetry of order 4.

7. a) 8   b) 8
  c) 45°

8. a) n b) n

  c) $\dfrac{360°}{n}$

Exercise 11b   Page 207

1. a)

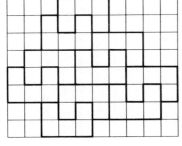

b)

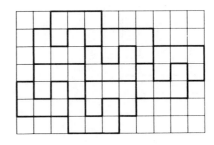

c)

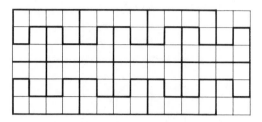

2. a)

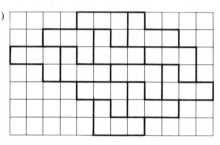

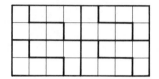

b)

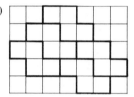

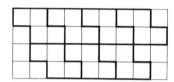

c)

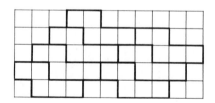

3. a) 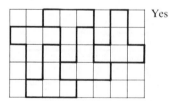   Yes

b) No

c) 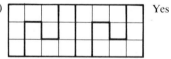   Yes

d)

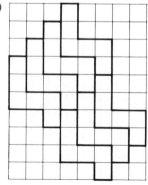

4. a) Equilateral triangle; 60°
    Square; 90°
    Pentagon; 108°
    Hexagon; 120°
    7-sided polygon; $128\frac{4}{7}$°
    Octagon; 135°
   b) Yes; Yes; No; Yes; No; No.
    The interior angle is a
    factor of 360°. There are
    no other polygons which
    tessellate. (The next
    factor of 360° is 180°.)

5. a)

b)

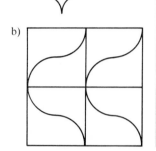

c)

For b) and c) there are
several other possibilities.

Exercise 11c   Page 210

1. a) Cuboid. I, K; H, L, N; A, C, G; D, F.
   b) Triangular prism. G, I; F, J; A, C, E.
   c) Pentagonal pyramid. B, D, F, H, J.
   d) Regular tetrahedron (triangular pyramid). A, C; F, D.

2.

	Cuboid	Pentagonal pyramid	Triangular prism	Tetrahedron
a) Edges	12	9	10	6
b) Vertices	8	6	6	4
c) Flaps	7	5	5	3
d) One possible net				

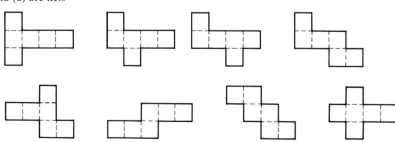

   e) All nets in one piece for a given solid need the same number of flaps.

3. (a) and (d) are nets

4. a)

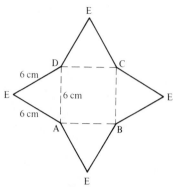

b) EM = 5.2 cm
c) Height = 4.24 cm
d) 50.9 cm³

5.

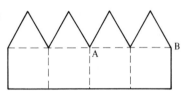

6. a)

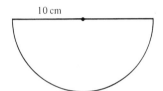

b)

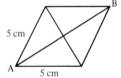

AB = 8.66 cm

7. i) a)

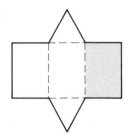

b) 4

ii) a)

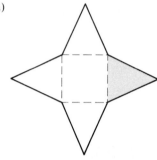

b) 3

iii) a)

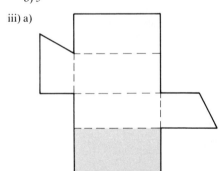

b) 4

Exercise 11d   Page 214

1. a)

	F	*E*	*V*
Cube	6	12	8
Tetrahedron	4	6	4
Pyramid	5	8	5
Prism	5	9	6

b) *E* = *F* + *V* − 2
d) 12 edges
e) No

2. a) 8 faces, 12 edges, 6 vertices
b) Yes

3. a) 20 faces, 30 edges,
   12 vertices
b) Yes

4. a) $L = 6, P = 4$

b)

	R	L	P
i)	4	6	4
ii)	2	4	4
iii)	3	5	4
iv)	5	8	5

d) $L = R + P - 2$

e) The plane figures could be pictures of solids, so lines are edges, points are vertices and regions are faces.

Exercise 11e   Page 217

1. a) 2   b) 4
   c) 2   d) 1

2. a) 9   b) Infinite
   c) 7   d) Infinite
   e) Infinite   f) 4

3. a) 1   b) 1
   c) Infinite   d) 1
   e) 1
   f) 0 (the pedals destroy any possible symmetry)
   g) 0   h) 6

Exercise 11f   Page 219

1. a)

b)
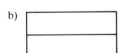

c)

d)
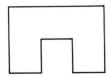

e)

f)

2. a) Sphere, cylinder, hemisphere

b) Cube, cuboid, wedge
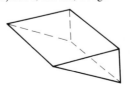

c) Either a square pyramid surmounting a cuboid

or a cuboid with a section in the shape of a square pyramid removed from it.

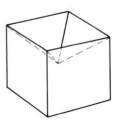

d) Triangular prism

3. a)   b)   c)

4. a)   i)   ii)   iii)

b)   i)   ii)   iii)

c)

d)   i)   ii)

5. a)   b)   c)

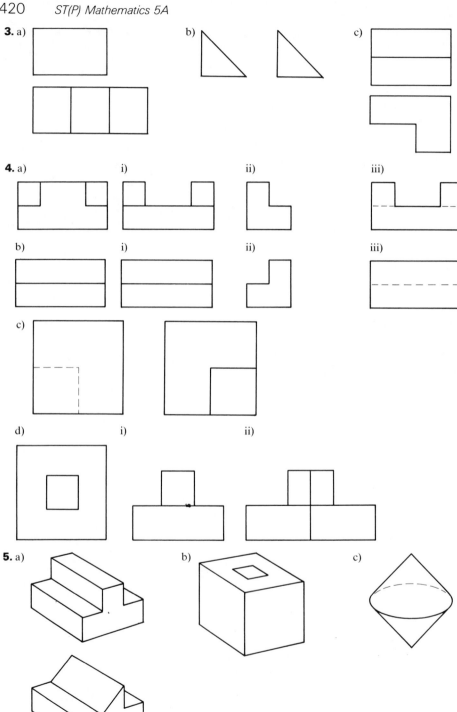

6.

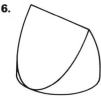

Exercise 11g   Page 224

---

1. a)

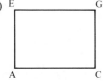

b) Yes

c)

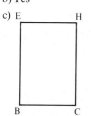

Congruent with
EACG

2. a)

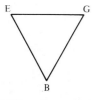

b) 8.49 cm
c) 31.2 cm²
d) 18 cm²
e) 36 cm³

3. a)

b) 17.7 cm²
c) 15.6 cm³, 109.4 cm³ or
$15\frac{5}{8}$, $109\frac{3}{8}$

4.

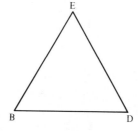

Isosceles; BD = 11.3 cm;
area = 56.6 cm²

b)

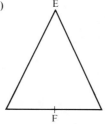

Area = 40 cm²

5. a) PQ = 3 cm
b)

Square; area = 9 cm²

c)

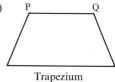

Trapezium

d) 1 : 7

6. a)

Area = 32 cm²

b)

Area = 24 cm²

c) 152 cm²
d) 96 cm³
e) 24 cm³
f) Cut (a)

1. a) Triangular prism or wedge
b) 6 vertices, 9 edges
c) ED = IH = GH = EF
   = 5 cm,
   AB = HE = GF = 8 cm,
   BC = AJ = JI = 3 cm
d) 108 cm³
e) 48 cm³
f)

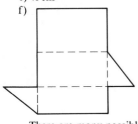

There are many possible nets
g) 10.6 cm
h) 9.43 cm
i) 12.0 cm

2. 20

3. a) Sphere
b)

c) Cone
d)   Cotton reel

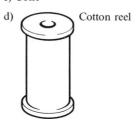

e)   Nail

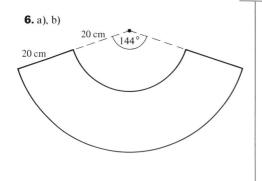

4. a) 14, 15, 16, 17, 18, 19.
b) 8, 9 or 10

5. a) 4
b)

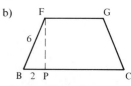

Trapezium

c) 5.66 cm
d) 317 cm²
e) FH = 8.48 cm,
   BD = 14.1 cm
f)

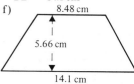

g)

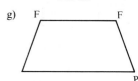

6. a), b)

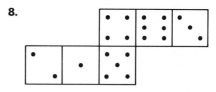

c) 48π cm²

7. a) No   b) F

8.

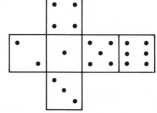

9. e.g.

CHAPTER 12

Exercise 12a   Page 233

1.

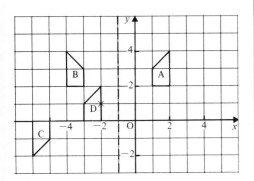

d) Rotation of $180°$ about $(-4, 0)$

e) C is turned over relative to B but there is no possible mirror line

2. a) $A \rightarrow B$, $B \rightarrow C$, $C \rightarrow A$

b) Reflection in line COR

c) Rotation of $120°$ clockwise about O.
$B \rightarrow A$

d) $Q \rightarrow C$, $R \rightarrow B$

3. a)

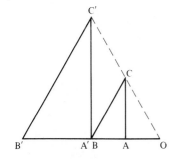

b) Trapezium BC ∥ B'C'

c) Isosceles $\triangle$ CC' = CB

4. i) Translations horizontally vertically, horizontally and diagonally in multiples of lengths of the pattern unit

ii) Rotations about corner and centre points of pattern unit through $90°$, $180°$, $270°$ anticlockwise

iii) Reflections in horizontal, vertical and diagonal lines

5. i) Reflection in line through midpoints of CS and BP. ABCD $\rightarrow$ QPSR

ii) Translation parallel to AB, distance 2AB. ABCD $\rightarrow$ PQRS

iii) Rotation of $180°$ about midpoint of BS. ABCD $\rightarrow$ RSPQ

iv)

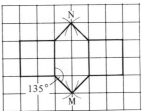

Rotation of $90°$ anticlockwise about N. ABCD $\rightarrow$ QRSP
Rotation of $90°$ clockwise about M. ABCD $\rightarrow$ SPQR

6.

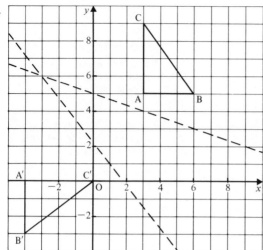

a) $(-3, 6)$
b) 90° clockwise

7. a) Invariant line is the *x*-axis. Scale factor is 4

b) Stretch parallel to the *x*-axis. Scale factor is 3.

c)

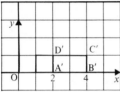

d)

8. a) Invariant line is the *x*-axis, $(1, 3) \rightarrow (3, 3)$

b) Shear, invariant line the *x*-axis, $(3, 3) \rightarrow (1, 3)$

c)

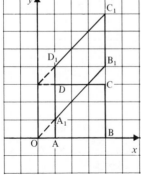

Exercise 12b   Page 236

1.

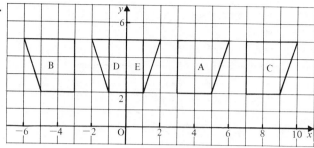

c) Translation $\begin{pmatrix} 4 \\ 0 \end{pmatrix}$

e) Translation $\begin{pmatrix} -4 \\ 0 \end{pmatrix}$

f) Two successive reflections in parallel mirror lines give a translation perpendicular to mirrors whose distance is twice the distance between mirrors

2.

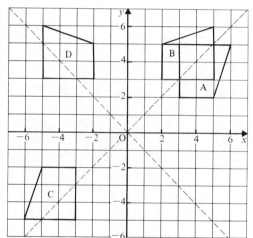

a) Rotation of 180° about O

e) Rotation of 90° anticlockwise about O

f) Two successive reflections in 2 non-parallel mirror lines give a rotation about O of twice the angle between the mirrors.

3.

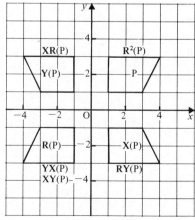

c) True

f) True

g) $X^2 = I$, $XY = YX = R$, $Y^2 = I$, $YR = RY = X$, etc.

4. a) $(0, 2)$, scale factor 2

b) Enlargement, centre $(0, 2)$, scale factor $\frac{1}{2}$

5.

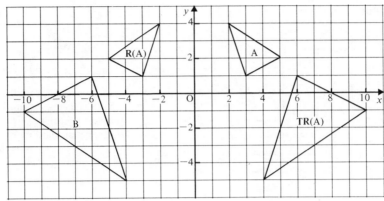

a) Enlargement, centre (0, 1), scale factor −2

b) Enlargement, centre (0, 1), scale factor −½

d) R

Exercise 12c   Page 238

1. 3×2 **2.** 4×3 **3.** 2×2 **4.** 4×1 **5.** 1×3 **6.** 2×2

Exercise 12d   Page 239

1. Not possible

2. $\begin{pmatrix} 1 & -3 \\ 7 & 3 \end{pmatrix}$

3. $\begin{pmatrix} 3 & -3 \\ 1 & -1 \end{pmatrix}$

4. $\begin{pmatrix} 1 & -3 \\ 7 & 3 \end{pmatrix}$

5. $\begin{pmatrix} -1 \\ 7 \end{pmatrix}$

6. $\begin{pmatrix} 4 & -6 \\ 8 & 2 \end{pmatrix}$

7. $\begin{pmatrix} 0 & -3 \\ 10 & 5 \end{pmatrix}$

8. $\begin{pmatrix} -11 & 12 \\ -7 & 2 \end{pmatrix}$

9. Not possible

10. $x = 2$ $y = -1$ $z = -4$

11. $p = 5$ $q = 3$ $r = -3$

Exercise 12e   Page 241

1. a) $\begin{pmatrix} 12 & 6 \\ 8 & 4 \end{pmatrix}$ b) (16)

c) Not possible   d) $\begin{pmatrix} 10 \\ 6 \end{pmatrix}$

e) Not possible   f) Not possible

g) Not possible   h) $\begin{pmatrix} 16 & -7 \\ 0 & 9 \end{pmatrix}$

i) $(12 \ 14 \ 8)$

2. $a = 1$ $b = 2$

3. $x = 2$ $y = 1$

4. $p = 2$, $q = 8$, $r = 0$

5. $\begin{pmatrix} 1 & 1 \\ 1 & 2 \end{pmatrix}$

Exercise 12f   Page 241

1. Not possible
2. Not possible
3. Not possible

4. $\begin{pmatrix} -6 & 5 \\ -4 & 9 \end{pmatrix}$

5. $\begin{pmatrix} -10 & 7 \\ -6 & 9 \end{pmatrix}$

6. $\begin{pmatrix} 6 & -1 \\ 2 & 0 \end{pmatrix}$

7. $\begin{pmatrix} 0 \\ 4 \end{pmatrix}$

8. Not possible

9. (4)

10. Not possible

11. $\begin{pmatrix} 4 & 1 & -2 \\ 8 & 2 & -4 \end{pmatrix}$

12. Not possible
13. D
14. C
15. A

Exercise 12g   Page 243

1. a) 2   b) -2
   c) 1   d) 2

2. a) $\begin{pmatrix} \frac{3}{5} & -\frac{1}{5} \\ -\frac{1}{5} & \frac{2}{5} \end{pmatrix}$ b) $\begin{pmatrix} 4 & -5 \\ -3 & 4 \end{pmatrix}$

   c) $\begin{pmatrix} -2 & 1 \\ \frac{5}{2} & -1 \end{pmatrix}$ d) $\begin{pmatrix} \frac{1}{2} & \frac{1}{2} \\ -\frac{1}{2} & \frac{1}{2} \end{pmatrix}$

3. a) $\begin{pmatrix} 1 & -1 \\ -1 & \frac{4}{3} \end{pmatrix}$ b) 2
   c) -4 d) 6
   e) -10 f) $\begin{pmatrix} -\frac{1}{2} & \frac{3}{2} \\ 0 & 1 \end{pmatrix}$
   g) $\begin{pmatrix} -2 & \frac{5}{2} \\ -1 & \frac{4}{3} \end{pmatrix}$ h) $\begin{pmatrix} -\frac{1}{2} & \frac{1}{2} \\ \frac{1}{2} & -\frac{1}{6} \end{pmatrix}$

4. a) $\begin{pmatrix} 4 & 7 \\ 3 & 5 \end{pmatrix}$ b) **A**

5. a) 2   b) 0
   c) $\begin{pmatrix} \frac{3}{2} & -2 \\ -2 & 3 \end{pmatrix}$ d) not possible
   e) Not possible

Exercise 12h   Page 244

1. a) $\begin{pmatrix} 10 & -14 \\ 8 & -11 \end{pmatrix}$ b) $\begin{pmatrix} -3 & -2 \\ 4 & 2 \end{pmatrix}$

   c) $\frac{1}{2}\begin{pmatrix} 2 & -2 \\ -5 & 6 \end{pmatrix}$ d) $\begin{pmatrix} 2 & 3 \\ 1 & 2 \end{pmatrix}$

   e) $\begin{pmatrix} 46 & 16 \\ 40 & 14 \end{pmatrix}$

   f) $\begin{pmatrix} 26 & -45 \\ -15 & 26 \end{pmatrix}$

   g) $\begin{pmatrix} 12 & 4 \\ 10 & 4 \end{pmatrix}$ h) $\begin{pmatrix} 6 & -9 \\ -3 & 6 \end{pmatrix}$

2. a) $\begin{pmatrix} 0 & 2 \\ 2 & 0 \end{pmatrix}$

   b) $\begin{pmatrix} 0 & 1 \\ 1 & 0 \end{pmatrix}$ i.e. **A**
   c) **I**   d) **A**
   e) **I**   f) **A**
   g) -1 h) 1
   i) **I**   j) **A**

3. $a = 23$, $b = -16$
   $c = 19$, $d = -6$

4. a) $\mathbf{AB} = \begin{pmatrix} 2 & 0 \\ 5 & k \end{pmatrix}$,

   $\mathbf{BA} = \begin{pmatrix} 2 & 0 \\ 4+3k & k \end{pmatrix}$
   b) $\frac{1}{3}$ c) $|\mathbf{B}| = 0$

5. $\begin{pmatrix} 16 & -4 \\ -6 & 2 \end{pmatrix}$

Exercise 12i   Page 245

1. a) Reflection in x-axis
   b) Enlargement centre O
      with scale factor 3
   c) Enlargement centre O
      with scale factor $\frac{1}{2}$
   d) Not possible to describe
      simply
   e) Rotation of 90°
      anticlockwise about O

2. a) Rotation of 180° about O
   b) Reflection in y-axis
   c) Reflection in x-axis
   d) Reflection in line $y = x$
   e) Reflection in line $y = -x$
   f) Rotation of 90°
      anticlockwise about O
   g) Identity transformation
   h) Rotation of 90° clockwise
      about O

3. a) Enlargement, centre O,
      with scale factor 2
   b) Enlargement, centre O,
      with scale factor -2
   c) Enlargement, centre O,
      with scale factor $\frac{1}{4}$
   d) Rotation of 180° about O,
      or enlargement, centre O,
      with scale factor -1

4. $a = 2$, $b = 0$, $c = 0$, $d = 3$

5. $\begin{pmatrix} 16 & -4 \\ -6 & 2 \end{pmatrix}$

Exercise 12j   Page 247

1. a) A$'(5,3)$, B$'(9,5)$,
      C$'(15,9)$, D$'(11,7)$
   b) $\begin{pmatrix} 2 & -3 \\ -1 & 2 \end{pmatrix}$ c) ABCD
   d) Inverse matrix gives
      inverse transformation

2. a) Square with vertices
      $(0,0)$, $(3,0)$, $(3,3)$, $(0,3)$
   b) Enlargement centre O
      with scale factor 3
   c) $\mathbf{M}^{-1} = \begin{pmatrix} \frac{1}{3} & 0 \\ 0 & \frac{1}{3} \end{pmatrix}$.
      Enlargement centre O,
      scale factor $\frac{1}{3}$

3. a) Rotation of 90° clockwise
      about O and reflection in
      $y = x$
   b) $\begin{pmatrix} 0 & -1 \\ 1 & 0 \end{pmatrix}$ and $\begin{pmatrix} 0 & 1 \\ 1 & 0 \end{pmatrix}$

4. a) $\begin{pmatrix} \frac{3}{2} & -2 \\ 1 & -1 \end{pmatrix}$ b) $(2,0)$
   c) $(1,3)$

Exercise 12k   Page 248

1. a) **P** gives a reflection in the x-axis, **Q** a rotation of $90°$ clockwise about O.

c) Reflection in the line $y = -x$

d) $\mathbf{R} = \begin{pmatrix} 0 & -1 \\ -1 & 0 \end{pmatrix}$

$\mathbf{S} = \begin{pmatrix} 0 & 1 \\ 1 & 0 \end{pmatrix}$

e) $A_3B_3C_3D_3$

f) Reflection in line $y = x$

g) (b) and (e) are the same. If we used **P** on $A_2B_2C_2D_2$ we would get (i)

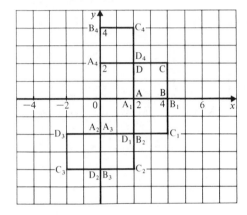

2. a)

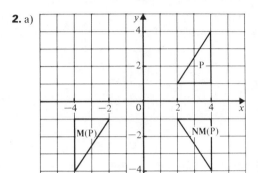

b) M is a rotation of $180°$ about O. NM is a reflection in the x-axis.

c) $\mathbf{L} = \begin{pmatrix} 1 & 0 \\ 0 & -1 \end{pmatrix}$

d) $\mathbf{L}(\mathbf{P}) = \mathbf{NM}(\mathbf{P})$, L is a reflection in the x-axis.

e) Yes.

3. a) $\begin{pmatrix} 0 & -1 \\ 1 & 0 \end{pmatrix}$ b) $\begin{pmatrix} 1 & 0 \\ 0 & -1 \end{pmatrix}$

c) $\begin{pmatrix} 0 & -1 \\ -1 & 0 \end{pmatrix}$ d) $\mathbf{R} = \mathbf{QP}$

4. a) $\begin{pmatrix} -y \\ x \end{pmatrix}$. Rotation of $90°$ anticlockwise about O.

b) $\begin{pmatrix} 1 & 0 \\ 0 & 1 \end{pmatrix}$.

Exercise 12l   Page 249

1.
$$\begin{array}{l} \text{Total cost in p} \\ \text{First week} \\ \text{Second week} \end{array} \begin{pmatrix} 166 \\ 279 \end{pmatrix}$$

2. a) $\mathbf{A} = \begin{pmatrix} 2 \\ 3 \\ 4 \end{pmatrix}$

b) $\mathbf{B} = \begin{pmatrix} 5 & 6 & 0 \\ 4 & 1 & 7 \\ 3 & 5 & 3 \end{pmatrix}$

c) $\mathbf{BA} = \begin{pmatrix} 28 \\ 39 \\ 33 \end{pmatrix}$

Distances in km travelled in first, second and third weeks

d) i) Number of visits over three weeks to each of the three areas, i.e. 12, 12 and 10

ii) Number of visits per week, i.e. 11, 12 and 11

3. a) $\mathbf{PQ} = \begin{pmatrix} 16 \\ 16 \end{pmatrix}$.

Sales of £16 are made each week.

b) $\mathbf{R} = \begin{pmatrix} 2000 \\ 500 \\ 1000 \end{pmatrix}$

$\mathbf{PR} = \begin{pmatrix} 8500 \\ 8000 \end{pmatrix}$.

Jigsaws sold in the first week had 8500 pieces between them and in the second week, 8000.

c) i) $\mathbf{PS} = \begin{pmatrix} 9 \\ 9 \end{pmatrix}$

9 jigsaws were sold each week.

ii) $\mathbf{TP} = (1 \ 5 \ 12)$. Over the two weeks, one A, 5 B and 12 C jigsaws were sold.

iii) $\mathbf{TPR} = (1650)$. The jigsaws contained 16 500 pieces altogether.

iv) $\mathbf{TPQ} = (32)$. The jigsaws sold for £32 altogether.

4. a) $\begin{pmatrix} 5 & 7 & 0 & 6 \\ 1 & 10 & 12 & 0 \end{pmatrix}$

b) $(6 \ 17 \ 12 \ 6)$ gives the number of bottles of milk delivered to each family.

c) **KM** gives the information (1×4).

d) $\mathbf{L} = \begin{pmatrix} 1 \\ 1 \\ 1 \\ 1 \end{pmatrix}$

ML gives the information (2×1).

c) i) **KML**   ii) **NML**

Exercise 12m   Page 252

1. From $\begin{array}{c} \\ A \\ B \end{array}\begin{array}{cc} \text{To} \\ A & B \\ \begin{pmatrix} 0 & 2 \\ 1 & 0 \end{pmatrix} \end{array}$

3. From $\begin{array}{c} \\ A \\ B \end{array}\begin{array}{cc} \text{To} \\ A & B \\ \begin{pmatrix} 2 & 1 \\ 1 & 2 \end{pmatrix} \end{array}$

2. From $\begin{array}{c} \\ A \\ B \\ C \end{array}\begin{array}{ccc} \text{To} \\ A & B & C \\ \begin{pmatrix} 0 & 2 & 1 \\ 2 & 0 & 1 \\ 1 & 1 & 0 \end{pmatrix} \end{array}$

4. From $\begin{array}{c} \\ A \\ B \\ C \end{array}\begin{array}{ccc} \text{To} \\ A & B & C \\ \begin{pmatrix} 0 & 1 & 0 \\ 0 & 0 & 1 \\ 1 & 0 & 0 \end{pmatrix} \end{array}$

5.

6.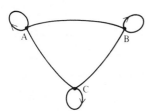

7.

8.

9. a) From $\begin{array}{c} \\ A \\ B \\ C \end{array}\begin{array}{ccc} A & B & C \\ \begin{pmatrix} 1 & 1 & 1 \\ 2 & 0 & 1 \\ 1 & 1 & 0 \end{pmatrix} \end{array}$

b) $\begin{pmatrix} 4 & 3 & 2 \\ 3 & 3 & 2 \\ 3 & 1 & 2 \end{pmatrix}$

c) 2   d) see b)
e) $\mathbf{R}^3$ gives three-stage routes (e.g. **AABC**) from A to C.

10. a) $\mathbf{R} = \begin{pmatrix} 0 & 2 & 1 \\ 2 & 0 & 2 \\ 1 & 2 & 0 \end{pmatrix}$

b) $\mathbf{R}^2 = \begin{pmatrix} 5 & 2 & 4 \\ 2 & 8 & 2 \\ 4 & 2 & 5 \end{pmatrix}$,

c) 4 routes
d) 9 routes (this includes routes such as ACBC.)

$\mathbf{R}^3 = \begin{pmatrix} 8 & 18 & 9 \\ 18 & 8 & 18 \\ 9 & 18 & 8 \end{pmatrix}$

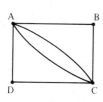
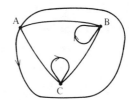

CHAPTER 13

Exercise 13a   Page 254 ──────────────────────────────────

1. b), c)

2. a)   b)   c) 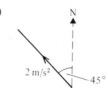   d)

Exercise 13b   Page 255 ──────────────────────────────────

1. $a = \begin{pmatrix} 3 \\ 4 \end{pmatrix}$, $b = \begin{pmatrix} -4 \\ -3 \end{pmatrix}$, $c = \begin{pmatrix} 0 \\ -3 \end{pmatrix}$, $d = \begin{pmatrix} 2 \\ -3 \end{pmatrix}$

2.

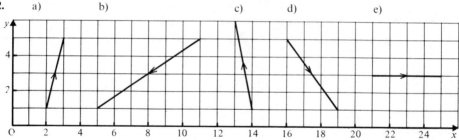

a)   b)   c)   d)   e)

3. $\overrightarrow{AB} = \begin{pmatrix} 3 \\ 6 \end{pmatrix}$, $\overrightarrow{BA} = \begin{pmatrix} -3 \\ -6 \end{pmatrix}$, $\overrightarrow{AC} = \begin{pmatrix} 1 \\ -3 \end{pmatrix}$

4. a) B(2, −2), C(3, 3)

   b)

5. (−4, 5)

6.

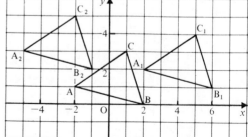

c) (i) $\begin{pmatrix} -7 \\ 1 \end{pmatrix}$ (ii) $\begin{pmatrix} 7 \\ -1 \end{pmatrix}$

Exercise 13c   Page 257

1. a) $a = \begin{pmatrix} -1 \\ 2 \end{pmatrix}$, $b = \begin{pmatrix} 1 \\ 2 \end{pmatrix}$,

$c = \begin{pmatrix} -2 \\ 1 \end{pmatrix}$, $d = \begin{pmatrix} 2 \\ 1 \end{pmatrix}$,

$e = \begin{pmatrix} 4 \\ 2 \end{pmatrix}$, $f = \begin{pmatrix} -1\frac{1}{2} \\ 3 \end{pmatrix}$,

$g = \begin{pmatrix} 4 \\ -2 \end{pmatrix}$, $h = \begin{pmatrix} 3 \\ 6 \end{pmatrix}$

b) $g = -2c$, $h = 3b$,
$f = 1\frac{1}{2}a$, $e = 2d$

4. a) $3:1$ b) $3:4$
c) $1:4$
5. a) $2:1$ b) $3:2$
c) $3:1$

2. a) $a = \begin{pmatrix} 3 \\ 0 \end{pmatrix}$, $b = \begin{pmatrix} 3 \\ 4 \end{pmatrix}$,

$c = \begin{pmatrix} 3 \\ 4 \end{pmatrix}$, $d = \begin{pmatrix} 0 \\ 6 \end{pmatrix}$,

$e = \begin{pmatrix} -3 \\ 4 \end{pmatrix}$, $f = \begin{pmatrix} 6 \\ 0 \end{pmatrix}$,

$g = \begin{pmatrix} -3 \\ -4 \end{pmatrix}$

b) $f = 2a$, $b = c = -g$

6. $\overrightarrow{AD} = \begin{pmatrix} -4 \\ 4 \end{pmatrix}$ D is $(-5,\ 4)$

3. b) $\overrightarrow{AC} = \begin{pmatrix} 4 \\ 3 \end{pmatrix} = \frac{1}{3}a$,

$\overrightarrow{CB} = \begin{pmatrix} 8 \\ 6 \end{pmatrix} = \frac{2}{3}a$,

c) $\overrightarrow{BA} = -a$
$\overrightarrow{CA} = -\frac{1}{3}a$,
$\overrightarrow{BC} = -\frac{2}{3}a$

7. $\frac{5}{3}a$

Exercise 13d   Page 259

1. $|a| = |b| = |c| = |d| = 2.24$,
$|e| = |g| = 4.47$,
$|f| = 3.35$, $|h| = 6.71$

2. $|a| = 3$,
$|b| = |c| = |e| = |g| = 5$,
$|d| = |f| = 6$

Exercise 13e   Page 261

4. $\begin{pmatrix} 3 \\ 6 \end{pmatrix}$

5. $\begin{pmatrix} 3 \\ 6 \end{pmatrix}$

6. $\begin{pmatrix} 9 \\ -2 \end{pmatrix}$

7. $\begin{pmatrix} 1 \\ -1 \end{pmatrix}$

8. $\begin{pmatrix} 10 \\ -3 \end{pmatrix}$

9. $\begin{pmatrix} 2 \\ 7 \end{pmatrix}$

10. $\begin{pmatrix} -2 \\ -7 \end{pmatrix}$

11. $\begin{pmatrix} 1 \\ -1 \end{pmatrix}$

12. $\begin{pmatrix} -9 \\ 2 \end{pmatrix}$

13. Yes
14. No
15. c) Yes   d) 6; no
16. c) The line representing
$2a + 2b$ is parallel to the
line representing $a + b$
and is twice its length

Exercise 13f   Page 262

1. a) Yes
b) i) $\overrightarrow{QS} = b - a$,
ii) $\overrightarrow{SR} = 2a - b$,
iii) $\overrightarrow{RS} = b - 2a$

2. a) $\overrightarrow{AB} = q - p$
b) $\overrightarrow{CD} = \frac{1}{2}q$
c) $\overrightarrow{DB} = \frac{1}{2}q$
d) $\overrightarrow{AD} = \frac{1}{2}q - p$

3. a) Trapezium
b) i) $\overrightarrow{BC} = a - b$,
ii) $\overrightarrow{BD} = a - 2b$,
iii) $\overrightarrow{AC} = a + b$
4. $\overrightarrow{BC} = \begin{pmatrix} -7 \\ 2 \end{pmatrix}$, $\overrightarrow{CB} = \begin{pmatrix} 7 \\ -2 \end{pmatrix}$
5. $h = 2, k = 3$

432   ST(P) Mathematics 5A

Exercise 13g   Page 264

1. a) $\overrightarrow{AP} = \frac{1}{3}b$
b) $\overrightarrow{AQ} = \frac{1}{2}c$
c) $\overrightarrow{BC} = c - b$
d) $\overrightarrow{PQ} = \frac{1}{2}c - \frac{1}{3}b$
e) $\overrightarrow{PQ} = \frac{1}{2}\overrightarrow{BC}$
f) $PQ = \frac{1}{2}BC$

2. a) $\overrightarrow{OP} = \frac{1}{2}a$
b) $\overrightarrow{AB} = b - a$
c) $\overrightarrow{AQ} = \frac{1}{2}(b - a)$
d) $\overrightarrow{PQ} = \frac{1}{2}b$
e) $\overrightarrow{SR} = \frac{1}{2}b$
g) Parallelogram

3. a) $\overrightarrow{BD} = b - a$
b) $\overrightarrow{BP} = \frac{1}{3}(b - a)$
c) $\overrightarrow{BQ} = \frac{2}{3}(b - a)$
d) $\overrightarrow{AP} = \frac{2}{3}a + \frac{1}{3}b$
e) $\overrightarrow{CQ} = \frac{2}{3}a + \frac{1}{3}b$

4. a) $\overrightarrow{AB} = b$
b) $\overrightarrow{AP} = \frac{2}{3}b$
c) $\overrightarrow{OP} = a + \frac{2}{3}b$
d) $\overrightarrow{OR} = \frac{1}{2}b$
e) $\overrightarrow{CQ} = \frac{3}{4}a$
f) $\overrightarrow{RQ} = \frac{3}{4}a + \frac{1}{2}b$
g) $\overrightarrow{RQ} = \frac{3}{4}\overrightarrow{OP}$
h) $RQ : OP = 3 : 4$

5. a) A, B and C lie in a
straight line
b) Parallelogram
c) Trapezium   d) ECBF

6. a) $\overrightarrow{BP} = 6a - 2b$
b) $\overrightarrow{BQ} = 3a - b$
c) $\overrightarrow{BP} = 2\overrightarrow{BQ}$
d) $1 : 2$

7. a) $3r - q$ b) $\frac{3}{2}r - \frac{1}{2}q$
c) $\frac{3}{2}r + \frac{1}{2}q$ d) $3r$
e) i) parallelogram,
ii) trapezium

Exercise 13h   Page 268

1. $\begin{pmatrix} 3\frac{1}{2} \\ 5 \end{pmatrix}$

2. a) $\begin{pmatrix} -1\frac{1}{2} \\ 6\frac{1}{2} \end{pmatrix}$ b) $\begin{pmatrix} 1\frac{1}{2} \\ 2\frac{1}{2} \end{pmatrix}$
c) $\begin{pmatrix} -3 \\ 4 \end{pmatrix}$

3. a) $\begin{pmatrix} 5 \\ 6 \end{pmatrix}$ b) $\begin{pmatrix} 2 \\ 1\frac{1}{2} \end{pmatrix}$
c) $\begin{pmatrix} 2 \\ 3 \end{pmatrix}$

4. a) $(-7, 7)$ b) $(-2, 3)$
c) $(-3, 1)$ i.e. B

5. a) $\overrightarrow{OC} = b - a$, $\overrightarrow{OD} = -a$,
$\overrightarrow{OE} = -b$, $\overrightarrow{OF} = a - b$
b) i) $\overrightarrow{AB} = b - a$,
ii) $\overrightarrow{BC} = -d$,
iii) $\overrightarrow{CD} = -b$,
iv) $\overrightarrow{DE} = a - b$
v) $\overrightarrow{EF} = a$
vi) $\overrightarrow{FA} = b$

6. $\overrightarrow{OA} = 3x - y$,
$\overrightarrow{OB} = -x - y$,
$\overrightarrow{OC} = -2x + 4y$

7. a) $\overrightarrow{BA} = a - b$
b) $\overrightarrow{AB} = b - a$
c) $\overrightarrow{BC} = \frac{1}{2}(a - b)$
d) $\overrightarrow{AC} = \frac{1}{2}(b - a)$
e) $\overrightarrow{OC} = \frac{1}{2}a + \frac{1}{2}b$; Yes

8. a) $\overrightarrow{AC} = y - x$
b) $\overrightarrow{AD} = \frac{1}{3}(y - x)$
c) $\overrightarrow{CA} = x - y$
d) $\overrightarrow{CD} = \frac{2}{3}(x - y)$
e) $\overrightarrow{BD} = \frac{2}{3}x + \frac{1}{3}y$

9. a) $\overrightarrow{AB} = b - a$
b) $\overrightarrow{BA} = (a - b)$
c) $\overrightarrow{AC} = \frac{1}{4}(b - a)$
d) $\overrightarrow{BC} = \frac{3}{4}(a - b)$
e) $\overrightarrow{OC} = \frac{3}{4}a + \frac{1}{4}b$
f) $\overrightarrow{OD} = \frac{1}{2}a + \frac{1}{6}b$
g) $\overrightarrow{DC} = \frac{1}{4}a + \frac{1}{12}b$

10. a) $\overrightarrow{PQ} = q - p$
b) $\overrightarrow{PR} = k(q - p)$
c) $\overrightarrow{RQ} = (l - k)q + (k - l)p$
d) $\overrightarrow{OR} = kq + (l - k)p$

Exercise 13i   Page 272

1. $h = -4, k = 5$
2. $h = 7, k = -6$
3. $h = 6, k = 12$
4. $h = 5, k = 7$
5. $h = 5, k = 6$

6. $h = \frac{1}{2}, k = -\frac{5}{6}$
7. $h = 2, k = 1$
8. $h = \frac{7}{5}, k = 1$
9. $h = 2, k = \frac{1}{2}$
10. $k = 2$

11. $k = -28$
12. $k = 4$
13. $k = \pm 6$
14. $k = 6$ or -4
15. $k = 5$ or -1

16. $k = 2$ or $\frac{1}{5}$

17. a) $\overrightarrow{OD} = k(\frac{2}{3}\mathbf{a} + \frac{1}{3}\mathbf{b})$,

$\overrightarrow{BD} = \frac{2}{3}k\mathbf{a} + (\frac{1}{3}k - 1)\mathbf{b}$

b) 3

c) $OC : CD = 1 : 2$

18. a) $\overrightarrow{BD} = k\mathbf{a} - \frac{1}{4}\mathbf{b}$

b) $k = \frac{1}{2}$ c) $\dfrac{BD}{OC} = \frac{1}{2}$

19. a) $\overrightarrow{RQ} = h\mathbf{p}$

b) $\overrightarrow{PR} = \mathbf{r} - \mathbf{p}$

c) $\overrightarrow{PQ} = (h - 1)\mathbf{p} + \mathbf{r}$

d) $\overrightarrow{PS} = k(\mathbf{r} - \mathbf{p})$

e) $\overrightarrow{OS} = (l - k)\mathbf{p} + k\mathbf{v}$

f) $h = \dfrac{1}{k}$

g) $h = 2$, $k = \frac{1}{2}$

20. a) $\overrightarrow{PQ} = \mathbf{q} - \mathbf{p}$

b) $\overrightarrow{PM} = \frac{1}{3}(\mathbf{q} - \mathbf{p})$

c) $\overrightarrow{OM} = \frac{2}{3}\mathbf{p} + \frac{1}{3}\mathbf{q}$

d) $\overrightarrow{ON} = \frac{1}{2}\mathbf{p}$

e) $\overrightarrow{QN} = \frac{1}{2}\mathbf{p} - \mathbf{q}$

f) $\overrightarrow{QL} = h(\frac{1}{2}\mathbf{p} - \mathbf{q})$

g) $\overrightarrow{OL} = \frac{1}{2}h\mathbf{p} + (1 - h)\mathbf{q}$

h) $\overrightarrow{OL} = k(\frac{2}{3}\mathbf{p} + \frac{1}{3}\mathbf{q})$

i) $h = \frac{4}{5}$, $k = \frac{3}{5}$

Exercise 13j   Page 277

1. a) $6\mathbf{i} + 8\mathbf{j}$ c) $\mathbf{i} - \mathbf{j}$ e) $-7\mathbf{i} - 4\mathbf{j}$ g) $6\mathbf{i} - 2\mathbf{j}$ i) $6\mathbf{i} + 15\mathbf{j}$

b) $2\mathbf{i} + 5\mathbf{j}$ d) $7\mathbf{i} + 4\mathbf{j}$ f) $5\mathbf{i} + 9\mathbf{j}$ h) $\mathbf{i} - 11\mathbf{j}$

2. a)

b)

c)

d)

e)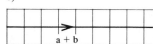

f)

g)

h)

i)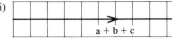

j) $|\mathbf{a}| = \sqrt{34}$, $|\mathbf{b}| = 5$, $|\mathbf{c}| = 2$

3. $2\mathbf{i} + 4\mathbf{j}$, $-3\mathbf{i} + 9\mathbf{j}$, $-6\mathbf{i} - 4\mathbf{j}$

4. a) $5\mathbf{i} + 4\mathbf{j}$, $-5\mathbf{i} + \mathbf{j}$, $\mathbf{i} - 3\mathbf{j}$

b) $\overrightarrow{PQ} = -10\mathbf{i} - 3\mathbf{j}$,

$\overrightarrow{QR} = 6\mathbf{i} - 4\mathbf{j}$,

$\overrightarrow{RQ} = -6\mathbf{i} + 4\mathbf{j}$

c) $2\frac{1}{2}\mathbf{i}$, $-2\mathbf{i} - \mathbf{j}$

d) $\overrightarrow{PS} = 3\mathbf{i} - 2\mathbf{j}$, $\overrightarrow{QS} = 13\mathbf{i} + \mathbf{j}$

e) $\overrightarrow{PS} = \frac{1}{2}\overrightarrow{QR}$, $\dfrac{PS}{QR} = \frac{1}{2}$

5. a) $2\mathbf{i} + 4\mathbf{j}$, $(2, 4)$

b) $\mathbf{i} + 2\mathbf{j}$

c) $\overrightarrow{EC} = 2\mathbf{i} + 1\frac{1}{2}\mathbf{j} = \frac{1}{2}\overrightarrow{OA}$

Exercise 13k   Page 279

1. a) $\overrightarrow{OR} = \frac{4}{5}\mathbf{p}$, $\overrightarrow{RP} = \frac{1}{5}\mathbf{p}$,

$\overrightarrow{PQ} = \mathbf{q} - \mathbf{p}$

b) $\overrightarrow{PS} = \frac{1}{5}(\mathbf{q} - \mathbf{p})$, $\overrightarrow{RS} = \frac{1}{5}\mathbf{q}$

c) RS is parallel to OQ and

$RS = \frac{1}{5}OQ$

d) trapezium   e) $120\ \text{cm}^2$

2. a) $\overrightarrow{AD} = 2\mathbf{b} - 2\mathbf{a}$

b) $\overrightarrow{BE} = 2\mathbf{b} - 4\mathbf{a}$

c) $\overrightarrow{EG} = 3\mathbf{a}$

d) $\overrightarrow{HG} = 2\mathbf{b} - 2\mathbf{a}$

e) $\overrightarrow{HG} = 2\overrightarrow{FE}$

f) parallelogram

3. a) $\overrightarrow{OD} = 2\mathbf{c}$, $\overrightarrow{AD} = 2\mathbf{c} - \mathbf{a}$

b) $\mathbf{b} = \frac{1}{2}\mathbf{a} + \mathbf{c}$

c) $k = \frac{3}{2}$

4. a) $(-1, 10)$

b) $h = 3$, $k = -2$

5. $A(2, -1)$, $B(5, 3)$,

$C(3, 8)$, $D(0, 4)$

CHAPTER 14

Exercise 14a   Page 281 ——————————————————

1. 25, 26
2. 15, 18
3. 35, 42
4. 10, 14
5. 19, 23
6. 15, 17
7. −9, −12
8. $\frac{1}{16}$, $\frac{1}{32}$
9. $1 + 3 + 5 + 7 + 9 + 11$,
   $1 + 3 + 5 + 7 + 9 + 11 + 13$
10. 13, 21
11. 720, 5040
12. 35, 48
13. 48, −96
14. 162, 486
15. 56, 72
16. 26, 37
17. $\frac{1}{81}$, $-\frac{1}{243}$
18. a) (5, 26)   b) (10, 101)
19.   1   5   10 10   5   1
      1   6   15 20 15   6   1
    1   7   21 35 35 21   7   1

20. 15, 21, 28, 36
21. 1, 4, 9, 16, 25, 36, 49, 64
22. 3, 12, 48, 192, 768, 3072
23. a)

b) 1, 5, 14, 30, 55

24. a) The sum of the squares of
     the first two is equal to
     square of the third.
    b) Any one of (9, 40, 41),
     (11, 60, 61), (13, 84, 85)…
     There is a pattern to these
     numbers:
     the first numbers go
     3, 5, 7, 9,…
     the second numbers go
     4, 4 + 8, 4 + 8 + 12,
     4 + 8 + 12 + 16,…
     the third number in each
     triple is the second + 1
25. $\frac{1}{2}$, $\frac{1}{4}$, $\frac{1}{8}$, $\frac{1}{16}$, $\frac{1}{32}$. In theory the
    bar will last for ever!
26. Each sequence ends in the
    form …, x, x, 0. The
    number x is the highest
    common factor of the
    integers that start the
    sequence.

Exercise 14b   Page 285 ——————————————————

1. 3, 5, 7, 9, …
2. 2, 4, 6, 8, …
3. 2, 4, 8, 16, ….
4. 1, 4, 9, 16
5. 0, 3, 8, 15

6. 5, 6, 7, 8
7. a) 1, 3, 6, 10, 15   b) 210
   c) $\frac{1}{2}(n-1)(n)$ or $\frac{1}{2}n(n-1)$
   d) n e) 2, 3, 4, 5
   f) 21

8. a) 1, 5, 14, 30, 55
   b) 2870
   c) $\frac{1}{6}(n-1)(n)(2n-1)$
   d) n^2
   e) 1, 4, 9, 16   f) 400

Exercise 14c   Page 287 ——————————————————

1. $2n$
2. $2n + 1$
3. $3(n-1)$
4. n^2
5. $n^2 + 1$
6. $n(n+1)$
7. n^3

8. 2^n
9. $\dfrac{1}{n+1}$
10. $(0.1)^n$
11. $4 - n$
12. $3 \times 2^{n-1}$

13. a)   1   6   15 20 15   6   1
        1   7   21 35 35 21   7   1
      1   8   28 56 70 56 28   8   1
   b) 1, 2, 4, 8, 16, 32, 64, 128, 256
   c) 2^{n-1}
14. a) 2 m   b) 20 m
   c) $n(n+1)$ metres

Exercise 14d   Page 289 ——————————————————

1. a) 0.$\dot{3}$ b) 0.0$\dot{3}$
   c) 0.01$\dot{6}$ d) 0.00$\dot{3}$
2. a) 0.0$\dot{9}$ b) 0.04$\dot{5}$
   c) 0.0$\dot{3}$ d) 0.1$\dot{8}$
3. a) 0.00$\dot{1}$ b) 0.00$\dot{2}$
   c) 0.00$\dot{9}$ d) 0.0$\dot{0}$$\dot{9}$
4. a) 0.$\dot{1}$ b) 0.$\dot{9}$

   c) $\frac{1}{9} \times 9 = 1$, but $\frac{1}{9} = 0.1111…$
   and $0.111… \times 9 = 0.999…$
   so $0.999… = 1$ (?)
   d) $1 - 0.999… = 0.000…$
   (0.$\dot{9}$ has no 'end' so
   0.0000… has no 'end')
   ∴ $1 - 0.\dot{9} = 0$, ie 0.$\dot{9}$ is
   equal to 1!

5. a) $\frac{1}{9}$ b) $\frac{1}{99}$ c) $\frac{1}{333}$
6. a) 0.058 823   b) 0.999 991
   c) 0.000 009
   d) 0.000 000 529 41
   e) 0.058 823 529 4
7. a) 0.076 923   b) 0.999 999
   c) 0.000 001   d) 0.000 000 076 923
   e) 0.076 923 076 92

Exercise 14e   Page 292

1. a) $-4, \sqrt{9}$
b) $-4, 0.5, -1.8, \frac{2}{7}, \sqrt{9}$
c) $\sqrt{5}, \pi$
$\left\{-4, -1.8, \frac{2}{7}, 0.5, \sqrt{5}, \sqrt{9}, \pi\right\}$

2. $-3 < -2 < 1.6 < 2.5 < 2.501$

3. a) true   b) true
c) false   d) false

4. $0.125 < 0.15 < 0.5 < 0.625$

5. a) 3   b) $-6, 3$
c) $-6, 0.5, \frac{2}{5}, 3, 1.25, -0.5$
$-6 < -0.5 < \frac{2}{3} < 0.5 < 1.25$
$< \sqrt{5} < 3$

6. a) yes   b) yes
c) yes
d) no, e.g. $4 \div 3$

7. $5 + 3$, and an infinite number of other possibilities.

8. a) yes   b) yes
c) yes   d) yes
e) no

9. a) $5 + 3$ b) $13 + 3$
c) $23 + 5$
d) $3 + 3$: if you assumed that the prime numbers had to be different you read more into the statement than was there: be careful.

10. 4.472

11. 1.703

12. 0.707

13. a) $1\frac{1}{2}$ b) $1\frac{2}{5}$
c) $1\frac{5}{12}$

14. a) $1 + \cfrac{1}{2 + \cfrac{1}{2 + \cfrac{1}{2 + \frac{1}{2}}}} = 1\frac{12}{29}$

b) $1 + \cfrac{1}{2 + \cfrac{1}{2 + \cfrac{1}{2 + \cfrac{1}{2 + \frac{1}{2}}}}} = 1\frac{29}{70}$

15. 2.25, 1.96, 2.006 944, 1.998 810, 2.000 204
The square of the fraction appears to be getting closer to 2, hence the fraction is getting closer to $\sqrt{2}$.

CHAPTER 15

Exercise 15a   Page 297

1.

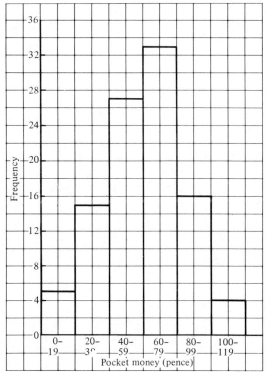

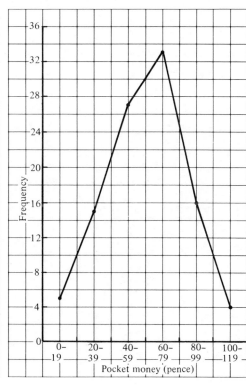

2.

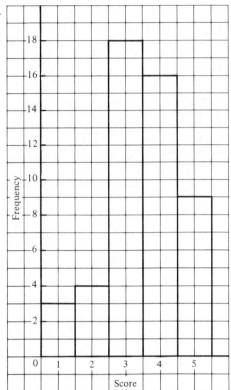

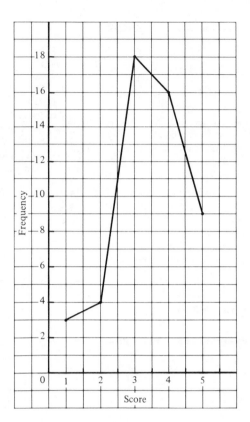

3.

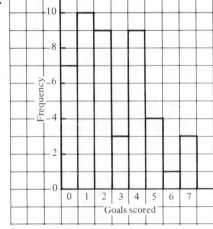

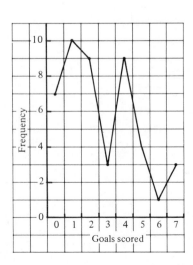

4.

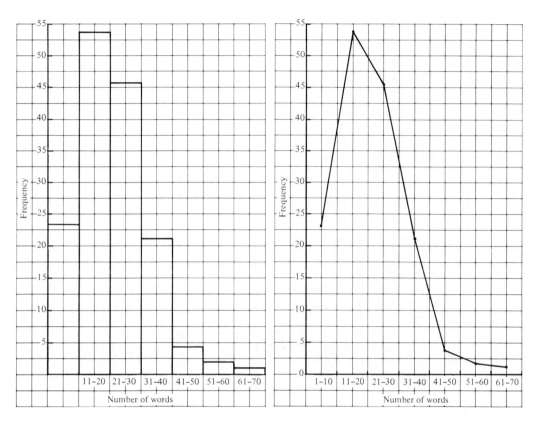

5.

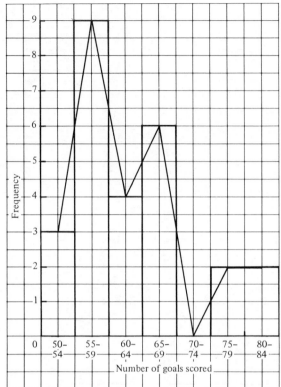

6.

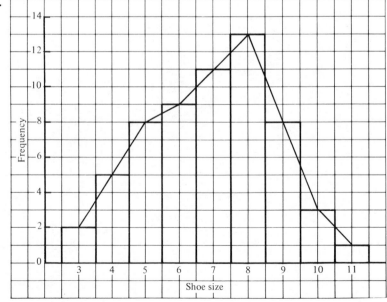

Exercise 15b   Page 298

1.
Weekday	Number of lunches served	Running total of lunches served
Monday	126	126
Tuesday	154	280
Wednesday	144	424
Thursday	175	599
Friday	118	717

2.
Place	Distance from Cardiff
Newport	10
Severn Bridge	26
Leigh Delamere	54
Swindon	72
Reading	111
London airport	139
Central London	154

3.
Day	Amount spent	Running total of expenditure
Monday	12	12
Tuesday	26	38
Wednesday	5	43
Thursday	8	51
Friday	32	83
Saturday	27	110
Sunday	4	114

Exercise 15c   Page 300

1.
Score	Frequency	Score	Cumulative frequency
0	3	$\leqslant 0$	3
1	8	$\leqslant 1$	$3 + 8 = 11$
2	4	$\leqslant 2$	$11 + 4 = 15$
3	3	$\leqslant 3$	$15 + 3 = 18$
4	5	$\leqslant 4$	$18 + 5 = 23$
5	2	$\leqslant 5$	$23 + 2 = 25$
6	1	$\leqslant 6$	$25 + 1 = 26$

2.
Mark	Frequency	Mark	Cumulative frequency
1–10	7	$\leqslant 10$	7
11–20	14	$\leqslant 20$	21
21–30	18	$\leqslant 30$	39
31–40	33	$\leqslant 40$	72
41–50	36	$\leqslant 50$	108
51–60	43	$\leqslant 60$	151
61–70	21	$\leqslant 70$	172
71–80	15	$\leqslant 80$	187
81–90	12	$\leqslant 90$	199
91–100	8	$\leqslant 100$	207

a) 207   b) 108   c) 56

3.

Score	≤ 19	≤ 39	≤ 59	≤ 79	≤ 99	≤ 119	≤ 139
Cumulative frequency	8	22	55	61	66	69	70

a) 70   b) 55   c) 48

4.

Score	67	68	69	70	71	72	73	74	75	76	77	78
Frequency	2	4	9	9	12	15	13	8	5	8	6	4

a) 13   b) 23

5.

Number of books sold	0–5	6–10	11–15	16–20	21–25
Frequency	77	124	182	228	164
Number of books sold	≤ 5	≤ 10	≤ 15	≤ 20	≤ 25
Cumulative frequency	77	201	383	611	775

Number of books sold	26–30	31–35	36–40	41–45	46–50
Frequency	92	73	32	22	9
Number of books sold	≤ 30	≤ 35	≤ 40	≤ 45	≤ 50
Cumulative frequency	867	940	972	994	1003

a) 136   b) 611   c) 666
It could have been shared

Exercise 15d   Page 303

1.

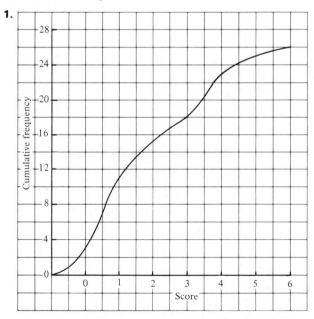

2.

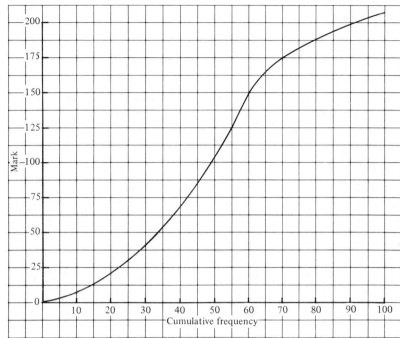

3.

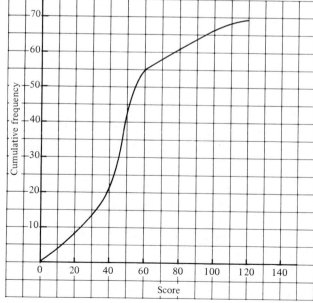

4.

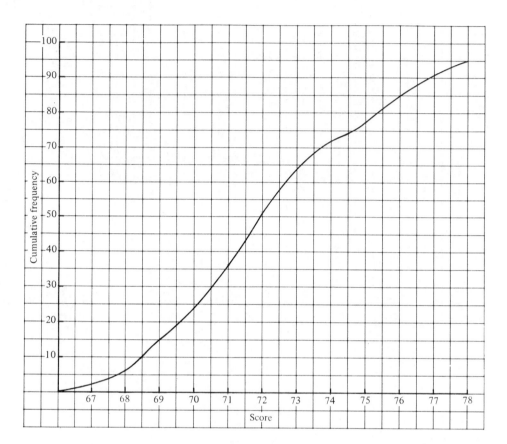

Exercise 15e   Page 304

1. 1.1, 49, 46, 71.9

2.

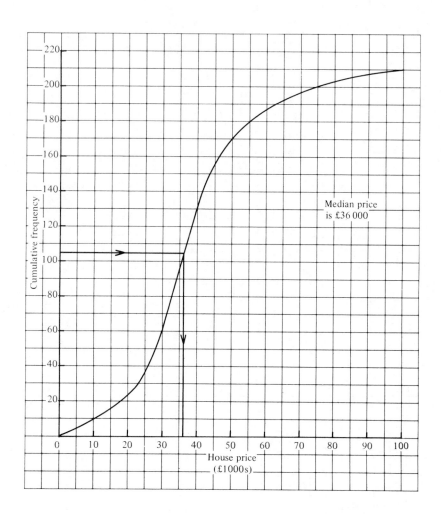

Median price
is £36 000

Cumulative frequency

House price
(£1000s)

3. a) 90   b) 56

4.

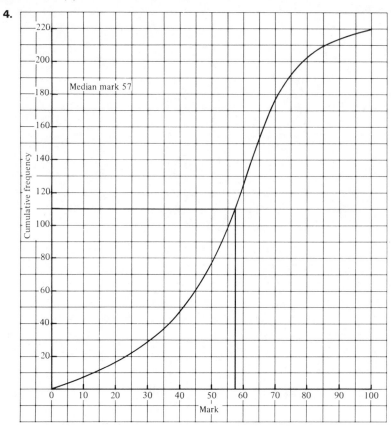

5.

Number of cars	≤ 40	≤ 41	≤ 42	≤ 43	≤ 44	≤ 45	≤ 46	≤ 47	≤ 48	≤ 49	≤ 50
Number of crossings	2	6	12	22	32	44	52	58	60	61	62

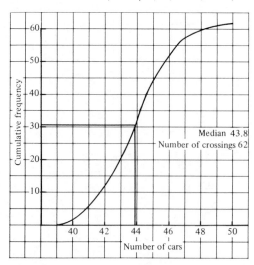

6.

Number of cars	0–5	6–10	11–15	16–20	21–25	26–30	31–35	36–40	41–45	46–50
Frequency	1	1	2	3	5	7	13	22	11	5

Number of cars	≤5	≤10	≤15	≤20	≤25	≤30	≤35	≤40	≤45	≤50
Cumulative frequency	1	2	4	7	12	19	32	54	65	70

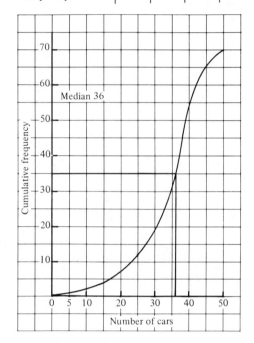

Exercise 15f   Page 308

1.

	Upper quartile	Lower quartile	Interquartile range
Question 4	67	42	25
5	45.4	42.5	2.9
6	40	29	11

2. a) 23
b) lower quartile is 16
upper quartile is 34.5
interquartile range is 18.5

3. a) 328   b) 62   c) 74, 52, 22

4. a) 29   b) 34.5, 23

5. a) 37   b) 50, 27, 23   c) 50

6. a)

Score	67	68	69	70	71	72	73	74
Frequency	4	7	9	9	6	3	1	1

b)

Score	≤67	≤68	≤69	≤70	≤71	≤72	≤73	≤74
Cumulative frequency	4	11	20	29	35	38	39	40

c) 20   d) 20   e) 70

7. a)

Score	≤ 10	≤ 20	≤ 30	≤ 40	≤ 50	≤ 60
Cumulative frequency	7	16	27	40	56	74

Score	≤ 70	≤ 80	≤ 90	≤ 100	≤ 110	≤ 120
Cumulative frequency	85	92	96	98	99	100

b) 56   c) 15   d) 46, 61, 28, 33

8. a) 11.7 cm
b) 12.4 cm and 10.8 cm

Exercise 15g   Page 312

1.

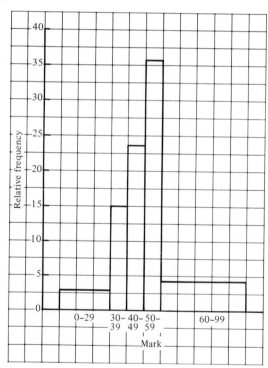

2.

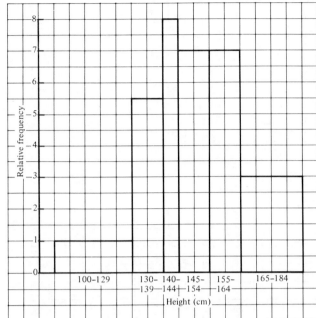

3.

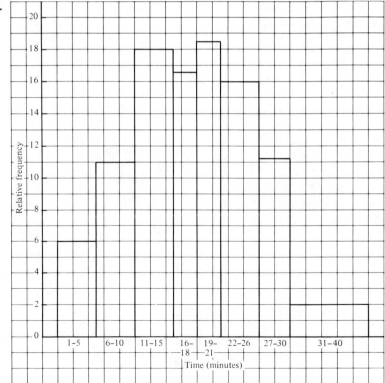

4.

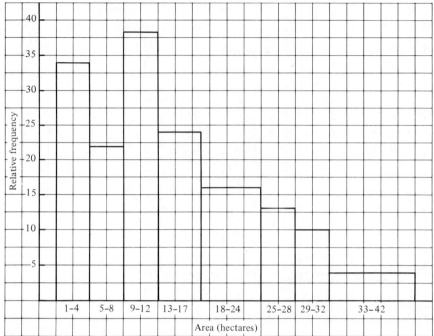

Exercise 15h   Page 314

1. a) $\frac{1}{3}$ b) $\frac{2}{3}$
c) 0   d) $\frac{13}{15}$

2. a) $\frac{21}{26}$ b) $\frac{2}{13}$
c) $\frac{4}{13}$

3. a) $\frac{1}{90}$ b) $\frac{1}{5}$
c) $\frac{1}{3}$

4. a) $\frac{3}{4}$ b) $\frac{19}{20}$
c) $\frac{1}{20}$

5. 4

6. a) P($\geqslant 7$) $= \frac{21}{36} = \frac{7}{12}$
b) P(odd number greater than 2 on both dice)
$= \frac{4}{36} = \frac{1}{9}$
c) P(product of the scores is even) $= \frac{27}{36} = \frac{3}{4}$

7. a) i) $\frac{1}{5}$,   ii) $\frac{37}{50}$
b) i) $\frac{9}{245}$,   ii) $\frac{12}{245}$

8. a) i) $\frac{3}{10}$,   ii) $\frac{1}{2}$
b) i) $\frac{29}{59}$,   ii) $\frac{17}{59}$

9. a) i) $\frac{9}{55}$,   ii) $\frac{18}{55}$
b) i) $\frac{9}{55} \times \frac{8}{54}$ i.e. $\frac{4}{165}$,   ii) $\frac{23}{33}$

10. a) $\frac{1}{100}$ b) $\frac{9}{50}$

11. a) $\frac{3}{5}$ b) $\frac{1}{20}$

12. $\frac{441}{500}$

13. a) $\frac{1}{8}$ b) $\frac{3}{8}$

14. a) $\frac{5}{36}$ b) $\frac{1}{36}$
c) $\frac{1}{18}$ d) $\frac{5}{12}$

15. a) $\frac{3}{51}$ b) $\frac{11}{221}$
c) $\frac{1}{221}$ d) $\frac{13}{102}$
e) $\frac{1}{2652}$

16. a) $\frac{16}{169}$
b) 0 (as only two letters have been taken)
c) $\frac{1}{169}$ d) $\frac{8}{169}$

17. a) $\frac{4}{15}$ b) $\frac{8}{27}$
c) $\frac{32}{225}$ d) $\frac{1331}{3375}$

18. a) 0   b) $\frac{1}{21}$
c) $\frac{2}{7}$ d) $\frac{5}{7}$

Exercise 15i   Page 319

1. a) 53   b) 90
c) graph   d) 58; 68, 43
e) $\frac{89}{250}$, $\frac{108}{250}$

2. a) 20   b) £2.60
c) One quarter had less than £1.50 per week
d) £4.20. A vertical line (CD) at £4.20 on the horizontal axis.

The range in the amount of pocket money received by the middle 50 % of the 240 pupils is £2.70.
e) £735   f) £3.06

3. a) 32   b) 38, 25
   c) 20   d) 27
   e) $\frac{3}{8}$

4. a) 22   b) 14
   c) graph
   d) median 74, $Q_3 = 80$,
     $Q_1 = 61$
   e) $\frac{3}{5}$

5. graph
   b) i) 3550,   ii) 56,
     iii) 21,   iv) 28 %
   c) i) $\frac{3}{16}$,   ii) $\frac{33}{80}$

6. a)

Age (years)	$\leqslant 25$	$\leqslant 29$	$\leqslant 33$	$\leqslant 37$	$\leqslant 45$
Cumulative frequency	3	7	15	27	40

   b) 7.3
   c) i) $\frac{1}{5}$,   ii) $\frac{5}{8}$
   d) i) $\frac{7}{195}$,   ii) $\frac{1}{15}$

CHAPTER 16

Exercise 16a   **Page 323**

1. b)

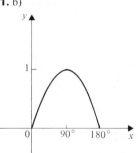

   c) $x = 90°$
   d) i) 53°, 127°,   ii) 37°, 143°,
     iii) 24°, 156°
     The sum of each pair
     is 180°

2. a) sin 30° = 0.5,
     sin 150° = 0.5,
     150° = 180° − 30°
   b) sin 40° = 0.6428,
     sin 140° = 0.6428,
     140° = 180° − 40°
   c) sin 72° = 0.9511,
     sin 108° = 0.9511,
     108° = 180° − 72°

Exercise 16b   **Page 324**

1. $\frac{5}{13}$ **3.** $\frac{2}{5}$ **5.** $\frac{7}{25}$ **7.** 15°   **9.** 28°   **11.** 5°

2. $\frac{1}{2}$ **4.** $\frac{3}{10}$ **6.** $\frac{40}{41}$ **8.** 40°   **10.** 80°   **12.** 89°

Exercise 16c   **Page 325**

1. b)

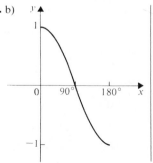

   c) 90°
   d) they are negative
   e) i) 37°,   ii) 143°
     They are supplementary
     (add up to 180°)

2. a) cos 30° = 0.8660
     cos 150° = −0.8660
     150° = 180° − 30°
   b) cos 50° = 0.6428
     cos 130° = −0.6428
     130° = 180° − 50°
   c) cos 84° = 0.1045
     cos 96° = −0.1045
     96° = 180° − 84°

Exercise 16d   Page 326

1. $-\frac{3}{5}$ **2.** $-\frac{2}{3}$ **3.** $-\frac{2}{5}$ **4.** $-\frac{7}{8}$ **5.** a) $\widehat{A} = 160\,°$ b) $\widehat{A} = 130\,°$

Exercise 16e   Page 328

1. $\cos A = \frac{24}{25}$, $\tan A = \frac{7}{24}$
2. $\sin A = \frac{12}{13}$, $\tan A = \frac{12}{5}$
3. $\sin P = \frac{3}{5}$, $\cos P = \frac{4}{5}$
4. $\tan D = \frac{4}{3}$, $\sin D = \frac{4}{5}$

5. $\cos X = \frac{40}{41}$, $\tan X = \frac{9}{40}$
6. $\sin A = \frac{1}{\sqrt{2}}$, $\cos A = \frac{1}{\sqrt{2}}$
7. $\frac{120}{169}$, 0.7101, 67.4 °, 0.7101;
$\sin 2A = 2 \sin A \cos A$

8. $\frac{7}{25}$, 0.28, 36.87 °, 0.2800,
$\cos 2A = \cos^2 A - \sin^2 A$
9. 1
10. $\frac{1320}{3479}$

Exercise 16f   Page 329

1. a) $-\frac{24}{25}$ b) $\frac{7}{25}$
2. 30 °, 150 °
3. −0.515
4. 108 °
5. a) $\frac{4}{5}$ b) $\frac{3}{5}$
   c) $-\frac{3}{5}$

6.
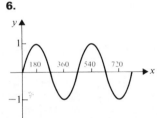

7.
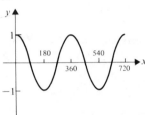

Exercise 16g   Page 331

1. 4.92 cm   **4.** 8.05 cm   **7.** 238 cm   **10.** 5.60 cm   **13.** 42.4 cm   **16.** 77.0 cm
2. 13.0 cm   **5.** 14.6 cm   **8.** 105 cm   **11.** 413 cm   **14.** 19.7 cm   **17.** 224 cm
3. 9.74 cm   **6.** 9.69 cm   **9.** 66.0 cm   **12.** 64.1 cm   **15.** 18.4 cm   **18.** 196 cm

19. QR = 7.57 cm, PQ = 7.84 cm
20. BC = 13.5 cm, AC = 9.08 cm
21. $p = 117$, $q = 175$
22. $a = 27$, $c = 35.4$

	AB	BC	AC	$\widehat{A}$	$\widehat{B}$	$\widehat{C}$
23.	23.7 cm	19 cm	19.5 cm	51 °	53 °	76 °
24.	217 cm	221 cm	146 cm	72 °	39 °	69 °
25.	81 cm	49.0 cm	69.8 cm	37 °	59 °	84 °
26.	114 cm	123 cm	97 cm	71 °	48 °	61 °
27.	12 cm	23.9 cm	28.9 cm	54 °	102 °	24 °
28.	7.97 cm	15.6 cm	9.9 cm	26 °	33 °	121 °

Exercise 16h   Page 337

1. 5.73
2. 7.95
3. 7.68 cm
4. 18.4 cm
5. 7.03
6. 14.2
7. 13.7
8. 11.1 cm
9. 8.30 cm

10. 35.3
11. 15.9 cm
12. 21.6 cm
13. 43.0 cm
14. 28.9
15. 4.75
16. 16.9
17. 96.4 cm

18. 9.05 cm
19. 24.1 °
20. 108.2 °
21. 92.9 °
22. 29.0 °
23. 95.7 °
24. 48.2 °
25. a) 13.1 cm   b) 72.3 °

Exercise 16i   Page 341

	a	*b*	*c*	$\widehat{A}$	$\widehat{B}$	$\widehat{C}$
1.		17.0				
2.	117					
3.		77.9				
4.			23.3			
5.			346			
6.			12.4			45 °
7.	29.7				38 °	
8.	25.9					
9.		28.6				
10.	21.6		29.3	47 °		

11. $b = 10.2\,\text{cm}$, $\widehat{C} = 54°$ $c = 8.62$

12. $\widehat{R} = 106.2°$, $\widehat{P} = 47.9°$, $\widehat{Q} = 25.9°$

13. $l = 17.7\,\text{cm}$, $\widehat{M} = 49.7°$, $\widehat{N} = 66.3°$

14. $\widehat{E} = 33.9°$, $\widehat{F} = 93.7°$, $\widehat{D} = 52.4°$

Exercise 16j   Page 344

1. $81.1\,\text{cm}^2$
2. $18\,900\,\text{sq units}$
3. $2610\,\text{cm}^2$
4. $572\,\text{sq units}$
5. $126\,\text{sq units}$

6. $22\,800\,\text{cm}^2$
7. $35.5\,\text{sq units}$
8. $6.86\,\text{m}^2$
9. $2030\,\text{cm}^2$
10. $148\,\text{sq units}$

11. $\widehat{C} = 83.3°$, $27.8\,\text{cm}^2$
12. $\text{PR} = 9.91\,\text{cm}$, $53.8\,\text{cm}^2$
13. $\widehat{M} = 48.9°$, $\text{LN} = 14.7\,\text{cm}$
14. $4.46\,\text{cm}$

Exercise 16k   Page 346

1. $62.3\,\text{km}$
2. $\text{A}\widehat{\text{P}}\text{B} = 18.1°$, $42.7\,\text{m}$
3. $88°$
4. a) $42\,\text{cm}^2$ b) $44.4°$
   c) $8.52\,\text{cm}$

5. $\text{AC} = 5.76\,\text{km}$,
   $\text{BC} = 4.29\,\text{km}$, $69.1\,\text{km/h}$
6. a) 0.2588 b) $97.6\,\text{m}$
7. a) $4.26\,\text{cm}$ b) $75.2°$
   c) $16.0\,\text{m}^2$
8. a) $3°$, $9°$ b) $1430\,\text{m}$
   c) $1080\,\text{m}$

9. a) $10.5\,\text{m}$ b) $6.42\,\text{m}$
   c) $81.0\,\text{m}^2$ d) $8.96\,\text{m}$
10. a) i) $4.1\,\text{cm}$,   ii) $8.89\,\text{cm}$
   b) $81.3\,\text{cm}^2$
11. a) $47.2°$ b) $17.3°$
   c) $10.7\,\text{cm}$ d) $51.5\,\text{cm}^2$

Exercise 16l   Page 351

1. $\frac{12}{13}$, $\frac{12}{5}$
2. $21.5°$, $158.5°$
3. -0.43

4. $196\,\text{cm}$
5. $10.9\,\text{cm}$

6. $38.0\,\text{cm}^2$
7. $13.5\,\text{km}$

Exercise 16m   Page 351

1. $143°$
2. $y = 180 - x$
3. a) $\cos(180° - A) = -\cos A$
   b) $\sin(180° - A) = \sin A$
   c) $-\cos A = \cos(180° - A)$

4. $16.7\,\text{cm}$
5. $\text{AB} = 8.94\,\text{cm}$, $\text{BC} = 6.90\,\text{cm}$
6. a) $42.6°$ b) $82.9°$

7. a) $70.9°$ b) $29.65\,\text{m}^2$
   c) $2.37\,\text{m}^2$

CHAPTER 17

Exercise 17a   Page 352

1. $\{-4, 0, 4\}$
2. $\{-2, -1, 0, 1\}$

3. $\{-5, -3, -1, 1, 3\}$
4. $\{\frac{1}{3}, \frac{1}{2}, 1, 2\}$

5. $\{2, 3, 7\}$
6. $\{0, 0.5, 0.8660, 1\}$

7.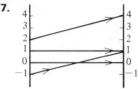

8.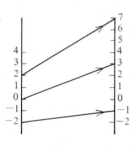

9.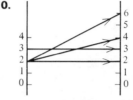

10.

11.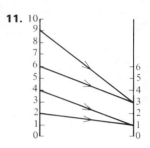

12. $x \mapsto x + 1$
13. $x \mapsto x - 1$
14. $x \mapsto 2x$
15. $x \mapsto \frac{1}{2}x$
16. $x \mapsto 2x + 1$
17. $x \mapsto -2x$
18. $x \mapsto \dfrac{1}{x}$

Exercise 17b   Page 356

1. yes
2. yes
3. no
4. yes
5. no
6. yes

7. yes
8. yes
9. yes
10. yes
11. no
12. yes

13. yes
14. yes
15. no
16. no (there are no values of x for which $\sqrt{x}$ exists if $x < 0$)

Exercise 17c   Page 358

1. a) -1 b) 7
   c) -3 d) -11
2. a) 8 b) 2
   c) $\frac{1}{2}$ d) -2
3. a) 2 b) 4
   c) -2 d) 6
4. a) -1 b) 8
   c) 5 d) -7
5. a) 0 b) 1
   c) 4 d) 9
6. a) 1 b) -3
   c) 3 d) -1
7. a) 5 b) 8
   c) 4 d) 13
8. a) 11 b) 3
   c) 6 d) 6

9. a) $\frac{1}{2}$ b) $\frac{2}{3}$
   c) 2 d) $1\frac{1}{2}$
10. a) 2 b) 7
   c) 2 d) $-15\frac{1}{2}$
11. a) $\frac{1}{2}$ b) 1
   c) 0 d) 0.9659
12. $x = 1\frac{1}{5}$
13. $x = \frac{1}{5}$
14. $x = 7$
15. $x = -5$
16. $x = 3$ or -3
17. $x = 1$ or -1
18. $x = 3$ or -1
19. $x = 1$
20. $x = -1$ or 2

21. $x = 30°$
22. a) -10 b) 18
   c) -1
23. a) -4 b) -6
   c) $x = 2$ or -3
24. $-8, -7, 2$
25. a) $k = \frac{4}{3}$ b) $k = 6$
26. a) $-4, -2, 2$ b) increasing
   c) 1.5 (actual value 1.56)
27. a) $8, 6, 2$ b) decreasing
   c) $x > 6$
28. a) $2, 6, 18, 102$
   b) increasing
   c) $2, 3, 11, 102$
   d) 2

Exercise 17d   Page 361

1.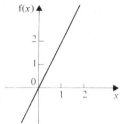

6.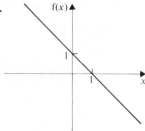

11.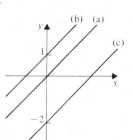

Translation of 1 unit upwards

i.e. defined by $\begin{pmatrix} 0 \\ 1 \end{pmatrix}$

2.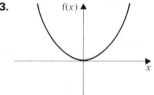

7.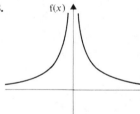

12.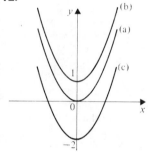

Translation of 2 units downwards

i.e. defined by $\begin{pmatrix} 0 \\ -2 \end{pmatrix}$

3.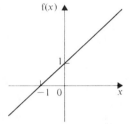

8.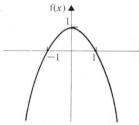

4.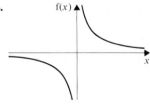

9.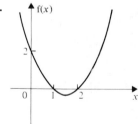

13.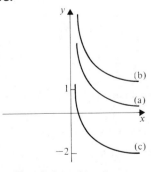

5.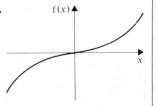

10.

Translation of 2 units downwards

14.

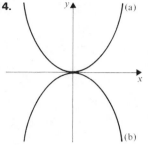

(a)

(b)

Reflection in *x*-axis

15.

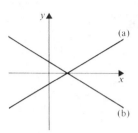

(a)

(b)

Reflection in *x*-axis

16.

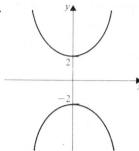

Reflection in *x*-axis

Exercise 17e   Page 363

1.

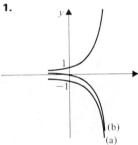

(b)
(a)

2. $-3, -1, 3; x \approx 1.3$
(actual value 1.303)

3. $-11, -4, 15; x \approx 2.2$
(actual value 2.289)

4.

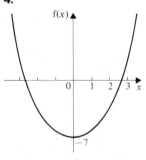

$\sqrt{7} \approx 2.7 \ (\sqrt{7} = 2.646)$

5. No

6.

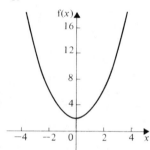

a) $x = 0$
b) $x = 2$ and -2
c) two values of x give the
   same value of $f(x)$
d) $f(2) = f(-2)$,
   $f(4) = f(-4)$,
   $f(-1) = f(1)$,
   $f(x) = f(-x)$,

7.

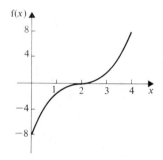

a) $(2, 0)$
b) $f(4) = -f(0)$
   $f(3) = -f(1)$
   $f(x) = -f(4-x)$

Exercise 17f   Page 366

1. $f^{-1}(x) = \frac{1}{2}x$

2. $f^{-1}(x) = \frac{1}{3}x$

3. $f^{-1}(x) = \frac{1}{x}, x \neq 0$

4. $f^{-1}(x) = 5x$

5. $f^{-1}(x) = x - 1$

6. $f^{-1}(x) = x + 2$

7. $f^{-1}(x) = 4 - x$

8. $f^{-1}(x) = \frac{1}{2}(x - 1)$

9. $f^{-1}(x) = \frac{12}{x}, x \neq 0$

10. $f^{-1}(x) = \frac{1}{2}(4 - x)$

11. $f^{-1}(x) = \sqrt[3]{x}$

12. $f^{-1}(x) = \frac{1}{3}(x + 5)$

13. no inverse

14. $f^{-1}(x) = \sqrt[3]{x}$

15. $f^{-1}(x) = \frac{1}{3}(x + 1)$

16. no inverse

17. $f^{-1}(x) = \frac{1}{2}x + 1$

18. no inverse

19. no inverse

20. $f^{-1}(x) = \frac{1}{x} - 1$

21. $f^{-1}(x) = \frac{1}{3}(7 - x)$

22. $f^{-1}(x) = \frac{12}{x}, x \neq 0$

23. $f^{-1}(x) = \frac{1}{x}, x \neq 0$

24. no inverse

25. $f^{-1}: x \mapsto \frac{1}{3}x, f^{-1}(2) = \frac{2}{3}$,
$f^{-1}(0) = 0, f^{-1}(-1) = -\frac{1}{3}$

26. a) $f^{-1}(x) = x - 4$
b) $f^{-1}(-2) = -6$
c) $x = 7$

27. a) 0   b) 1

28. −4

Exercise 17g   Page 369

1.
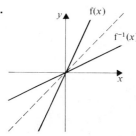

2.
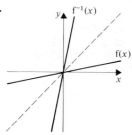

3.
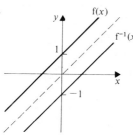

4.
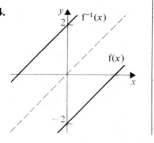

5.
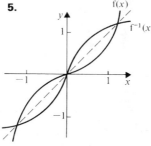

6.
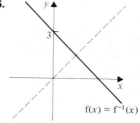

7.
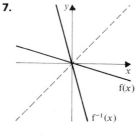

8.
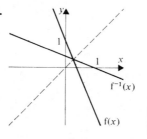

9.
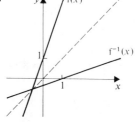

10.
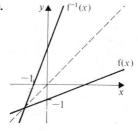

11. Any function of the form
$f(x) = k - x, f(x) = \frac{k}{x},$ or
any other function whose
graph has $y = x$ as a line
of symmetry.

Exercise 17h   Page 370

1. $j(4) = 16\frac{1}{4}$, $j(-1) = -5$
2. $k(2) = 24$, $k(-3) = 24$
3. $m(4) = 3\frac{1}{5}$, $m(-2) = 8$,
   $m(-1)$ has no meaning
   (we cannot divide by zero).
4. $j : x \mapsto 4x + \dfrac{1}{x}$; $x = \frac{1}{4}$ and 1
5. $k : x \mapsto 4x(x+1)$;
   $x = -1\frac{1}{2}$ and $\frac{1}{2}$

6. a) $1\frac{1}{2}$,
   b) $q(x) = \dfrac{x+1}{x}$
   c) -2 d) No
7. $m(x) = 5x + 1$,
   $m^{-1}(x) = \frac{1}{5}(x-1)$
8. j^{-1} does not exist;
   $q^{-1}(x) = \dfrac{1}{x-1}$

9. $x = 0$; for j, x can take all
   real values except 0.
   for k, x can take all
   real values.
   for q^{-1}, x can take all
   real values except 1
10. $\frac{1}{3}$.
11. a) i) 5,   ii) 0
   b) -2 c) yes, $x = 0$
   d) $a = 2$

Exercise 17i   Page 373

1. $fg(x) = 2x + 1$,
   $gf(x) = 2x + 2$
2. $fg(x) = 12 - 4x$,
   $gf(x) = 3 - 4x$
3. $fg(x) = 16 - 3x$,
   $gf(x) = 4 - 3x$
4. $fg(x) = 5 - 8x$,
   $gf(x) = 11 - 8x$
5. $fg(x) = x$, $gf(x) = x$
   f and g are inverses of each
   other.
6. $fg(x) \mapsto 3(x-1)$
7. $gf(x) \mapsto 3x - 1$

8. $fh(x) \mapsto \dfrac{6}{x}$
9. $hf(x) \mapsto \dfrac{2}{3x}$
10. $gh(x) \mapsto \dfrac{2}{x} - 1$
11. $hg(x) \mapsto \dfrac{2}{x-1}$
12. $fi(x) \mapsto 3x^2$
13. $ig(x) \mapsto (x-1)^2$
14. $hi(x) \mapsto \dfrac{2}{x^2}$
15. $gi(x) \mapsto x^2 - 1$

16. $fg(2) = 1$, $gf(2) = 3$
17. $fg(3) = -\frac{5}{6}$, $gf(-1) = \frac{1}{32}$
18. $fg(-1) = \frac{1}{3}$, $gf(0) = \frac{1}{4}$
19. $gf(-2) = -3$, $fg(-2) = -5$
20. $gf(3) = \frac{1}{36}$, $fg(3) = \frac{1}{18}$
21. a) $fgh(4) = 49$,
   b) $gfh(1) = 6$
   c) $hgf : x \mapsto 9(x+1)^2$
22. $x = 1$ and -3
23. $x = -2$ and 2
24. a) 1   b) $\frac{1}{2}$
25. 1 and -1

Exercise 17j   Page 376

The graphs for this exercise are on p. 377.

Exercise 17k   Page 378

1.

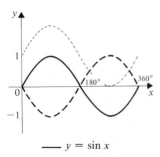

—— $y = \sin x$
- - - $y = -\sin x$
- - - - - $y = 1 + \sin x$

2.

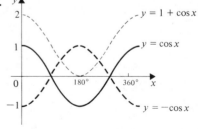

3. b) 0, 180 °, 360 °

c)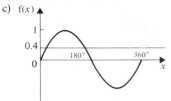

4. b) 90 °, 270 °

c)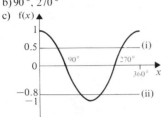

5. a) $-1 \leqslant \sin x \leqslant 1$ b) $-1 \leqslant \cos x \leqslant 1$

6.

7.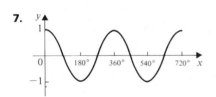

The cosine curve is the sine wave translated 90 ° to the left.

8. a) $fg(x) = \sin 2x$ b) 0.342
c) $gf(x) = 2 \sin x$ d) 1.97

9. $hf(x) = (\sin x)^2$,
$hf(15) = 0.0670$

10. $gf(x) = \sin x - 1$

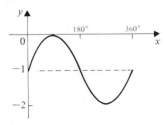

11. a) $g^{-1}(x) = x + 1$
b) $g^{-1}f(x) = \sin x + 1$

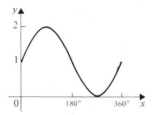

12. $fh(x) = \cos 3x$,
$fh(50) = -0.866$

13. $gf(x) = 1 + \cos x$,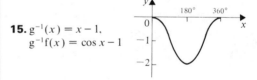

14. $hf(x) = 3 \cos x$
a) 2.60   b) 1.5

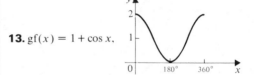

15. $g^{-1}(x) = x - 1$,
$g^{-1}f(x) = \cos x - 1$

16. $g(x) = 1 - \cos x$

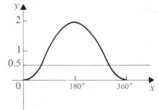

17.

$x = 45$ ° and 225 °

18. $gh(x) = \cos 4x$
1, 0.5, 0, −0.5, −1, −0.5, 0.5, 1

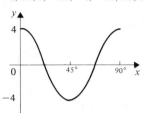

20.

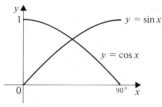

a) $x \approx 45°$
b) $\sin x = \cos x$

19. $hf(x) = 4 \sin x$

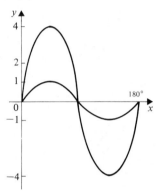

21.

22. c) the finished curve is the sine wave.

Exercise 17I   Page 381

1. A   **2.** B   **3.** A   **4.** D   **5.** C   **6.** C   **7.** A   **8.** B   **9.** D

Exercise 17m   Page 382

1. $\{0, 2, 6\}$

2. a) yes   b) yes
c) yes

3. a) −2   b) 28
c) −22

4. a) 1   b) 0
c) 4

5. a) 6   b) 5
c) 4

6. $k = 6$

7. a) $f^{-1}(x) = x − 1$
b) $f^{-1}(x) = 2 − x$
c) $f^{-1}(x) = \frac{1}{3}(x + 2)$

8. a) $f^{-1}(x) = \frac{1}{5}x$
b) $f^{-1}(10) = 2$
c) 10

9. a) $f(x)g(x) = 6x(x + 2)$
b) $g(x) + h(x) = x + 2 + \frac{1}{x}$,
$x \neq 0$
c) $fg(x) = 6(x + 2)$
d) $hg(x) = \frac{1}{x + 2}, x \neq −2$

10. a) 0   b) 0
c) −2   d) 4

11. a) $fg(x) = 2(3 − x)^2$,
$gh(x) = 3 − 6x$
b) $g^{-1}(x) = 3 − x$,
$h^{-1}(x) = \frac{1}{6}x$
c) $g^{-1}f(x) = 3 − 2x^2$,
$gh^{-1} = 3 − \frac{1}{6}x$
d) $x = 2\frac{4}{7}$

12. a) $\frac{1}{2}$ b) $\frac{1}{2}$
c) 2 and $−\frac{2}{3}$
d) $h^{-1}(x) = \frac{2}{3}\left(2 + \frac{1}{x}\right)$

13. a) 10   b) 4
c) −14

14. a) $x = 0$ and $2\frac{1}{2}$
b) $h^{-1}(x) = \frac{1}{20}(x + 5)$
c) 350